Experimental Life

Experimental Life

Vitalism in Romantic Science and Literature

ROBERT MITCHELL

Johns Hopkins University Press
Baltimore

© 2013 Johns Hopkins University Press
All rights reserved. Published 2013
Printed in the United States of America on acid-free paper
2 4 6 8 9 7 5 3 1

Johns Hopkins University Press
2715 North Charles Street
Baltimore, Maryland 21218-4363
www.press.jhu.edu

Library of Congress Cataloging-in-Publication Data

Mitchell, Robert, 1969–
Experimental life ; vitalism in Romantic science and
literature / Robert Mitchell.
pages cm.
Includes bibliographical references and index.
ISBN 978-1-4214-1088-3 (hardcover : acid-free paper) —
ISBN 978-1-4214-1089-0 (electronic) — ISBN 1-4214-1088-5 (hardcover :
acid-free paper) — ISBN 1-4214-1089-3 (electronic)
1. English literature—19th century—History and criticism.   2. Literature
and science—England—History—19th century.   3. Romanticism—
England.   4. Vitalism in literature.   5. Life in literature.   I. Title.
PR585.S33M58 2013
820.9'36—dc23   2013004819

A catalog record for this book is available from the British Library.

*Special discounts are available for bulk purchases of this book. For more
information, please contact Special Sales at 410-516-6936 or
specialsales@press.jhu.edu.*

Johns Hopkins University Press uses environmentally friendly book
materials, including recycled text paper that is composed of at least 30
percent post-consumer waste, whenever possible.

Contents

Acknowledgments

This book is the result of the intersection of two lines of research I initially thought of as quite separate. It had its origins in research that I conducted for *Sympathy and the State in the Romantic Era* (2007). That book focused on eighteenth- and nineteenth-century theories of sympathy, a term that then referred not only to moral relationships among individuals but also to a physiological action-at-a-distance that occurred within a living body. Although I was not able to include discussion of the latter, medical sense of sympathy in that earlier project, my research on this topic put me on the path of those Romantic concepts of life that are the focus of this book.

The book is also the result of a quite different line of research, one focused on the cultural logic and economics of our own, contemporary systems of biological research and healthcare. Relatively little of my work in this area has engaged literature, and I have focused primarily on technologies and techniques developed since the 1970s (e.g., genetic engineering, organ transplantation, and biobanking). Yet with some degree of regularity, peculiar resonances emerged between my literary critical work on the Romantic era and my more sociological work on contemporary biotechnology. I became fascinated, for example, with the final stages of the process by means of which contemporary biologists create immortal cell lines, as cells were lowered into waist-high cylindrical liquid nitrogen tanks to be frozen. I saw these cylinders as urns that quite literally produced what Keats had called frozen time, as the liquid nitrogen suspended the living processes of this matter so that the biologist could return to these same cells later, as though no time at all had passed for the cells. When, at about the same time, I read John Hunter's late-eighteenth-century reflections on the newly coined term "suspended animation," and reflected on his hopes that suspended animation might allow for a new kind of human existence and a new mode of scientific observation, I became convinced that it was worth pursuing further these resonances between the Romantic era and the present—and this volume is the result of that endeavor (though early versions of chapters 2 and 6 appeared in *PMLA* and *European Romantic Review*, respectively).

Because this project has been under way for a relatively long time, and because I have pursued its concepts, questions, and problems across several research communities, I have been assisted by a quite considerable list of people, and I especially wish to thank the following: Lindsey Andrews; Alan Bewell; James Bono; Andrew Burkett; David Clark (to whom I owe the phrase "experimenting with experimentation"); Matthew Cohen; David Collings; Allison Dushane; Michael Eberle-Sinatra and the members of the Technologies, Media, and Representations in Nineteenth-Century France and England Research Group; Tim Fulford; Denise Gigante and the members of the Seminar on Enlightenment and Revolution; Nick Halmi; Orit Halpern; Tim Lenoir; Tracie Matysik, Sam Baker, and Phillipa Levine and the members of the Nineteenth-Century Workshop; Sean Metzger; Colin Milburn; Thomas Pfau; Michelle Puetz; Phillip Thurtle; Charles Rzepka; Alexander Schlutz; Barbara Herrnstein Smith; Alan Vardy, Priscilla Wald; and Nancy Yousef.

I also wish to thank Colleen M. Weum and Cheyenne Maria Roduin for assistance with textual resources in the Health Sciences Rare Book Room at the University of Washington; Duke University's Arts and Sciences Research Council, which provided vital financial support for this project at various points; and the National Humanities Center and the Duke Endowment Fellowship, which made it possible for me to complete the manuscript of the book in the Center's convivial surroundings. I am also grateful to Meghan O'Neil, Chris Catanese, and Glenn Perkins for their able proofreading assistance, and to Matthew McAdam, my editor at the Johns Hopkins University Press, with whom it was a joy to work.

And last, but definitely not least, my heartfelt thanks go to Inga Pollmann, who not only patiently read and commented on many drafts of each of these chapters, and whose work on the life of film has been an inspiration for my own project here, but who has, in addition, always been a vitalizing pneuma for me.

Experimental Life

Three Eras of Experimental Vitalism

A vitalist turn seems to be under way in contemporary natural sciences, social sciences, and humanities. While biologists have, for much of the twentieth century, generally understood vitalism as an outdated and invalidated nineteenth-century approach to living beings—an atavism from the irrational prehistory of biology proper, when natural philosophers believed that they could explain vitality by means of a principle of life that exceeded the laws of chemistry and physics—an understanding of "life itself" as a source of mystery and provocation for thought and experimentation nevertheless seems to drive several recent attempts to reorient and renew explicitly scientific biological practice in the hard sciences. In attempting to understand the implications of the genomic revolution, for example, one respected molecular biologist has sought to distinguish between "conscious life," of which we ourselves are examples, and "life-in-general," a more expansive and mysterious force of vitality that constantly evolves new living forms.[1] Interest in the mysteries and provocations of life extends beyond the field of biology. A prominent geochemist, for example, has urged us to understand the earth itself as a living entity and life as fundamentally an "activity" and "process" that produces "increasing complexity," while some computer scientists, arguing that vitality can be instantiated in silicon as well as carbon, have developed forms of artificial life that they offer as proof of their claims.[2]

Scholars in the social sciences and humanities have also taken this vital turn. Apparently dissatisfied with the exclusive emphasis of poststructuralist thought on representation and signs, an ever-increasing number of literary critics, philosophers, and cultural critics have coined and adopted terms such as "non-organic life," "bare life," and "precarious life" to denote ontological dimensions of vitality that exceed, or stand as the conditions of possibility of, semiotics and representation.[3] Others, using less precise but no less earnest terminology, have urged us to attend to what they describe as the "life of images" and the "life of art."[4] What we can discern in the work of many in the sciences and the humanities, in short, is a sense of life as *provocation*; for these various authors, life is not a self-evident fact that can be taken for granted but rather a source of perplexity that demands new modes of conceptual and practical experimentation.

Such intense multidisciplinary interest in life itself is not unprecedented, and I suggest that current fascination with vitality represents a third wave of experimental vitalism. As I use the term, "experimental vitalism" denotes a submovement—a sort of transverse flow—within broader vitalist movements. The first era of experimental vitalism occurred at the end of the eighteenth and start of the nineteenth centuries, during what we now think of as the Romantic period in literature. This first era of experimental vitalism was a transnational European affair, as British, French, and German physicians, surgeons, philosophers, and literary authors struggled to understand the relationship of a "principle of life" that seemed to animate and connect living beings to the concrete matter of which these bodies were composed. The second era of experimental vitalism was also a fin-de-siècle phenomenon, this time taking place at the end of the nineteenth and early part of the twentieth centuries, during what we tend to think of as the modernist period in literature and art. In Germany, biologists such as Hans Driesch and Johannes Reinke developed "neo-vitalisms" on the basis of embryological and botanical experiments, while "life-philosophers" such as Friedrich Nietzsche and Max Scheler speculated more broadly about the metaphysical status of life. In France, philosopher Henri Bergson developed a wildly popular vitalist philosophy premised on an élan vital that enabled "creative evolution," and his philosophy served as the basis for artistic experiments in multiple media, ranging from painting to sculpture to literature, in France, England, and the United States.[5]

This book is an attempt to understand what is at stake in our own, current fascination with vitality in the light of these two earlier eras of experimental vitalism, especially in relation to its first, Romantic, era. There are, of course, significant differences between the methods, objects of research, and premises at play in these three different eras. Scientific understandings of living beings have changed significantly from the late eighteenth to the early twenty-first centuries, as have aesthetic norms and artistic practices in what eventually came to be called the humanities. Yet despite these differences, a distinctive and singular style of thought and practice traverses all three eras of experimental vitalism. This style of thought and practice begins with an active search for what is perplexing and inscrutable in vital phenomena. As Romantic-era poet P. B. Shelley noted in his unfinished "Essay on Life," the rush of our everyday concerns tends to produce a "mist of familiarity" that encourages us to take life for granted, with the consequence that life is simultaneously "certain and unfathomable" (172). Experimental vitalism begins with an attempt to locate phenomena most likely to dispel this mist of familiarity and thus encourage perplexity about the being of life. This often involves a search for liminal entities or "altered states" that seem to confuse the line between life and death. In the Romantic era, for example, scientists and artists were fascinated by apparently "non-organic" living beings such as polyps and fertilized eggs, which clearly lived but at the same time lacked the organs that characterized all other living things. They were also fascinated by physiological processes, such as digestion and embryological development, that transformed inanimate into living matter, and by states of "trance" and "suspended animation," which seemed to place vitality in abeyance, at least for brief periods.

Experimental vitalism seeks out and isolates these liminal beings and states not simply because they produce astonishment but also, and more fundamentally, because they encourage invention on the part of artists and scientific researchers. For scientists and artists, invention takes the form of a search for new concepts and experiences, which in turn enable new ways to solicit the potentials of life. Scientists often focus their experimental inventiveness on nonhuman living beings, while artists seek to create new states of sensation and embodied perception in humans. Yet experimental vitalism also tends to confuse these apparently firm distinctions, soliciting modes of self-experimentation from scientists and encouraging artists to link human sensation and perception to nonhuman forms of vitality. Experimental vitalism confuses the categories "scientist" and "artist" in part because it is a project of tracing, and thereby encouraging, the self-reflexivity of life, producing new possibilities *for individual and social life* through new sensations and perceptions *of* life.

In this book, I focus most extensively on the first era of experimental vitalism, exploring at length Romantic-era engagements with liminal vital objects and states. My emphasis on the Romantic era is in part a function of my own professional training as a literary critic of Romanticism. However, it is also the result of my belief that an analysis of this first era of experimental vitalism can be especially useful in helping us to understand the later modernist and postmodern eras, for these latter often reanimate potentials from the Romantics. Equally important, it was in the Romantic era that literary authors first explicitly appropriated the term "experiment," as well as a set of experimental practices, from the sciences in order to alter both the nature and function of the arts. While twentieth- and twenty-first-century literary and art critics have adopted the terminology of experiments to describe both Romantic and post-Romantic art, scholars have tended to use such terminology simply as a synonym for innovation in general or for formal "difficulty" (i.e., works of art that depart from the conventions of their genre or medium to such an extent that spectators or auditors have to work hard to understand what questions the work might be posing or what experiences the work seeks to facilitate). Romantic authors, by contrast, understood literary experiments much more broadly and as much further-reaching in their impact. Writing in the wake and shadow of what Edmund Burke had famously described as an unprecedented "experiment" in politics—the French Revolution—the Romantics sought in their art to reformulate the very concept of an experiment and thereby channel the revolutionary energies unfolding around them to make possible new, and more equitable, social and political forms of life.[6]

Romantic authors were able to appropriate the term "experiment" from the sciences in large part because the meaning of the word was not then as firmly fixed as it would become by the early twentieth century. Though by the Romantic era "experiment" was firmly linked to, and often functioned as a synecdoche for, a fundamentally modern vision of science—that is, science as a communal investigation of nature by means of discrete, documented experiments, which were then communicated to peers through periodicals—many of the modern sciences that we now take for granted, such as chemistry and biology, were only just emerging in this period. Moreover, borders between scientists and nonscientists were sufficiently porous that J. W. von Goethe could write as both scientist and literary author, G. W. F. Hegel could synoptically interpret vast amounts of contemporary science in his philosophical *Encyclopedia*, and—perhaps most famously—Mary Shelley could draw upon contemporary biological research to make the creature of *Frankenstein* seem scientifically plausible. And while the term "experiment" had by the Romantic era increasingly come to refer to what historian of science Peter Dear has called "event experiments"—dated observations of nature as it acted in a specific place—it still bore an earlier, Aristotelian sense of "experience," that is, commonsense understandings of nature as it manifested itself habitually.[7] Romantic authors took advantage of this ambiguity to experiment with the term "experiment" itself, rather than simply applying a scientific concept of experiment to the field of art.

VITALISM, EXPERIMENT, AND EXPERIMENTAL VITALISM

"Life" served as the focus for the Romantics' hybridization of scientific and literary experimentalism in large part because, as historian of science Robert J.

Richards has stressed in *The Romantic Conception of Life*, terms such as "life" and "vitality" referred both to the biology of living beings as well as to those social and political forms of life by means of which groups of people were bound to one another. As other scholars have documented, Romantic authors were especially attracted to "vitalist" claims that life had laws and principles separate from those of physics and chemistry (a plausible claim at the time, given that chemical laws were themselves just then being separated from the laws of physics).[8] While I draw on these discussions of late eighteenth- and early nineteenth-century interest in life, my account is premised on the claim that we need to begin our investigation of Romantic vitality with a distinction between *theoretical* vitalism and *experimental* vitalism.

In order to understand what is at stake in this distinction, it is important to recognize that for much of the twentieth and into the twenty-first century, the term "vitalism" has functioned within the life sciences primarily as a gatekeeping term. To label someone a vitalist is to question his or her scientific legitimacy, and thus implicitly to defend one approach to living beings (generally, a mechanist approach) by attacking the legitimacy of another. The charge of vitalism thus also generally implies a historical narrative: the vitalist is someone who has fallen prey to the hazy concepts and thinking that characterized earlier, less rational, and ultimately theological approaches to life; as a consequence, the vitalist, far from working in the present or even future of a scientific or medical field, rather represents a danger to the progress of scientific research. The charge of vitalism, moreover, is designed to serve both as a warning and to encourage self-monitoring, for it is generally implied that the vitalist's collapse into the irrational prehistory of biology is the result of some psychological failing. The vitalist, that is, is simply not willing or able to accept the truths that science has to offer, preferring instead the emotional consolations provided by concepts such as a "principle of life" or "intuition" that cannot be verified by scientific means. "Materialism" and "mechanism" often function as the purported sources of the vitalist's fear, with vitalism understood as a spirited, though irrational, reaction to claims that life can be fully explained either through its material structures or by the mechanical—that is, automatic—operations of the laws of physics and chemistry.

Yet if it is clear to many that vitalism belongs to the past of science—and, moreover, ought to remain there—we find far less consensus among historians of medicine and science about precisely which past authors ought to be counted among the "vitalists." Some historians of vitalism have defined the term quite broadly, as any belief that the vital processes of living beings are in some way autonomous from the rest of the world of matter. From this perspective, the history of vitalism is a very long one, reaching back to Aristotle at least.[9] Others, defining vitalism more narrowly as the belief that matter is "infused" with "the power of reason and self-motion," discern a more limited history of this mode of thought, tracing its origins back to the sixteenth and seventeenth centuries.[10] Still others have understood vitalism in an even more limited sense, employing the term to denote the claim that the actions of living matter cannot be fully understood in terms of the laws of modern—that is, post-seventeenth-century—physics or chemistry, a definition that implies a history of vitalism that begins no earlier than the eighteenth or nineteenth centuries. Michel Foucault has articulated one of the most provocative and compelling versions of this latter claim, arguing that "up to the end of the eighteenth century . . . life does not exist: only living beings."[11] Foucault's claim is that prior to the end of the eighteenth century, "life does not constitute an obvious threshold beyond which entirely new forms of knowledge are required" (161), and it is thus only from the late eighteenth century on—that is, following the epistemic reorganization that Foucault documents in *The Order of Things*—that life becomes a sort of "transcendental," in the sense that it makes "possible the objective knowledge of living beings." From the late eighteenth century on, "life" is therefore "outside knowledge, but by that very fact [it is a] conditio[n] of knowledge" (244). From this perspective, Foucault argues, nineteenth-century vitalism cannot be understood as a "reaction against" an earlier mechanistic science; rather, "vitalism and its attempt to define the specificity of life are merely the surface effects" of a much more fundamental "archaeological even[t]" that affected knowledge-production as a whole (232).

To confuse matters even further, some historians have questioned whether vitalism can even be understood as a unified doctrine. Some historians of science have suggested that the term in fact denotes different positions that may share some sort of Wittgensteinian family resemblance among themselves but manifest no real essential unity. E. Benton, for example, suggests that we ought to understand vitalism as "the belief that forces, properties, powers, or 'principles' which are neither physical nor chemical are at work in, or are possessed by living organisms, and that any explanation of the distinctive features of living organisms which did not make reference to such properties, forces, powers or principles would be incomplete."[12] Nevertheless, Benton contends, it would be "a mistake to regard vitalism as one doctrine, or set of doctrines," for often "the issues which divided . . . vitalists from one another can be shown to be more significant than those which divided vitalists from non-vitalists" (17).[13] Historians of science have also urged us to be wary of too quickly categorizing all critiques of mechanism and materialism as vitalist and have proposed further categories, such as "teleo-mechanism."[14] Finally, to introduce yet another twist, scholars from the humanities—particularly those involved in literary criticism and art history—tend to think of vitalism not primarily as a medical or biological theory but rather as a nineteenth-century philosophical and artistic movement centered around the French philosopher Henri Bergson.

I have introduced the phrase "experimental vitalism" in part to avoid both the ambiguity of, and the automatic gatekeeping function associated with, the more generic term "vitalism"; "experimental vitalism," in other words, denotes something much more historically and conceptually specific than the more general term "vitalism." "Experimental vitalism" implies, for example, a specific history, for "experimental" generally refers not to any and all attempts to interrogate nature by means of objects or protocols, but rather to that much more specific set of social and technical practices—the so-called experimental life—that emerged in the late seventeenth century around figures such as Robert Boyle and Isaac Newton and institutions such as the Royal Society. Where an older generation of scholars tended to attribute this scientific revolution of the seventeenth century to courageous decisions on the part of individual natural philosophers to turn away from the authority of ancient authors and toward experiments, more recent historians of science have documented the extent to which the emergence of experimental science depended on the construction of complicated social systems. These social systems certainly encouraged natural philosophers to conduct experiments, but also, and equally importantly, they enabled experimentation to become a *collective* endeavor, regulating the style of communications between researchers, the mode in which disagreements would be conducted, and the criteria by means of which valid and invalid "witnesses" of experiments could be distinguished.[15] As I document in this book, experimental vitalism also assumes this collective sense of experiment, for experimental vitalists too orient their claims, concepts, and theories toward groups and social constellations. As a consequence, though, no matter how much the *content* of theoretical claims made by experimental vitalists might seem to echo those of earlier authors, experimental vitalism itself became possible only after the late seventeenth century.

In emphasizing an experimental mode of vitalism, I am distinguishing it from "theoretical" vitalism, which is, I suggest, generally what scholars have meant when they have discussed vitalism. The difference between these two modes of vitalism hinges on the temporal relationship between experimentation and theoretical concepts. Experimental vitalists begin with a *sense* that life cannot be fully explained by current scientific concepts and assumptions and then develop experiments in order to provoke new questions and concepts about life and living beings. Theoretical vitalists, by contrast, begin with a theory about life—that, for example, there is an immaterial "principle of life" or "life-forces"—and then marshal experiments already performed (often by others) as evidence of the theory. The molecular biologist François Jacob makes the same point from a slightly different perspective, noting that eighteenth- and nineteenth-century theoretical vitalists did not *design* experiments in order to prove vitalist claims but rather sought out earlier experiments or observations as proof for their claims; as a consequence, while theoretical vitalists do not necessarily "produc[e] fewer observations than mechanists . . . their observations were very rarely made *because* of vitalism or in order to demonstrate a vital force. Vitalism generally came into the picture *after* observation, as an aid not to investigation but to interpretation."[16] For experimental vitalists, by contrast, experimentation is a means for making more complex and nuanced our understanding of the nature of living beings' potentials.

The career of modernist vitalist Hans Driesch exemplifies the movement from experimental to theoretical vitalism. Driesch began his career as a zoologist working under the Darwinian embryologist Ernst Haeckel, but in the wake of a series of vexing experiments that he conducted on sea urchin embryos in the 1880s and 1890s, he began to suspect that there must be a nonmaterial "life force" that regulated and supervised embryological development. However, when Driesch was unable to replicate these embryological anomalies in other organisms, he essentially gave up experimentation and spent the rest of his career working out the logical structure and implications of the vitalism that he claimed had been definitely established by his earlier series of experiments with sea urchins.[17] Driesch did not begin his scientific career as an experimental vitalist but rather became one because his experiments with sea urchin embryos encouraged him to produce new questions and concepts about living matter and embryological development. He did not feel such questions could be well posed and framed if one assumed at the outset that all movements of matter could be explained mechanistically. Yet he was not able to remain an experimental vitalist for long, opting instead to settle for a set of dogmatic a priori principles about life.

In describing experiments as mechanisms experimental vitalists use to produce new questions and concepts about life, I am drawing on a particular approach to understanding the nature and role of experiments in science, one associated with historians of science such as Gaston Bachelard, Georges Canguilhem, Ludwik Fleck, and Hans-Jörg Rheinberger, and with literary and cultural critics such as Richard Doyle and Elizabeth Wilson. It may seem strange to describe experiments as mechanisms for enabling *new* questions and concepts, for the popular view of experiments is that they answer existing questions either by confirming or invalidating concepts or theories that have already been constituted.[18] However, for the authors in the tradition of science studies from which I draw inspiration, this understanding of experiments as oriented toward answers, confirmation, and invalidation misses the forest for the trees, confusing a local function of experiments with their more general function within a particular science. Experiments, of course, often do provide answers to particular questions, and sometimes function to consolidate or invalidate existing theories or concepts.[19] However, as Rheinberger notes, "a research system [as a whole] is organized in such a way that *the production of differences* becomes the organizing principle of its reproduction."[20] Thus, while any particular experiment may aim at specific answers and conceptual clarification, experimental *systems* persist only to the extent that they create *new* questions and concepts. I consider an example of such an experimental system in chapter 2, describing a series of vitalist experiments with animal temperature undertaken by the eighteenth-century surgeon and experimentalist John Hunter. Though Hunter began with the specific goal of establishing whether fish and snakes could remain alive after being frozen, as "had been confidently asserted" by earlier natural philosophers, his quick disproof of this claim proved to be simply the start of a series of experiments in which he froze and then thawed plants, animals, animal parts, and eggs, a process that in turn led him to invent the concept of "simple life" and to refine existing concepts of "suspended animation."[21]

This tradition of understanding experiments as mechanisms for producing differences also grounds the understanding of the links among science, medicine, and literature upon which this book is premised. I have found particularly useful historian of science Georges Canguilhem's description of medicine and the life sciences as vitalist ways of being-in-the-world—that is, as "attempts by life to understand life."[22] For Canguilhem, medicine and the life sciences—and, by extension, the history and philosophy of medicine and the life sciences—should not be understood simply as observations about a specific "object" (i.e., living beings). In addition, insofar as these fields of research constitute activities undertaken by communities of living beings, they must be understood as a reflexive return of life upon itself—that is, a method by means of which living beings seek to alter, by means of understanding, the nature of living. Experimental engagements with life by living beings should thus be understood less in terms of "objective observers" and "isolated objects" than as ways particular kinds of living beings—for example, humans and the other living beings to which they bind themselves through experiments—end up creating new experiences, concepts, and objects of knowledge by means of reflexive procedures. From this perspective, investigations into the ontology of life (that is, the kind of being of life) always requires reconceptualizations of epistemology (the conditions under which knowledge—in this case, of life—is possible).[23]

THE POLITICS OF VITALISM

My emphasis on the production of differences as characteristic of experimental vitalism bears on a topic that is frequently linked to discussions of vitalism, experimental or otherwise—namely, the question of the politics of vitalism. The fact that it is difficult to discuss vitalism without also discussing politics is in large part a function of the fact that in both its Romantic and modernist moments, vitalism was frequently tied—occasionally even by its advocates—to conservative, repressive, and violent political positions. Romantic-era surgeon John Abernethy, for example, explicitly linked his brand of vitalism to support of the British crown and encouraged the political persecution of advocates of alternative "materialist" philosophies of life, while the conservative political leanings of modernist vitalist fellow traveler Jakob von Uexküll encouraged the German novelist Thomas Mann to conclude that "an interest in biological questions, even of the new, less mechanistic, anti-Darwinian sort, disposes one to be conservative and rigid in political manners."[24] Nor is this link between vitalism and political conservatism necessarily restricted to these earlier moments: as Melinda Cooper notes in a survey of recent vitalist work in geology and computer science, "the philosophy of *life as such* runs the risk of celebrating *life as it is,*" and modern-day vitalisms often come "dangerously close to equating the evolution of life with that of capital."[25]

No doubt in part as a response to these perceived dangers, recent historical and literary critical accounts have tended to emphasize the political implications of vitalist theories. Thus, where earlier generations of scholars tended to write histories of vitalism in whiggish terms (by showing how vitalism had been either invalidated or revealed the wisdom of the past), more recent historians and literary critics have stressed ways in which the purportedly scientific claims of vitalists can be aligned with particular political and professional battles occurring within national polities. Abernethy's vitalism, for example, has been interpreted both as a response to tensions engendered by Britain's war with France and as an attempt to protect the reputation of the newly formed Royal College of Surgeons, while Driesch's vitalism and Uexküll's quasi-vitalism have been described as expressions of a German desire for unity and wholeness that could not be achieved in political terms prior to the rise of Hitler.[26] Within these accounts, scientific debates about the nature of life are decoded as battles for control over social and political institutions.

I have some sympathy with this approach, and I draw on work undertaken from such a perspective throughout this book. At the same time, though, my fundamental goal is not to demonstrate "the politics of life," if by that one understands an exposure of the vested political or professional interests involved in the production of knowledge and texts deemed to be scientific or literary. This is in part because I find the national frame most of these analyses assume unnecessarily limiting. As I note in chapter 4, which focuses on Romantic-era theories of digestion, while conceptions of vital processes were often linked in some way to cultural anxieties, these anxieties were frequently tied to *global*, rather than strictly national, concerns. Moreover, an attention to vital processes, rather than limiting debate and policing borders, served most of the authors that I consider here as a means for encouraging perceptions of *increasing* social complexity. I am also a bit suspicious of the narrative structure of many accounts of "the politics of vitalism," for they seem to assume precisely that (questionable) narrative of progress that I outlined above in my account of the gatekeeping function of the term "vitalism." By presenting vitalism as simply an emotional outlet for cultural anxieties produced by the traumas of modernity, we risk obscuring the commitment to change and transformation that I argue is central to experimental vitalism. Finally, I am not certain that experimental vitalism is linked—either structurally or historically—to *any* particular politics. Some of the authors I consider in this book simply seemed to have avoided an explicit engagement with politics entirely: this was the case, for example, for the eighteenth-century surgeon and ur-experimental vitalist John Hunter. Other experimental vitalists, such as the poet P. B. Shelley, were more explicit about their political commitments, yet these commitments were far more radical—or at least liberal—than one would expect from the usual narratives of vitalism as a conservative reaction to modernity. This difficulty in pinning down the politics of experimental vitalism is no doubt in large part a function of its intrinsically liminal status: less a theory than a practice of invention, it is hard, if not impossible, to align it with a specific interest group. While—as I discuss in the book's conclusion—this interest in invention does have a structural relationship to the capitalist practices of innovation and obsolescence characteristic of modernity, it cannot be taken as identical to these.

OVERVIEW OF THE BOOK

I begin with a discussion of the term "experimental." As I noted above, art first began to be called experimental in the Romantic era, and there are now long traditions of "experimental" work in poetry, music, film, video, and other arts. Yet it is not, for all that, always clear what these invocations of experiment and experimentalism mean in the context of art, and whether such descriptions have much, if anything, to do with *scientific* experimentation. I attempt to clarify these issues in part by drawing on several key discussions of scientific experiments developed within the field of science studies, and in part by reminding us of John Cage's and Theodor Adorno's especially astute claims about artistic experimentation. Combining these two traditions of reflection on experimentation illuminates the fact that the emergence of a singular conception of "art" in the Romantic era was facilitated by artists' appropriation of the term "experiment" from the sciences. I outline one example of such appropriation through a case study of Wordsworth's and Coleridge's self-described experiments in *Lyrical Ballads*, and I conclude chapter 1 with a more general reflection on the ways the concept of artistic "experimentation" has transformed our understanding of the temporality of art, focusing especially on the question of whether the movement of experiment into the arts implies that art (like the sciences) demonstrates "progress" (or its contrary, "decline").

Chapters 2 through 6 analyze crossing points between Romantic art and science by focusing on what we might call "vital sensations"—that is, sensations or experiences that both emphasize the opacity of life and encourage an experimental approach to social and political forms of life. These five chapters collectively trace a sort of vital spiral. I begin in chapter 2 with a state that hovers at the edge of vitality (suspended animation); in chapter 3, I consider experiences of orientation and disorientation within an environment; I focus in chapter 4 on experimental interest in digestion, which was often understood as the question of how living matter drawn from one's surroundings could be incorporated into one's own body. Chapter 5 considers what we might describe as the converse of digestive assimilation—namely, an awareness of becoming oneself a medium for something else—and I conclude in chapter 6 with the human experience of being solicited by a nonhuman form of life.

While each chapter focuses on a concept—suspended animation, disorientation, digestion, media, and plant vitality—indigenous to the Romantic era, and while my analyses focus almost exclusively on Romantic-era texts, these concepts and analyses are nevertheless intended to establish relationships of resonance with aspects of our contemporary interest in vitality. My discussion of Romantic-era suspended animation, for example, helps us think further about the twentieth- and twenty-first-century fascination with suspended animation and with what Giorgio Agamben has described as "bare life," as well as with what Gilles Deleuze calls "non-organic life." The chapter emphasizes that there are, beginning in the Romantic era, two quite different—and in fundamental ways opposed—traditions of thinking about both the sites and the potentials of vital suspension, and we thus ought to read the differences between (for example) concepts of bare life and nonorganic life as new attempts to work out the conflict between these two traditions in our own contemporary context. To take another example, my chapters on digestion and media both explore the Romantic-era origins of a kind of "materialist paranoid logic" that has been replicated, in significant ways, in both contemporary food activism (which often seeks to change patterns of food consumption by producing nausea around the facts of food production) and in contemporary evolutionary psychological and neurological accounts of human consciousness and actions (which seek to convince us that our most seemingly selfless acts and even our sense of free will are simply media for the operation of evolutionary forces). The Romantics can help us think about these phenomena in part because they were often far more willing than our contemporaries to think these topics through to the end. In their discussions of digestion (chapter 4), for example, it was precisely the *coarseness* of the materialism of authors such as novelist and political theorist William Godwin and poet and antislavery advocate Coleridge that forced them to take literally the question of the relationship between digestion and thought, while it was the truly cosmological aspirations of Romantic-era attempts to explain media and mediation (chapter 5) that allow us to see the thorny problems that result if one simultaneously assumes that one element of nature (the natural sciences) can explain all aspects of human biological and intellectual existence but neglects to account, in that explanation, for the sciences that are doing this explaining.

While discussions of these resonances between Romantic-era and contemporary experimental vitalism appear primarily by way of asides and in chapter conclusions, I end the book with a brief discussion of contemporary criticisms of the "neo-vitalisms" of our own era. Focusing on two trenchant critiques of contemporary appeals to "life"—on the one hand, the charge that such appeals inevitably obscure importance differences between different kinds of living beings; on the other hand, the suggestion that contemporary fascination with "life itself" should be decoded as in fact a fascination with capital—I seek to draw out the implications of my account of Romantic experimental vitalism for contemporary biopolitics. I suggest that what both criticisms reveal is the need for greater understanding of the relationship between *experimentation* and life. This is, admittedly, a hard link to think, but I hope my book convinces its readers that beginning with the Romantic era would not be bad a place to start.

Romanticism, Art, and Experiments

It is best for the *Philosophers* of this Age to imitate the *Antients* as
their *Children*; to have their Blood derived down to them; but to add
a new Complexion, and Life of their own: While those, that endeavour to come near them in every Line, and Feature, may rather be
called their dead *Pictures* or *Statues*, than their *genuine Off-spring*.

Thomas Sprat, The History of the Royal Society of London, *51*

Experimental, in the legitimate sense, means nothing other than
art's self-conscious power of resistance against what is conventionally
forced upon it from the outside, by consensus.

Theodor Adorno, "Difficulties," 651

Can art really ever engage in *experiments*? Consider the following two quite
different—even contradictory—assertions about the relationship between experiments and art. In "The Aporias of the Avant-Garde," media theorist Hans Magnus Enzensberger contended that insofar as the term "experiment" "designates
a scientific procedure for the verification of theories or hypotheses through me-
thodical observation of natural phenomena," works of art *cannot* truly engage in
experiments: "the experiment is a procedure for bringing about scientific insights,
not for bringing about art."[1] Enzensberger does not deny that artists and critics
often describe works as experiments, and he archly notes that "the obligatory
modifier is *bold*, but the choice of ennobling epithet *courageous* is also permitted"
(35). However, he contends that applying this term to the arts is simply a kind of
"bluff," a way of "flirting with scientific methods and their demands" without
"ever seriously getting involved" with these latter while at the same time seeking
to adopt the "moral immunity" associated with the sciences.

Frankfurt School critic Theodor Adorno, by contrast, argued that "art is now
scarcely possible unless it does experiment."[2] Adorno defended this claim with

the assertion that "the idea of the experimental" is one means by which twentieth-century artists have resisted the demands of a capitalist culture industry that has come to control and administer almost every aspect of the social world. The idea of experiment provides art with resources for resistance in part by displacing what had become an increasingly nostalgic, and hence problematic, "image of the artist's unconscious organic labor" with an image of creation drawn from science: namely, the image of "conscious control over materials." The idea of experiment also enabled artists and audiences to appropriate and transform into a form of activity the passivity that has become part and parcel of modern life: rather than simply passively accepting an administered, and hence completely alien, social world, the experimental work of art allows individuals to "integrat[e what is alien] into subjectivity's own undertaking as an element of the process of production [of the work of art]." It is difficult to imagine a starker contrast: Enzensberger denies that art can ever *really* be experimental; Adorno insists not only that modern art *is* experimental but that art cannot persist if it ceases to be so. Which side to take?

It seems fair to say that subsequent literary, art, and media criticism has generally been less interested in taking sides than in hoping for a possible happy middle ground between the two positions. Thus, while "experimental" often seems to play a descriptive or classificatory role in anthologies, essays, and criticism devoted to various "experimental" arts—"experimental literature" and "experimental film," for example, presumably mark out kinds of literature and film that differ from other modes or genres of those arts—there is often little or no engagement with the term "experimental" or, alternatively, an open admission that the denotation of the term is a bit hazy.[3] The primary function of the term in contemporary criticism seems to be less denotation than approbation: that is, describing a piece of writing, a film, or a musical composition as experimental is less a means of identifying specific qualities than of encouraging renewed inspection and appreciation of that work. Works described and valued as experimental frequently question the principles of earlier works of art or in some other way present viewers, listeners, or readers with interpretive difficulties, and it is almost invariably these aspects that are valued in such works; experimental art, in other words, is good *because* it opens new horizons or possibilities for art. Perhaps not surprisingly, then, "experimental" is thus often used in literary and art criticism as a synonym for other terms that connote transgression or the expansion of boundaries, such as "radical" or "avant-garde." The implication seems to be that even if Enzensberger is correct that the arts cannot *truly* experiment, they can at any rate engage in important activities such as risk-taking and transgression.[4]

One of the problems inherent in employing "experimental" as primarily an evaluative term is that such usage tends to obscure the actual history of the term experiment in the arts. It is common to point to Émile Zola's 1880 essay "The Experimental Novel" as the point at which the terminology of experiment explicitly made its way over into the arts, a chronology that reinforces a sense of intrinsic connection between the experimental and the avant-garde.[5] Yet in fact literature first became "experimental" in the Romantic era. William Wordsworth and Samuel Taylor Coleridge, for example, explicitly described many of the poems in their 1798 volume *Lyrical Ballads* as "experiments" that were to help readers "ascertain how far the language of conversation in the middle and lower classes of society is adapted to the purposes of poetic pleasure," and this terminology was taken up by early reviewers of the volume, many of whom claimed to assess the success or failure of these experiments.[6] The terminology of the "experiment" was subsequently adopted by, or applied to, other artistic movements, including the late nineteenth-century "naturalist" novel (especially as developed by Zola); the series of late-nineteenth- and early-twentieth-century art and anti-art movements collectively grouped under the rubric of the avant-garde; Gertrude Stein's writing; abstract and minimalist strands of music practice (some of which explicitly positioned themselves against the avant-garde); the Fluxus and Oulipo projects; and various forms of contemporary film, video, and digital imaging practices.[7] While some of these post-Romantic artists were explicitly interested in the Romantics' inaugural interest in (artistic) experiments, this was not always the case. As a consequence, tempting as it is to assume that there must be *some* family resemblance among the various "experimental" works and movements, it is—to put it bluntly—hard to find much of a unified genealogy or common thread in the rather heterogeneous list of works of art to which this term has been applied.

What hinders the effort to determine in what ways—and, more fundamentally, *if*—art can be experimental are not only a tendency to confuse experimental with other terms such as radical and the avant-garde but also an ambiguity or uncertainty around the very term "experiment." Drawing on just the invocations of experiment I quoted above, we can already see a rather wide diversity of meanings: for Wordsworth and Coleridge, experiments are bound up with measuring; for Enzensberger, with "scientific methods and their demands"; for Adorno, with an image of creation and a corresponding relationship of activity to passivity. Each of these references to experiments is an engagement with an image of science and an image of the scientific experiment, yet it is not clear that these authors each have in mind the *same* image of science and its experiments. This in turn suggests that the question of whether or not art can really experiment begs an anterior question: namely, what does it mean to experiment in the sciences?

This chapter takes this question of what it means to experiment in the sciences as its starting point by outlining, in broad strokes, three different twentieth-century approaches to scientific experiments. Such a summary helps us to recognize that, first, there is no unified understanding of "the scientific experiment" (or, for that matter, "the scientific method") and, second, that differing accounts of scientific experiments are based on differing assessments of whether experiments are best understood in terms of epistemology, sociology, or ontology. The second part of the chapter focuses on two important discussions of *artistic* experiments, taking up the composer John Cage's account of his own experimental practice and returning to Adorno's discussion of the role of experimentalism in twentieth-century art. The third and fourth sections employ Wordsworth and Coleridge's *Lyrical Ballads* as a bridge for bringing together the discussions of scientific experiments, outlined in part one, with the accounts of artistic experiments discussed in the second part. I stress, though, that the "experiments" of *Lyrical Ballads* can be fully understood *neither* solely in terms of accounts of scientific experiments nor—and this is perhaps more surprising—solely in terms of Cage's or Adorno's accounts of artistic experiments. *Lyrical Ballads* requires us, instead, to hybridize these two contemporary modes of reflection upon experiments and in this way understand Wordsworth and Coleridge's volume of poems as part of the emergence of a new concept of *artistic* experimentation. In the chapter's final two sections, I discuss the implications of this new concept of artistic experimentation for our understanding of both the nature and the temporality of art. I argue that the concept of artistic experiment both unified and fractured the arts: unified, in the sense that what had been the separate "arts" became the singular concept of "Art," but also fractured, in the sense that the connections among past, present, and future works of art could no longer be understood in terms of a linear series—it has for a long time been difficult to understand art as something that progresses or declines—but must instead be understood through the form of an increasingly ramified network.

FROM EPISTEMOLOGY TO ONTOGENETIC SYSTEMS: SCIENCE STUDIES AND SCIENTIFIC EXPERIMENTS

As I document later in this chapter, musicians and theorists such as Cage and Adorno, as well as recent scholars of Romanticism, have generally been attentive to the need to define terms such as "experiment" and "experimentalism" before using them to explain aspects of literature and other arts. Yet as we will also see, even when these terms are explicitly defined, definitions tend to rely on commonsense understandings of scientific experiments (most frequently, an understanding of the scientific experiment as the testing of a hypothesis). They tend as well to assume that the concept of experiment is "proper" to the modern sciences, and thus purported artistic experiments should be judged according to whether they fulfill criteria and standards set by the sciences; if they do not, they are not *really* experiments. Both of these tendencies are problematic, for scholars working in the loosely defined field of "science studies" have revealed that scientific experiments can be understood in multiple ways; as a consequence, there is no one set of criteria or standards according to which one could assess if artistic experiments live up to their peers in the sciences. We can distinguish, in fact, three quite different science studies approaches to scientific experiments, each developed in a relatively distinct period: an *epistemological* approach that emerged in the early twentieth century and focuses on the contribution of experiments to true accounts of the world; a *sociological* approach that emerged in the 1970s and emphasizes the role of experiments in both facilitating and resolving social conflicts; and a relatively recent *ontogenetic* approach that focuses on the roles of experiments in creating new assemblages of inanimate matter, animate matter, and social groups.[8]

Bookended by Karl Popper's *The Logic of Scientific Discovery* (1934) and Thomas Kuhn's *The Structure of Scientific Revolutions* (1962), the *epistemological approach* to scientific experiments is likely already familiar to many readers. It is to this approach, for example, that we owe the lay understanding of the scientific experiment as something that cannot actually verify true theories but rather only has the power to falsify inaccurate theories (the so-called falsificationist understanding of experiments). The key to the epistemological approach is that it considers experiments primarily in terms of their contribution to true descriptions of the world. This approach is frequently underwritten by the belief that the fundamental activity of science is to develop *theories* about the natural world; from this perspective, experiments are a means by which to distinguish more accurate from less accurate theories. This in turn implies both a clear separation and hierarchy of activities (and, often, people) within science, such as the distinction between "theory development" and "experimental testing." In Karl Popper's classic account of experiments, theorists are fundamentally in charge of science, for they develop the clear theories experimentalists are then asked to test by means of well-defined experiments: "the theoretician puts certain definite questions to the experimenter, and the latter, by his experiments, tries to elicit a decisive answer to these questions, and to no others. All other questions he tries hard to exclude" (89). Experimental results are relayed back to the theorist, who determines whether the theory has been falsified, or is worth further pursuing. In the actual history and practice of science, the same individual may serve as both theorist and experimentalist, but Popper's account seeks to clarify the hierarchal relationship of these tasks.

I have included Thomas Kuhn's well-known work under the epistemological approach to emphasize that it is possible to take an interest in sociological aspects of scientific communities while nevertheless retaining a fundamentally epistemological approach to experiments. Kuhn is best known for his account of the history of the sciences as a series of shifts between discontinuous and incommensurable "paradigms," with a paradigm understood as both a theoretical and a practical orientation toward problem-identification and -solving that helps unify a specific scientific community (e.g., "physicists"). As Isabelle Stengers stresses, a Kuhnian paradigm is not just a

> vision of the world but [also] a *way of doing*, a way not only of judging phenomena, of giving them a theoretical signification, but also of *intervening*, of submitting them to unexpected [experimental] stagings, of exploiting the slightest implied consequence or effect in order to create a new experimentation situation.[9]

As the theoretical and practical means for unifying a specific community of scientists, a paradigm necessarily encourages certain forms of blindness that have no place in Popper's account of the rational theorist who passes on precise questions to a clear-eyed experimentalist. Kuhn's concept of the paradigm stresses that, even if we accept Popper's hierarchical distinction between theory and experiment, a theorist will nevertheless be unable to consider certain kinds of questions about the world and still remain within his or her paradigm, and, by the same token, experimentalists will also (and necessarily) tend to ignore experimental results that do not conform to an existing paradigm. Kuhn's account suggests, moreover, that the shift from one paradigm to another depends upon factors more sociological than epistemological in nature. Kuhn notes, for example, that even extraordinarily capacious scientists such as Joseph Priestley and Lord Kelvin were unable to make the shift to the paradigms that ended up supplanting their own research projects: "Priestley never accepted the oxygen theory, nor Lord Kelvin the electromagnetic theory."[10] Kuhn sees in this fact a lesson articulated by the physicist Max Planck: "a new scientific truth does not triumph by convincing its opponents and making them see the light, but rather because its opponents eventually die, and a new generation grows up that is familiar with it."[11]

Yet despite Kuhn's interest in these sociological dimensions of science and experimentation, his account is nevertheless fundamentally, like Popper's, an account of *how* science is able to develop true descriptions of the world.[12] Though

Kuhn stresses forms of theoretical and experimental blindness that are absent in Popper's account, these limitations are not, for Kuhn, anti- or a-scientific factors but rather means by which science is able to produce truth. "Lifelong resistance" to a paradigm on the part of a scientist, Kuhn writes, "is not a violation of scientific standards but an index to the nature of scientific research itself"; without such resistance, scientists could not cohere into communities of researchers and thus would not be able to produce true accounts of the world. Nor are claims about the incommensurability between paradigms evidence that Kuhn believed that scientific truths were "socially constructed"; rather, incommensurability simply meant for Kuhn that science produces many true accounts of the world that cannot be coordinated into one unified account.[13]

We can distinguish this epistemological interest in how experiments lead to true accounts of the world from a second, *sociological* approach to scientific experiments. Initially developed in the 1970s and 1980s by historians and sociologists such as David Bloor, Steven Shapin, Simon Schaffer, and Peter Galison, this approach emphasizes how experiments facilitate or engender social distinctions, such as that between "scientific experts" and "laypeople."[14] In order to distinguish its focus from the epistemological question of how experiments contribute to true accounts of the world, many proponents of the sociological approach advocate a "symmetry" principle. The symmetry principle proposes that when historians and sociologists discuss conflicts between rival accounts of the natural world, they should not take either side of the conflict but should instead treat each side equally (i.e., symmetrically). If a historian, for example, wishes to discuss the Romantic-era debate between Priestley and Antoine Laurent Lavoisier about the proper way to describe the chemical composition of air, the goal ought not to be to explain why Lavoisier's "true" account of the world—that is, the account closest to the chemical view we now accept—defeated Priestley's erroneous account. The symmetry principle brackets this question of true and false scientific claims about the world in order to focus attention on a different realm of dynamics: the political and economic conflicts between groups fighting for the prestige (and often funding) associated with terms such as "science" and "truth." So, for example, in *Leviathan and the Air-Pump*—a text to which I shall return below—Shapin and Schaffer detail how the seemingly value-free "experimental life" associated with the Royal Society in late seventeenth-century England depended upon social and political strategies that created distinctions between trustworthy and untrustworthy witnesses of experiments and distinguished between valid and invalid prose styles of communicating scientific truths. The purpose of Shapin and Schaffer's account is not to explain why, for example, Robert Boyle's commitment to experiments enabled him to develop a true account of the existence of vacuums while his opponent Thomas Hobbes's disinterest in experimentation prevented him from seeing the truth. Rather, Shapin and Schaffer emphasize that our modern understanding of experimentation as essential to the practice of science was itself an outcome of a fundamentally political battle between Boyle and Hobbes over the very nature of what would count as "knowledge." For advocates of the sociological approach, then, experiments are to be understood first and foremost as "persuasion technologies," and the important question is that of the political and economic ends for which people are being persuaded by experiments.[15]

The third approach, which I describe as *ontogenetic* accounts of experiments, emerged in the late 1980s in the work of authors such as Bruno Latour, Donna Haraway, and Hans-Jörg Rheinberger. Where the epistemological approach aims to explain how experiments contribute to true accounts of the world, and the sociological approach seeks to account for the role of experiments as technologies of persuasion in political and economic battles, the ontogenetic approach focuses on ways in which experiments bring *new* entities and assemblages into being. In his now classic account of Louis Pasteur's endeavors to locate the cause of anthrax in cattle, for example, Latour argues that Pasteur did not so much discover as help *create* the "anthrax bacillus."[16] Latour's counterintuitive claim that the anthrax microbe came into being around 1870 emphasizes that "anthrax bacillus" is something that only appears in the space of the laboratory (even if one then attributes past cattle deaths to this newly isolated biological entity). Latour contends that the bacillus could appear in the laboratory *as* a cattle-causing disease agent only because Pasteur forged a series of "alliances" with different human and nonhuman entities: an alliance with an unknown microorganism (which Pasteur facilitated by developing laboratory biological media within which the bacillus would grow); an alliance with hygenicists, who had their own, alternate explanations of the causes of cattle disease; an alliance with farmers, who were not initially eager to allow their farms to become "part" of the laboratory; and so on. Though many of these alliances might be interpreted in terms of the sociological approach to experiments outlined above, Latour distinguishes his approach by his insistence that science always involves constructing novel alliances between human and *nonhuman* "actants," resulting in what he calls a "parliament of things."[17]

Where Latour's concepts of actants and a parliament of things emphasize the priority of ontogenetic over sociological concerns, Hans-Jörg Rheinberger's concept of "epistemic things" stresses the priority of ontogenetic over (narrowly) epistemological questions. *Pace* Popper, Rheinberger asserts that scientists do not judge an experiment "successful" when it answers a single, well-formulated theoretical question; rather, they judge an experiment successful when it leads to further experiments. Scholars of science thus risk misunderstanding the real function of experiments if they focus on the relationship of a singular experiment to a specific theoretical claim; their object of study ought instead to be the way in which a *system* of experiments produces new experiments, objects of research, and concepts. Rheinberger claims that the real work of a successful experimental system is less the testing of previously generated theories than the project of enabling the emergence of new "epistemic things"—new kinds of entities or concepts, such as Pasteur's "anthrax bacillus"—within the experimental apparatus. Drawing on Jacques Derrida's concept of *différance*, Rheinberger contends that "a research system is organized in such a way that *the production of differences* becomes the organizing principle of its reproduction."[18]

The epistemological, sociological, and ontogenetic approaches to scientific experimentation I have outlined here are not necessarily disjunctive, and arguably some of the most interesting work in science studies results from hybridization of the central concerns of each approach.[19] However, in their usual instantiations, each kind of account tends to enforce a different hierarchy of considerations vis-à-vis explanations of experiments. The distinctions among these three approaches will help us identify what is stake in different accounts of artistic experiments, though developing a theory of experimental art will ultimately require that we bring together multiple elements of these approaches.

CAGE AND ADORNO ON ART AND EXPERIMENTS

We can make a first step toward a theory of experimental art by drawing on a distinction between two modes of artistic experimentation initially proposed by minimalist composer John Cage and further developed by Adorno in his account of the role of experiment in modern art. In his writings on his work as a composer, Cage distinguished between two senses in which a musical composition could be experimental. In the case of many works, he claimed, the process of experimentation occurs during the *composition* of a musical work, and thus that which is presented to the public is simply the result of an experiment, but not the experiment itself: "the experiments that had been made had taken place prior to the finished work, just as sketches are made before paintings and rehearsals precede performances."[20] Cage was more interested, however, in experiments that occur during the *performance* of the work; in these cases, the "results" of the experiment cannot be known before the actual performance, and the audience itself becomes part of the experiment. A well-known example of this kind of experiment is Cage's 4′ 33″, a musical composition in which performers remain silent during each of the three short movements of the piece (for a total of roughly four minutes and thirty-three seconds of silence). By combining the frame of musical performance with silent performers, Cage sought to enable the audience to adopt an experimental approach to their environment, listening to sounds—coughing, the muffled sound of automobiles driving by outside the performance hall, and so forth—that would otherwise be ignored as incidental or as noise.

Considered from the perspective of the science studies approaches to experiments that I outlined above, Cage's reflections seem closest in spirit and interests to the ontogenetic approach. For Cage, a musical work can be experimental not in the sense that it leads to "truths" about the world but rather in the sense that it enables the emergence of new kinds of "musical things." Phenomena such as coughing or the rustling of clothes become, as a consequence of the experimental frame of the performance, new musical sounds. The function of an "experiment" in music, Cage wrote, is to open "the doors of the music to the sounds that happen to be in the environment" (8). An experiment is thus for Cage not an "act to be later judged in terms of success or failure, but simply . . . an act the outcome of which is unknown" (13). While Cage's account has a sociological dimension— the kind of musical experimentation that he values happens only when an audience is assembled—this is a rather bare-bones sociology, as Cage is decidedly not interested in the social conflicts, allegiances, and rivalries that are the focus of the sociological approach to scientific experiments.

Adorno's reflections on experimental music, by contrast, link the ontogenetic dimension of the artistic experimentation Cage emphasized to a much more focused interest in the social conflicts that are revealed by practices of experiment. Adorno implicitly accepted Cage's distinction between two modes of artistic experimentation—a mode in which experimentation happens before a work is performed and a mode in which the performance is the experiment—but he plotted this distinction along a historical axis. Adorno stressed that there is a huge gulf between Cage's two kinds of artistic experimentation: while *we* can use the term "experimentation" to describe the efforts of, for example, ancient Greek sculptors to expand the techniques of their individual art, this has almost no relationship to the recent use of the term by artists themselves to describe their artworks as a whole. When we talk about artistic experimentation in the first sense, we mean a willing, conscious "test[ing of] unknown or unsanctioned technical procedures" that aims to expand the resources of a particular art, such as sculpture, poetry, or painting.[21] This use of "experimentation" is premised on the principle that a new work of art that employs a technique discovered through experiment is continuous with past instances of that art and that, as a consequence, the success or failure of experimentation can be judged in terms of whether a given art now has more resources with which subjects can express themselves. Yet when art becomes experimental in Cage's second sense—that is, when "the artistic subject employs methods whose objective results cannot be foreseen" (24) by either the artist or the audience—then the meaning of "experimental," and both the artist's and audience's comportments toward art, have shifted radically. Artistic experiments such as 4′ 33″ no longer aimed at expanding the powers of self-expression within a given art but instead were means by which subjects sought to submit themselves to an "object" (for example, the ambient sounds of an environment).

Adorno suggested that artistic experimentation in this second sense was possible only after "art" had become understood as a field of human endeavor distinct from other realms of human action, such as religion, work, and politics. Adorno stressed that art "did not become a unified whole until a very late stage" of human history, and it achieved this unity only by taking on a contestatory role: in the face of a world that was becoming increasingly subordinated to capitalism, "art" became a field of human action capable of "denounc[ing] the particular essence of a *ratio* that pursues means rather than ends."[22] However, insofar as art contests instrumental rationality and capitalism, it is also bound to them; it bears a "constitutive relation . . . to what it itself is not" (326). As a consequence, none of the elements of modern art, including the formal procedures employed by artists, the themes depicted in works of art, or the nature and function of audiences, can be understood as invariants or constants but are instead always bound up with specific historical contexts. This is true as well for explicit invocations of "experimentation" by artists such as Cage; thus, one can only understand Cage's interest in expanding the realm of musical sounds in connection with the increasingly instrumentalized world of modern capitalism.

Though Cage understood this second mode of artistic experimentalism as a method that enabled music to *expand* its tools and field—such experimentalism, he claimed, "opened the doors of the music to the sounds that happen to be in the environment"—Adorno suggested that these new experimental techniques did not in fact increase the possibilities of music but rather introduced a series of ruptures into the history of twentieth-century music.[23] After Cage, Adorno claimed, it becomes difficult to hear pre-experimental musical compositions as anything but the desiccated past of music, rather than its vital present: "There is no going back. If, in contrast to the twelve-tone technique, the serial principle, and aleatory music, one simply sought to get a fresh grasp on subjective freedom, i.e., free atonality in the sense of Schoenberg's expectation, one would almost necessarily recapitulate to reaction."[24]

Adorno did not blame experimentalism for draining past art of its vitality but rather saw in artistic experimentalism testimony to the fact that conscious and deliberate "submission to the object" is the only way the subject can seek to regain some control over a world that has become increasingly administered by technologies and institutions of power. Adorno emphasized that when one attends a performance of $4'$ $33''$, one is *choosing* to submit oneself to an "object":

> The subject, conscious of the loss of power that it has suffered as a result of the technology unleashed by himself, raised this powerlessness to the level of a program and did so in response to an unconscious impulse to tame the threatening heteronomy by integrating it into subjectivity's own undertaking as an element of the process of production.[25]

Where Cage saw in his second mode of experimentalism a Buddhist-inspired means for becoming free of the self, Adorno saw a last-ditch effort by artists and audiences to retain an admittedly paradoxical form of subjective autonomy: *one chooses* to submit to what is nonsubjective rather than passively accepting such submission. However, in a world in which even minimalist artistic experiments can become commodified—"ambient music," for example, is now packaged and sold as soothing background sounds for commercial environments—further experimentalism remains the only horizon of possibility for art: "art is now scarcely possible unless it does experiment" (37).

If Cage's focus on the new musical entities enabled by experimental music seems most consonant with the concerns of the ontogenetic approach to scientific experiments, Adorno's approach seems uneasily suspended between the sociological and the epistemological approaches. As in the case of authors such as Shapin and Schaffer, Adorno was fundamentally interested in what artistic experimentation reveals about the nature of social conflict, though Adorno's understanding of social conflict is cast in much more abstract form than in the close historical studies of Shapin and Schaffer; where the latter are interested in the role of scientific experiments in a very particular (even if especially significant) late-seventeenth-century battle between Thomas Hobbes and members of the Royal Society, Adorno was interested in the place of artistic experimentation vis-à-vis a much more general historical tendency toward the "homogenization" of all aspects of life. At the same time, though, Adorno does not employ anything like a "symmetry principle" in his analysis of artistic experiments. For him, the truth of the artistic experiment should not be bracketed, but rather must be understood from within. He was, as a consequence, far more willing than Shapin and Schaffer to *judge* and evaluate experiments. Cage's experimentalism, for example, was for Adorno a central event in twentieth-century music and art, but it was also an event he approached with some degree of suspicion. As one of a series of musical practices that allow "externally imposed rules" to determine what shall happen in a musical composition and thus afford both composer and audience "relief" (*Entlastung*) from the need to process each aspect of music subjectively, Cage's experimentalism worried Adorno, and he explicitly urged his readers to "hold fast to the ideal" of what he called "informal music" rather than placing all their hopes in "the techniques of relief."[26]

EXPERIMENTALISM AND SCIENCE
IN THE ERA OF *LYRICAL BALLADS*

Cage's distinction between two modes of artistic experimentation and Adorno's historicization and sociologization of that distinction emphasize several points essential to a theory of artistic experimentation. These include the question of whether experimentalism depends upon a subjective comportment or instead allows freedom from subjectivity, the importance of collectives for artistic experimentation (e.g., the role of the audience for Cage's 4' 33"), the relationship between experiment and tradition, and the impact of artistic experimentalism on the concept of art itself. Yet the example of Wordsworth and Coleridge's *Lyrical Ballads* suggests that Cage's and Adorno's accounts of experimentalism must also be supplemented with several further considerations. The experiments of *Lyrical Ballads* complicate, for example, Cage's distinction between two kinds of experimental art. On the one hand, early reviewers of *Lyrical Ballads* generally described the collection as the *result* of experiments previously undertaken by the volume's authors, a reading that aligns the volume with Cage's first category of art experiments. Yet Wordsworth and Coleridge's "Advertisement" for their volume suggested that the poems were to be understood as technologies for experiments that would take place only in those "performances" of the poems that each reader makes for him- or herself, which aligns the volume with Cage's second category of experimental art.

   Lyrical Ballads also complicates Adorno's historical account of the dialectical transformation of the experimental comportment in art. At first glance, *Lyrical Ballads* seems like an example of Adorno's first mode of artistic experimentalism: "test[ing of] unknown or unsanctioned technical procedures" that aimed to expand the field of a particular art (in this case, poetry). Insofar as ballads were traditionally understood as a low form of song, for example, simply writing down ballads can be seen as an attempt to expand the field of poetry, while the wager of the volume's title was that a new genre of poetry could be created by binding together two apparently opposed kinds of voice (the collective voice of the ballad and the singular voice of the lyric). The varied subject matter of the poems in the volume, which ranged from the supernatural events described in *The Rime of the Ancyent Marinere* to the mundane conversation between father and son in "Anecdote for Fathers," also seems like an attempt to capture a greater field for poetry. (Even contemporary reviewers who contended that Wordsworth and Coleridge had not succeeded in enlarging the field of poetic topics—Robert Southey, for example, described the subjects of some of the poems as too "bald in story" or "uninteresting" for verse—by the same token implicitly granted that the authors had *sought* to expand what would count as poetry.)[27] Yet *Lyrical Ballads* contested precisely that premise of artistic continuity that Adorno saw as definitional of pre-twentieth-century experimental approaches to art, for the "Advertisement" used the language of the experiment to attempt to introduce a violent break into poetic tradition. The volume aimed at a *reform* of taste, and Wordsworth and Coleridge anticipated that readers committed to a more traditional understanding of poetry would "have to struggle with feelings of strangeness and awkwardness" and would perhaps fail to find any poetry at all in the book.[28] For Wordsworth and Coleridge, in other words, the criterion of success was not whether their experimentation enabled an expansion of the existing field of verse, but rather whether these experiments enabled readers to reject a great deal of what had, to that point, counted as poetry. From the perspective of Adorno's history of experimentalism in art, *Lyrical Ballads* is an untimely volume, mixing the various aspects of his two quite distinct historical stages of experimentalism in art.

Part of the difficulty of applying either Cage's or Adorno's theories of experimentation in art to *Lyrical Ballads* is that of anachronism, of trying to apply to Romantic-era artistic experiments reflections that were intended to explain twentieth-century artistic movements such as minimalism and the avant-garde. Yet this difficulty is also a function of the fact that both Cage and Adorno implicitly grant early-twentieth-century sciences (and philosophy of science) too much "ownership" of the term "experiment" and thus miss the chance not only to engage with the long history of multiple and shifting meanings of "experiment" in the sciences but also to consider the related possibility that an artist might draw the term "experiment" from this history of the sciences precisely in order to transform its meaning. We thus need to supplement Cage's and Adorno's accounts of purely artistic experimentalism with an understanding of the history of the term "experiment" in the century leading up to Wordsworth's and Coleridge's volume. Such an understanding will allow us to see that Wordsworth and Coleridge did not precisely take over a stable, ready-made, and "properly" scientific concept of experimentation, but rather experimented with the term in an attempt to create a new kind of social collective, one that borrowed something from both the sciences and the arts.

Though "experiment" was firmly linked to "science" by the time Wordsworth and Coleridge wrote *Lyrical Ballads*, it was nevertheless far from clear precisely where the boundaries of properly scientific experimentation began and ended. There was widespread and trans-European agreement that the proper method, aspirations, and rigor of science had been embodied in the figure of Isaac Newton, and that science required institutions such as the English Royal Society and the French Académie Royal des Sciences (or, after the French Revolution, the Institut de France) and the journals these societies used to disseminate the results of individual experiments to communities of experimentalists. Yet historians of science have established that the extraordinary valorization of science over the course of the eighteenth century depended upon its more widespread popularization among members of social and economic elites who were not—and never aspired to be—natural philosophers or scientists. In order to solicit this wider public interest in science, scientific popularizers developed public forms of experimentation, such as lecture series on Newton's discoveries, public demonstrations of electricity (often rendered particularly spectacular by the use of human bodies as electrical conductors), or hot-air balloon ascents.[29] The latter were a particularly public form of experiment: purportedly half the population of Paris (400,000 people) watched Charles and Nicolas Robert's manned hydrogen balloon ascent from the Jardin des Tuileries in 1783.[30]

Were these public demonstrations scientific "experiments" or simply exhibitions of the results of scientific experiments? Popularizers sought to convince their audiences and readers that these were properly scientific forms of experimentation that could be distinguished from the merely sensational, or "experiential," practices of nonscientists. The popular French lecturer Jean Antoine Nollet, for example, wrote in his *Leçons de physique experimentale* that in both his public performances and writings, his "intention has always been that one would find a course of experimental physics [*un Cours de Physique Expérimentale*], and not [simply] a course of experiences [*un Cours d'Expériences*]."[31] In the late eighteenth century, in other words, "experiment," did not have one stable meaning but was rather the site of a continuing contest over *which* collectives and *which* practices were to be bound together, and what relationship such collectives and practices would have to the term "science." And as Wordsworth's and Coleridge's "Advertisement" emphasizes, using the term "experiment" could enable authors to appropriate prestige, even practices, from the sciences for not-quite-properly scientific collective practices.

The key dimensions of Wordsworth's and Coleridge's appropriation of the term "experiment" have been partially mapped by scholars of Romanticism such as Alan Bewell and Clifford Siskin. Bewell has made a convincing case, for example, that Wordsworth was influenced by the way eighteenth-century moral philosophers understood "experiment," namely, as a mental activity by means of which a philosopher discovered laws of human nature by imagining an original state of humanity and its subsequent development. Bewell argues that some of the poems in *Lyrical Ballads* should thus be read as attempts to "reshap[e moral philosophy's] experimental language" by developing a new, lyrical protocol for reimagining human origins.[32] Siskin comes to a similar conclusion by arguing that the experiments of *Lyrical Ballads*—and Romantic-era lyrical poetry more generally—must be understood in terms of Francis Bacon's seventeenth-century approach to experiments. Siskin argues that Bacon understood experimentation as a method that "alternated empirical practice, proceeding from sense experience, with the formulation of general propositions" derived from that practice.[33] For Bacon, an individual could generate general propositions about the world only by *acting* on the world in controlled ways, rather than (as in the earlier Aristotelian approach to knowledge) by developing deductive chains of reasoning that began with commonsense axioms. Siskin contends that Romantic-era authors sought to instantiate Bacon's protocol in the realm of creative writing, and they did so by "coupling" the lyric to "personal, subjective feeling—the *I* expressing itself." This subjectification allowed the lyric to take on what Siskin calls its "experimental role," for a reader could draw general conclusions from a lyric on the basis of a subjective experience by engaging in the same kind of disciplined, experimental alternation between concrete experiences and general propositions that Bacon had promoted.[34] Siskin suggests that if Wordsworth's and Coleridge's approach to experimentation indeed has a Baconian origin, it indicates that Romantic poetry is less a rebellion against the "conservativism" of eighteenth-century literature than a "disciplinary venture" that "naturaliz[ed]" and "narrow[ed]" a certain kind of writing into the valorized form of "Literature" (135, 12).

Valuable as Bewell's and Siskin's historical contextualizations are, it is nevertheless striking that both emphasize understandings of experiments focused on isolated subjective activities rather than on communal practices. For both Bewell and Siskin, in other words, an experiment is a way in which an isolated individual thinks or feels (about) something, rather than an activity that requires tools and other practitioners. In this sense, both adopt what we might describe as a "skeptical Popperism": though Bewell and Siskin are skeptical of the "truths" purportedly produced by (for example) *Lyrical Ballads*, both nevertheless approach the volume primarily through the lens of epistemology, focusing on the ways Coleridge and Wordsworth's literary experimentation facilitated truth claims about some aspect of the world (in this case, the nature of "humanity"). Moreover, the overtly critical tone of Siskin's account depends itself upon the (implicit) claim that Romantic experimental literature did not—perhaps even constitutively could not—deliver on the truth claims that it promised; that is, the power of Siskin's account depends upon our sense that he is exposing the ways in which Romantic-era literary experimentation produced falsehoods rather than truths. Insofar as both Bewell's and Siskin's accounts imply an epistemological understanding of experimentation, it is unclear how their image of literary experimentation relates to that eighteenth-century world of scientific experimental devices and communities described by historians such as Schaffer, Golinski, Stewart, and Lynn, and which fascinated most major Romantic authors. Thus, without contesting the contexts proposed by Bewell and Siskin, we can nevertheless expand our understanding of what was at stake in Wordsworth's and Coleridge's appropriation of "experiment" by emphasizing how their usage bore upon *sociological* aspects of seventeenth- and eighteenth-century science.

An emphasis on the collective aspects of science suggests, for example, that Siskin's account of the Baconian experiment as a movement between empirical practice and generalizations is a bit too schematic, for it does not allow us to distinguish between seventeenth-century neo-Aristotelian approaches to experience and experiment, on the one hand, and Baconian approaches to experience and experiment, on the other. As historian of science Peter Dear notes, seventeenth-century neo-Aristotelian natural philosophers also employed and discussed the importance of experiments. However, neo-Aristotelians treated experiments simply as occasions for manifesting or illustrating "experience," that is, "*how things happen* in nature" in general.[35] Because the neo-Aristotelian experiment was an illustration of how things by their nature happen, there was no need for a neo-Aristotelian to date precisely *when* an experiment happened or provide details of its context. Moreover, for the neo-Aristotelian, the experience that the experiment illustrated was to serve as the starting point for a deductive, syllogistic argument about the *causes* of phenomena; that is, an answer to the question of "Why thus and not otherwise?"[36] The neo-Aristotelian approach to experiment and experience is fundamentally distinct from a more modern approach in which experiments are focused on unusual events, and in which an experiment is intended to lead to "a statement of *how something happened* on a particular occasion" (4). This latter approach to experiment — the experiment as discrete event, or, as Dear puts it, the "event-experiment" — owed something to Bacon's texts, but it owed much more to experimental protocols and systems of publication established in the late seventeenth century by natural philosophers such as Christopher Wren, Robert Boyle, and John Wilkins, who went on to establish the Royal Society.[37]

Something important is at stake in the question of whether Wordsworth and Coleridge were drawing on a Baconian or a neo-Aristotelian approach to experiment and experience. Siskin's Baconian interpretation lends support to his claim that *Lyrical Ballads* was part of a modern disciplinary project intended to elevate "Literature" above other kinds of writing by connecting it to the procedures and prestige of the modern sciences. An interpretation of Wordsworth's and Coleridge's experimental approach as neo-Aristotelian would lead us to more or less precisely the opposite conclusion, for it would suggest that these two authors sought to revive an approach to experiment and experience that aimed to illuminate causes (that is, answers to the question of "why?"), which by the late eighteenth century had been almost entirely eclipsed by the scientific event-experiments encouraged by the scientific revolution.

Yet *Lyrical Ballads* complicates even this apparent choice between Baconian and neo-Aristotelian approaches to experiment and experience, for Wordsworth and Coleridge embraced *both* approaches and, by crossing and hybridizing them, created an unprecedented new mode of experimentalism. Some of Wordsworth's contributions to the volume seemed to draw on the requirement of event-experiments for relatively precise dating and geographical localization; one thinks, for example, of a poem title such as "Lines written a few miles above Tintern Abbey, on revisiting the banks of the Wye during a tour, July 13, 1798." Both Wordsworth and Coleridge, moreover, contributed poems that seemed to follow the demand of the event-experiment for unusual events: for example, the superstitious events and narrator of "The Thorn," the mentally challenged protagonist of "The Idiot Boy," or the apparently efficacious curse described in "Goody Blake and Harry Gill: A True Story." And Wordsworth and Coleridge's interest in the language of "the middle and lower classes of society" echoes the preference of the early Royal Society for a "close, naked, natural way of Speaking" — "the Language of Artizans, Countrymen, and Merchants," rather than "that of Wits, or Scholars."[38] Yet the poems in *Lyrical Ballads* nevertheless ultimately seem oriented more toward an Aristotelian than a Baconian sun. Most serve as occasions for reminding us of our experience of things as they tend to happen, rather than seeking to ground a universal claim in the veracity of a specific unusual event.

The poems, moreover, seem concerned with an analysis of *why* things happen rather than simply a delineation of *what* happens to happen.

Wordsworth's "The Thorn," for example, relates an unusual event, but it also features a narrator who hybridizes late medieval with late-seventeenth-century investigative approaches. Like late-seventeenth-century proponents of the event-experiment, the narrator carefully describes the particulars of the thorn and its surroundings, often in quantitative terms: the thorn itself is "Not higher than a two-years' child" (line 5) and it stands "Not five yards from the mountain-path" (line 27), while "to the left, three yards beyond, / You see a little muddy pond / Of water, never dry; / I've measured it from side to side: 'Tis three feet long, and two feet wide" (lines 29–33). The narrator also distances himself from the explanations that unnamed "others" give for the peculiarities of the thorn—namely, that Martha Ray killed her child and buried it there—noting that though "all do still aver / The little babe is buried there," "I cannot tell how this may be" (lines 241–43). Yet unlike a late-seventeenth-century natural philosopher, the narrator seems willing to consider a wide variety of causes, including preternatural or supernatural causes, which by the late seventeenth century had become forms of "superstition."[39] "The Thorn" is thus not an appropriation of the Baconian experimental protocol for determining laws but rather an occasion to reflect on multiple understandings of what it means to account for causes. This in turn emphasizes that the "experiments" of *Lyrical Ballads* are neither fully Baconian nor fully Aristotelian, but rather a composite of these two modes.

EXPERIMENTS AND THE VIRTUAL WITNESS

Shapin and Schaffer's discussion of the importance of "virtual witnesses" for late-seventeenth-century scientific experiments helps us better understand Wordsworth and Coleridge's hybridization of Baconian and Aristotelian approaches to experiments, for the concept of the virtual witness encourages us to think more closely about *where* that experimental activity of measurement Wordsworth and Coleridge described in their "Advertisement" to *Lyrical Ballads* was to take place. In their discussion of Robert Boyle's late-seventeenth-century air-pump experiments, Shapin and Schaffer emphasize that Boyle's ability to produce belief in others about his experiments was not, in general, a consequence of the fact that readers replicated what Boyle had done.[40] Air pumps were extraordinarily expensive to build in the seventeenth century and those that were built tended not to work very well; as a consequence, air-pump experiments were rarely reproduced. Boyle's ability to engender belief in readers instead depended on what Shapin and Schaffer call a "literary technology" that allowed readers to engage in a kind of "virtual witnessing." The journal of the *Philosophical Transactions of the Royal Society* provided a large and geographically distributed readership with polite, descriptive, and authorized testimony of gentlemen who *had* witnessed Boyle's experiments, which in turn allowed readers to become "virtual" witnesses of these experiments. Shapin and Schaffer thus illuminate a model of science that employs a specific genre of literary technology as a means for binding a certain approach to experiment (event-experiment) to a certain conception of community (trustworthy witnesses who, via print, transmit experience to a wider body of virtual witnesses).

Shapin and Schaffer's account allows us to see that it was precisely the possibility of virtual witnessing that was at stake in many of the early reviews of *Lyrical Ballads*. Exploiting Wordsworth and Coleridge's use of the term "experiment," professional reviewers of *Lyrical Ballads* replicated the relationship between actual and virtual witnesses of scientific experiments by positioning themselves as trustworthy witnesses who passed on the results of these literary experiments to a wider public of virtual witnesses: the reviewer, in other words, would explain to readers whether Wordsworth and Coleridge's experiments had succeeded. Yet the "Advertisement" to *Lyrical Ballads* suggests that Wordsworth and Coleridge were unwilling to grant professional reviewers this distinction between actual and virtual witnesses. Wordsworth and Coleridge qualified carefully the kind of reader able to perform successfully the experiments in the volume, contending that only readers who: (1) were familiar with "our elder writers"; (2) were familiar with writers "in modern times who have been the most successful in painting manners and passions"; and (3) had engaged in "severe thought, and a long continued intercourse with the best models of composition" could determine the success of their literary experiments (47–48). Yet Wordsworth and Coleridge did *not* suggest that these qualified witnesses could then communicate the results of the experiments to others by means of reviews or descriptions of the poems. The experiments in *Lyrical Ballads* were, rather, premised upon a different relationship among experiment, literary technology, and community than that upon which the Royal Society relied. Where the Royal Society relied on print distribution networks to enable virtual witnessing of experiments that had taken place in a limited number of physical locations, *Lyrical Ballads* used print distribution networks to communicate the experimental apparatus itself. Copies of *Lyrical Ballads* served to distribute experimental activity both spatially and temporally: spatially, because every qualified reader could perform these experiments at the same time; temporally, because the determination of "how far" the language of conversation could accommodate the purposes of poetic pleasure need never actually conclude but could go on perpetually.

The temporal distribution of experimental activity enabled by *Lyrical Ballads* thus implied a different model of historicity, and of historical community, than that of the modern natural sciences from which Wordsworth and Coleridge had partially drawn the term "experiment." The experiments of the Royal Society were intended to function as events, in the sense that each experiment would contribute to, and mark, the progress of a science. However, the goal of "progress" meant that once an experiment had gained acceptance by the relevant scientific community, the experiment itself was for all intents and purposes a part of the past of a science. Rather than *repeating* past experiments, future generations could *build upon the results* of past scientific experiments. By contrast, though the experiments of *Lyrical Ballads* were also intended to create an event—in this case, the event of enabling the future of poetry to be better than its recent past—the sense of historicity upon which *Lyrical Ballads* relied was not progress, at least in its scientific modality. Unlike scientific experiments, the experiments of *Lyrical Ballads* did not become "past" as soon as they had succeeded in producing an event in the history of poetry. The poetry of *Lyrical Ballads* is, of course, now undeniably part of the past of English-language poetry. Yet rather than ceding their vitality to subsequent developments in poetry, the experiments of *Lyrical Ballads* were intended to retain a form of suspended animation that perpetually offers new generations the opportunity to become part of the experimental community of art-witnesses that began in 1798.[41]

EXPERIMENTING WITH THE EXPERIMENTAL

Drawing on Shapin's and Schaffer's account of virtual witnessing in the sciences emphasizes that Wordsworth and Coleridge did not simply *apply* an existing scientific concept of experiment to their poetry but rather *experimented* with the concept and practice of experimentation itself. The notion of "applying something" from one realm to another depends upon the premise of distinct realms, each of which has its own proper elements and borders. Applying Aristotelian, Baconian, Royal Society, or moral philosophical modes of experimentalism to art, for example, would be a matter of appropriating a procedure proper to science or philosophy and then transforming this properly scientific or philosophical procedure so that it conforms to the proprieties of another distinct realm, that of art (or a specific art). Applying a scientific concept of experiment to art thus relies on what Adorno described as the premise of artistic continuity (namely, the premise that a specific art has its own proprieties and a continuous history). Experimenting with concepts and practices of experiment, by contrast, means aspiring to a new mode of experimentalism that will be proper to neither science nor art, at least as each term has been understood in the past. Such an endeavor is always attended by the possibility that one will end up abandoning art—or, at the very least, that the word "art" will come to connote something other than what it did prior to this crossing. Wordsworth and Coleridge recognized, and even embraced, this risk, urging readers of their volume not to allow "the solitary word Poetry, a word of very disputed meaning, to stand in the way of their gratification" as they perused *Lyrical Ballads*.[42]

This willingness to abandon the label of "poetry"—or "literature" or "art"—in favor of an activity that perhaps has yet to receive its proper name occurs repeatedly in nineteenth- and twentieth-century experiments with art and science, from Zola's attempt to create an "experimental novel," to the Fluxus emphasis on confusing distinctions between art and life, to more recent bioart in which biological protocols, materials, and living matter are transported into the art gallery. Though individual experimental artists may set for themselves particular goals for their hybridizing activity—Wordsworth and Coleridge, for example, aimed to enable readerly pleasure—there is nevertheless no one telos that guides all experimentations with the concepts and practices of scientific experiments. Experimenting with experimentation is instead an eccentric, excessive exercise, one that aims— if it can be said to "aim" at anything—only at facilitating new forms of thought and sensation.

The concept of "experimenting with experimentation" is admittedly difficult to think, since it involves thinking at least two, and often three, different senses of the term "experiment" at the same time. Understanding the nature of the "experiments" in *Lyrical Ballads*, for example, requires us to consider "experiment" in: (1) an Aristotelian sense, as common experiences that exemplify the way things habitually happen; (2) a Baconian sense that had, in the sciences, almost entirely displaced the older Aristotelian sense, and which relied not on common occurrences, but unusual and often "forced" experiences; (3) a new sense, as a kind of activity that is able to join the central concerns (1) and (2) (by, for example, using time- and date-specific unusual events not to produce general laws but rather to encourage open-ended reflection on what it means to "explain" something). This new sense of experimentation cannot be reduced to either (1) or (2) but is instead a fundamentally synthetic activity that produces something new by linking elements of both present and past domains or practices.[43]

The eccentric desire that motivates the artistic experiment in the Romantic era is thus not truly proper to either art or science but is rather a desire to transform art by means of the sciences, to make art something other than what it has properly been to date. As I have suggested above, something like this same eccentric desire can be found at the origins of modern science, for what Shapin and Schaffer call the modern scientific "experimental life" emerged when authors such as Bacon and Boyle transformed natural philosophy by reconfiguring the relationship between Aristotelian concepts of experiment and experience. The sciences continue to rely, moreover, on an attenuated form of this eccentric desire insofar as, in the sciences, "a research system [as a whole] is organized in such a way that *the production of differences* becomes the organizing principle of its reproduction."[44] While any particular scientific experiment may aim at specific answers and conceptual clarification, experimental *systems* persist only to the extent that they continue to facilitate new questions, concepts, and even modes of science.[45] From this perspective, Wordsworth and Coleridge's invocation of the language of experiment was the moment when this principle of the production of differences traversed the space between the sciences and the arts and was thus the point at which the arts began actively to solicit this power of differentiation from the sciences.

At least one of Wordsworth and Coleridge's early critics recognized in the experiments of *Lyrical Ballads* precisely this possibility of a differential and differentiating system of artistic production. Francis Jeffrey, editor of the *Edinburgh Review*, saw in the poetry of Wordsworth, Coleridge, Southey, and other members of what he called the "Lake School" the formation of a "new system," which he understood as oriented toward the production of differences within the field of artistic activity. Jeffrey seems to have intuited that simply by naming the poems in *Lyrical Ballads* "experiments," Wordsworth and Coleridge threatened to transform art into something other than, for example, a canon or storehouse of tradition, wisdom, and beauty. Arguing that the "standards" of poetry "were fixed long ago," Jeffrey charged that the Lake Schoolers were "dissenters from the established system"; this meant, in practice, the "positive and *bona fide* rejection of art altogether," and Jeffrey clearly hoped that mockery and censure could put an end to the Lake School experiment.[46] We can agree with Jeffrey's analysis even as we evaluate the risks to art differently, seeing in Wordsworth and Coleridge's wager a new future for art rather than simply its rejection. As Jeffrey intuited, when an art experiments with concepts and practices of experiment, that art necessarily risks becoming something other than it was in the past. Nor could this be simply a movement from one stable concept of art (e.g., art as storehouse of wisdom) to another stable concept: insofar as the artistic experiment aims neither to mirror nor simply "apply" philosophical or scientific experimental systems, it opens up a system of differences, each of which can serve as the starting point for new artistic experiments and new images of art.

In this sense, scientific experimentation functions like a virtual dimension for the artistic experiment: that is, as sets of ideals and protocols that can never be fully instantiated in art but which nevertheless regulate judgments and actions. Alternatively, we might think of the scientific experiment as the "quasi-object" of desire of the artistic experiment, as that which guides the artistic experiment—or which the artistic experiment desires to be or to possess—but which cannot be grasped directly. (This latter interpretation is partially captured in Enzensberger's suggestion that the arts "flirt" [*kokettieren*] with the sciences, though flirtation is probably not a strong enough term for this relationship.) In his 1802 preface to *Lyrical Ballads*, Wordsworth suggested something along these same lines, noting,

> If the labours of men of Science should ever create any material revolution, direct or indirect, in our condition, and in the impressions which we habitually receive, the Poet will sleep then no more than at present, but he will be ready to follow the steps of the man of Science, not only in those general indirect effects, but he will be at his side, carrying sensation into the midst of the objects of the Science itself. The remotest discoveries of the Chemist, the Botanist, or Mineralogist, will be as proper objects of the Poet's art as any upon which it can be employed, if the time should ever come when these things shall be familiar to us, and the relations under which they are contemplated by the followers of these respective Sciences shall be manifestly and palpably material to us as enjoying and suffering beings. If the time should ever come when what is now called Science, thus familiarized to men, shall be ready to put on, as it were, a form of flesh and blood, the Poet will lend his divine spirit to aid the transfiguration, and will welcome the Being thus produced, as a dear and genuine inmate of the household of man.[47]

Thinking of the scientific experiment as the quasi-object of desire of the artistic experiment also emphasizes that what counts as a scientific experiment has changed over the course of the past three centuries, and that what a given artist understands under the rubric of "scientific experiments" is even more variable. Artistic experiments can orient themselves toward the scientific experiment in a variety of different ways: by seeking to produce phenomena with the apparent constancy and givenness characteristic of scientific objects, for example, or by working directly with scientists, or by aspiring to the formation of communities of witnesses (and perhaps even virtual witnesses). No matter which path it takes, though, the artistic experiment cannot become the same as the scientific experiment or the former will cease to be art. This is not a tragic narrative, for it is precisely through their non-coincidence with scientific experiments that artistic experiments continue to produce differences, and thus continue to make possible new concepts and sensations.

EXPERIMENTS AND THE TEMPORALITY OF "ART"

I want to conclude by returning to Jeffrey's fear that the experimental approach of Lake Schoolers such as Wordsworth and Coleridge—their "new system" of poetry—posed a threat to traditional conceptions of the arts, for not only was Jeffrey's fear well-founded, but it also helps us understand the particular, and particularly complex, temporality the concept of the experiment introduced into subsequent discussions of art and the arts. Admittedly, when Jeffrey wrote that the simple language of Lake School poetry constituted a "positive and *bona fide* rejection of art altogether," he likely meant the term "art" to be understood as a synonym for "elevated technique" rather than as a general term that encompassed all of the individual arts. Yet the ambiguity of Jeffrey's phrase—an ambiguity that makes it possible to sense in his critique a fear that a *specific kind* of art (Lake School poetry) could bear upon *art-in-general*—is telling, and can be read as an index of the more widespread shift, in the nineteenth century, from a fundamentally plural conception of "the arts" to a conception of "Art" as a general rubric that encompasses all of the individual arts in both their historical and future instantiations.[48] Such a shift cannot be attributed solely to Wordsworth's and Coleridge's introduction of the term "experiment" into the practice of poetry. Yet the conception of artistic practices or works as potentially experimental facilitated this shift from "the arts" to "Art," for the concept of artistic experimentation depends upon a concept of "Art" that has a perpetually provisional and open denotation; conversely, the concept of Art depends upon the premise that experimentation is what enables new forms of art. The concept of Art means that one can neither predict nor foreclose what will count as an instance of Art: one cannot limit Art, for example, to existing categories such as poetry, sculpture, and painting, but one must instead always remain open to the addition of new practices and objects (e.g., Pop art, performance art, and so on), which latter are produced by experimentation.

Yet as Jeffrey seems to have intuited, when the arts become Art, this does not involve simply an additive logic—it is not just a matter of adding new objects or arts to an existing "stock" of past art—but it also involves a continual reconfiguration of the relationships among the set of things that have in the past counted as the arts. So for example, to allow the simple language of the Lake Schoolers' verse to count as "art"—with the term here used in the sense of poetic technique—would be, necessarily, to redefine retrospectively the borders of poetic technique, allowing a kind of language formerly deemed art-less to count as Art. Yet this would mean, in turn, redefining the distinction between poetry and non-poetry, and between Art and non-art. Hence Jeffrey's insistence that the Lake School experiments are not *bad* Art but rather not an instance of art at all. Jeffrey recognized, in short, that to allow their experiments to wear the mantle of *any* kind of art would be to open up the prospect of perpetual retrospective reassessments of the borders separating the individual arts from one another and from non-art.

Jeffrey, of course, lost this battle, and we now indeed inhabit a world of Art, rather than the arts. We can perhaps best capture the peculiar shape and dynamics of Art through the figure of a *heterogeneous network*, for the concept of Art names not only parallel series of media-specific practices that had existed for centuries (poetry, painting, sculpture, etc.) but also the virtual links between elements of these series that enable new series (abstract art, conceptual art, etc.), experiments in "multimedia" art (e.g., painting-sculptures), and series yet to be invented by means of future experiments. The Art-network retrospectively encompasses past artworks: thus, Egyptian architecture and Sophoclean tragedy are as much a part of the Art-network as *Lyrical Ballads* or Cage's *4' 33"*. Yet because this is an Art-*network* (rather than, for example, a single or even a determinate number of series), there are no longer any scales or criteria by means of which the different nodes in this network can be valued against one another. One can *link* Coleridge's *Rime of the Ancyent Marinere*, for example, to the work of "elder poets," but Coleridge's *Rime* does not represent an improvement on (or falling away from) this earlier verse; it is simply a different link. Nor are links restricted to a given medium: one can link Wordsworth and Coleridge's *Lyrical Ballads* to Émile Zola's *Germinal* to John Cage's *4' 33"*.

The network-image of Art renders unbelievable earlier claims about hierarchies among the arts and linear historical accounts that purport to encompass all artworks. The supposed "rivalry" between the ancients and the moderns that obsessed critics in the seventeenth and eighteenth centuries, for example, has for us only historical interest, while G. W. F. Hegel's narration of the spiritual ascent of the arts from architecture to poetry now reads like a last-ditch effort to force into narrative form a cultural phenomenon that had already begun to manifest itself in the shape of a network. G. E. Lessing's distinction between the time-based and the spaced-based arts, by contrast, seems much more modern: for us, the importance of Lessing is that he clarifies that painting and poetry are not rivals but rather fundamentally different means for producing aesthetic experience, a position that accords well with our sense of artistic media as simply different from, rather than better or worse than, one another.[49] One can, of course, always seek to reestablish continuity in the realm of the psychology of artists, rather than the ontology of art: Harold Bloom, for example, emphasized the agonistic rivalries of artists with their predecessors within individual series within this network (e.g., within the series of "poetry," Wordsworth felt oppressed by, and sought to outdo, Milton).[50] Yet even if we grant the legitimacy of this account, such rivalries never affect the whole of the Art-network, and artistic experimentation always makes it possible to negate or escape such local rivalries simply by creating a new kind of art into which these Oedipal dynamics fail to extend. From the perspective of the Art-network, artists' names now function primarily as local maps that allow us to group nodes within a vast network of links unified only by the term—or more accurately, the framing gesture implied by the term—Art.

The image of the Art-network captures both the totality implied by the modern concept of Art while at the same time it emphasizes the insularity of each individual artwork within the network. The image of the network underscores, in other words, that however many formal, biographical, and even affective links we can establish between individual artworks (i.e., individual nodes), each node nevertheless remains simply different from—and in this sense, incomparable to—the nodes that have come before and those that will follow. The links between artwork-nodes remain resonances rather than architectural relationships of foundation or scalar relationships of evaluation (good/better/best). The Art-network is the *form* that art-in-general takes in what Jacques Rancière has called the "aesthetic regime of the arts," that is, "the regime that strictly identifies art in the singular and frees it from any specific rule, from any hierarchy of the arts, subject matter, and genres."[51]

As the activity that produces new nodes within the Art-network, (artistic) experimentation thus bears a fundamentally different relationship to Art than scientific experimentation bears to the sciences. Both Ludwik Fleck and Thomas Kuhn stressed that scientific experiments do *not* enable one continuous, linear line of "scientific progress" but rather facilitate occasional paradigmatic shifts within the sciences. Yet it is nevertheless always possible, from within a given scientific paradigm, to describe the work of earlier scientists as simply limited and special cases of the new paradigm: one can see Newton's formulas, for example, as applying only to a limited aspect of the Einsteinian universe.[52] The temporality of science and its experiments, as Gilles Deleuze and Félix Guattari note, is thus "a peculiarly serial, ramified time, in which the before (the previous) always designates bifurcations and ruptures to come, and the after designates retrospective reconnections . . . Scientists' proper names are written in this other time, this other element, marking points of rupture and points of reconnection."[53] Understanding the temporality of science and its experiments as serial yet ramified allows us to recognize the importance of paradigmatic shifts—the Newtonian universe of gravitation is simply not the same as the Einsteinian universe of mass-energy—while clarifying how scientists also, and simultaneously, tend to see contemporary experimental science as "building upon" past experiments.

The artistic experiment, by contrast, creates new nodes within the Art-network: that is, ramification without any overall seriality. As a consequence, artistic experiments prevent, rather than enable, any sort of transgenerational "progress" (or "decline") of Art, or even of the individual arts, and instead simply facilitate further ramification of the Art-network. One can at best belong to a movement or school (e.g., the Lake School of poetry; nineteenth-century naturalism; the "first" avant-garde), yet such groupings rarely last beyond a single generation and never seem to cohere into "progress" in the fashion of the sciences, no matter how experimental the school or movement might be.[54] Even Clement Greenberg's powerful interpretation of modern art as a collective attempt by artists to locate the pure form of each medium by means of abstraction—to reduce painting, for example, to the bare minimum of conditions necessary for an object still to count as a "painting"—is fundamentally a narrative of the diaspora of the separate arts (and is, in any case, a narrative that has been undone by the rise of multimedia art).[55] In place of the serial "progress" that occurs in the sciences, the artistic experiment instead facilitates the emergence of new nodes and links within the Art-network.

The concept of the Art-network allows us to see in a new light Adorno's claim that, since at least the 1930s, "art is now scarcely possible unless it does experiment." For Adorno, the explicit turn to experiment emphasized the difficulty that artists faced in employing means that were not already occupied in advance and rendered homogenous by capitalism; the only way to oppose such a world was to design works that allowed audiences to submit themselves to sounds or experiences that could not be known in advance. Yet for Adorno, this turn to experimentation simultaneously drained past art of its vitality; rather than expanding the realm of art, experimentalism transformed art into a moving "front" that left nothing vital behind: "expansion appears as contraction."[56] Thus, though in principle "artworks . . . have a life [Leben] sui generis" (4), experimentalism is part and parcel of a world in which it "is self-evident that nothing concerning art is self-evident anymore, not its inner life, not its relation to the world, not even its right to exist" (1). The concept of the Art-network allows us to understand better this ambivalent "life" of art to which Adorno alluded. Though the art-nodes marked as "past" art indeed no longer appear vital, they are not dead but rather in a state of suspended animation. Past art-nodes can, as a consequence, always be revived—that is, made to resonate with a new node—and artistic experimentation is in large part a matter of precisely such combinatory reanimations. From this perspective, then, to return to Romantic-era experimental art is not to catalogue or describe the dead past but rather to create resonance among nodes by articulating the importance of experiment for the Art-network as a whole.

CHAPTER 2

Suspended Animation
and the Poetics of Trance

There is an interval . . . of suspended animation, a kind of psychological shock or paralysis . . . Experienced interrogators recognize this effect when it appears and know that at this moment the source is far more open to suggestion.

Central Intelligence Agency, quoted in Klein, The Shock Doctrine, 16

I think I'll will my heirs to revive me one day a century. That way I can observe the fate of all mankind.

Dick, Ubik, 6

Among the more enduring legacies of Romantic-era literary criticism is Samuel Taylor Coleridge's suggestion that literary works demand from readers a "willing suspension of disbelief." Coleridge employed this phrase in his famous discussion of the reception of poetry, recalling in *Biographia Literaria* (1817) that his contribution to the coauthored *Lyrical Ballads* (1798) was to describe "persons and characters supernatural, or at least romantic" such that readers would be encouraged "to transfer from our inward nature a human interest and a semblance of truth sufficient to procure for these shadows of imagination that willing suspension of disbelief for the moment, which constitutes poetic faith."[1] Since the Romantic era, the concept of "willing suspension of disbelief" has been applied to the reception of various artistic media, leading Thomas McFarland to describe Coleridge's phrase as "undoubtedly the single most famous critical formulation in all of English literature."[2]

Though "disbelief" is the most well-known object of Coleridge's thoughts on suspension, he was equally interested in another modality of abeyance, suspended animation. "Suspended animation" emerged as a concept in the late eighteenth century around the efforts of the newly founded Royal Humane Society to convince lay and medical readers that individuals who had apparently drowned might still be alive, albeit in states of "suspended animation" (a state that we would now likely describe as a coma). The phrase was quickly taken up by medical and literary authors. Coleridge, for example, used the concept to describe processes associated with literary production and reception: in the introduction to *Christabel* (1816), he excused his lack of literary productivity with the claim that his "poetic powers [had] been, till very lately, in a state of suspended animation" (v), while elsewhere, he contended that some forms of mass print media induced a "morbid Trance, or '*suspended Animation*'" in readers.[3] Mary Shelley drew implicitly upon the concept in *Frankenstein* (1818), in which she described her creature's efforts to "restore animation" (165) to a small girl who had apparently drowned in a river and made explicit use of the term in "Roger Dodsworth: The Reanimated Englishman," a short story inspired by reports that a man from the seventeenth century had been revived after having spent more than a hundred years frozen in ice.[4] P. B. Shelley, for his part, sought to demythologize the purported miracle of Jesus's resurrection by noting in *Queen Mab* (1813) that the Royal Humane Society frequently "restores drowned persons," but "because it makes no mystery of the method it employs, its members are not mistaken for the sons of God" (201).

That Coleridge's brief comments on the willing suspension of disbelief have been of greater interest to literary critics than his claims about suspended animation is no doubt a function of the clearly demarcated psychological domain of the former. As psychological acts, "willing" and "disbelief" are phenomena that literary critics feel competent to assess; suspended animation, by contrast, seems to concern physiological and ontological dimensions about which literary critics are, perhaps justifiably, more skittish. Thus, while it has seemed reasonable to take Coleridge's psychological claims about the willing suspension of disbelief literally, it has appeared more prudent to treat his comments on suspended animation as "simply metaphorical," hence, of limited use in discussions of the creation and reception of art.

This chapter moves against the grain of these assumptions, drawing on Romantic-era approaches to suspended animation as a means for better understanding the reception and formal structures of creative literature, especially what Jonathan Culler has described as the "rhythmical shaping and phonological patterning" of verse.[5] Romantic-era discussions of suspended animation are useful in part because—even when used as a metaphor—the concept links epistemology to ontology, emphasizing embodied modes of trance produced by printed media.[6] These discussions thus help us to articulate further the specific ways in which Romantic-era authors understood literary experience as embodied, and point to the importance of rhythmic aspects of literature that enable, yet operate beneath, the judgments readers make about relationships between "the rhyme and reason" of a work. Moreover, focusing on the concept of suspended animation helps us better understand and articulate the close, and often counterintuitive, links that Romantic-era authors established between "altered states" and drug-induced experiences, on the one hand, and what Orrin Wang has called "Romantic sobriety," on the other.[7]

Early-nineteenth-century interest in suspended animation also has contemporary relevance, for Romantic-era authors presciently understood suspended animation to be bound up with the question of modernity, linking the concept to institutions, like mass print media, that seemed to accelerate the temporality of modern life. As I discuss below, Romantic-era authors understood the link between suspended animation and modernity in two quite different ways. For authors such as Coleridge, suspended animation was a dangerous condition, and the concept was employed to describe a loss of subjective autonomy produced by the distractions of modernity. John Keats and the Shelleys, by contrast, drew on a more idiosyncratic medical understanding of suspended animation—one advanced by, for example, the surgeon John Hunter—as a potentially desirable state that could *regulate* the otherwise relentlessly swift and automatized "animations" of modern life. Where Coleridge feared suspended animation, seeing in it only a narcosis of the will, Keats and the Shelleys aimed at a poetics of trance, deploying literary form as a technology that could vitalize a reader's will and understanding precisely by suspending animation. And where Coleridge's negative assessment of suspended animation survives within twentieth- and twenty-first century discussions of the narcosis purportedly produced by mass-media technologies, Keats's and the Shelleys' more positive assessment resonates with, for example, Gilles Deleuze and Félix Guattari's recent attempt to rethink sensation, and it helps us as well to articulate a rationale for the continuing importance of poetry and other "slow" media within our contemporary new media landscape.

SUSPENDED ANIMATION AND SIMPLE LIFE

Suspended animation first came to prominence in Britain in the 1770s as part of the publicity efforts of the newly founded Humane Society. Following the lead of the Dutch Society for the Drowned, established in 1773, William Hawes and Thomas Cogan founded the Humane Society in 1774—it became a "Royal" society a decade later—and set two central goals. First, the society sought to convince laypeople and physicians that it was possible for individuals to *appear* dead, without actually being dead. The society focused especially on cases in which individuals were unintentionally submerged in cold water for extended periods (for example, when an individual fell into a canal). Society members argued that in such cases, a vital principle persisted in the body even though organ functions such as circulation, respiration, and excretion had ceased. Second, the society argued that it was often possible to rescue people from these states of "trance" or "suspended animation," and they established an award for physicians who invented and perfected methods for resuscitating individuals from these states.[8] The society acknowledged in its *Transactions* that "suspended animation" was a modern discovery, sadly noting the "unnecessary" loss of life in earlier times: "can we reflect on the vast numbers of the human beings, that have been sacrificed *in all ages and in all countries*, and not feel the utmost remorse, and the most poignant regret?"[9]

In different printed media, including the society's *Transactions*, individual pamphlets, and articles in the *Philosophical Transactions of the Royal Society of London*, medical authors proposed various techniques for restoring animation to afflicted individuals. These ranged from vigorous rubbing of the victim's body to electrical shocks and the introduction of tobacco smoke into the patient's bowels.[10] Yet some authors also wondered whether suspended animation might be employed as a therapeutic technology. Could the "awful pause" produced by suspended animation, for example, be "safely imitated by art" in order to "produce more lasting and salutary changes in certain highly obstinate affections of the brain and nerves than can be accomplished by any ordinary means?"[11]

Whether understood as threat or possible therapy, suspended animation begged theoretical explanation. If, as the Royal Humane Society maintained, this state was entirely unknown to earlier ages, what modern facts and theories explained how a body could appear dead yet remain alive, or how a state of quasi-death might prove salutary? Of the authors who took up these questions, the well-known surgeon and experimentalist John Hunter was among the most innovative in his approach. The society had solicited Hunter's support for their cause, and Hunter obliged with a short set of "Proposals for the Recovery of People Apparently Drowned." However, even prior to the society's solicitation, Hunter had been fascinated by states in which vitality persisted in the absence of animation. In the 1760s, Hunter had pursued a series of experiments in which he measured changes in the internal temperature of a variety of living beings—including bean plants, vipers, snails, and humans—as each was brought to the point of near freezing. He then compared the velocity of temperature change of these living beings with that of dead instances of the same animals and plants, and observed that living creatures cooled more slowly than the materially identical corpses. He also found that the living instances were extremely difficult to cool below a certain temperature. These experiments suggested to Hunter that even when all the "actions" of life—

for example, blood circulation, respiration, and cognition—had been suspended by extreme cold, the living matter of plants and animals continued to respond to their environments, resisting the effects of the increasingly cold medium that surrounded the plant or animal.

Hunter's essay was part of a more general attempt to determine the ways in which the principle of life reacted to changes in the temperature of a medium surrounding a living body, and "Proposals for the Recovery of People Apparently Drowned" was in part a response to several earlier papers by physician Charles Blagden that had appeared in the *Philosophical Transactions of the Royal Society of London*.[12] Blagden had described several experiments conducted by Hunter's friend Dr. George Fordyce in which Fordyce had measured his own body temperature, as well as the body temperatures of several associates (including Blagden), as they stood in rooms that were heated to progressively hotter temperatures (Blagden reported that the final experiment was conducted in a room heated to the impressive temperature of 260° F). Blagden noted that standing in extremely hot air produced peculiar, and sometimes uncomfortable, phenomenological experiences of intensity. Blagden wrote, for example, that "every expiration, particularly if made with any degree of violence, gave a very pleasant sensation of coolness to our nostrils, scorched just before by the hot air rushing against them when we inspired," but he also noted that when he touched the side of his body, "it felt cold like a corpse."[13] Blagden also stressed that the increased temperature of the medium, despite its tendency to produce phenomenological novelty in the experimentalists, did not seem to affect the actual internal and external temperatures of the experimentalists' bodies, which stayed essentially constant, between 98° and 100° F (114–18). Blagden suggested that his experiments thus "explode[d] the common theories of the generation of heat in animals"—namely, theories of chemical or mechanical cooling—and proved instead that the power of cooling could be attributed only "to the principle of life itself" (122). He contended that chemical and mechanical explanations of cooling were incapable of explaining the *intentionality* of the body's response to heat—that is, its power of "producing or destroying heat, just as the circumstances of the situation require" (122). Such hypersensitive regulation, Blagden concluded, could only be produced by "the principle of life," which was able to generate heat "in those parts of our bodies in which life seems particularly to reside" and then "readily communicat[e] [that heat] to every particle of inanimate matter that enters into our composition."[14]

Where Fordyce's and Blagden's work considered the effects of heat on living bodies, Hunter's work engaged the other side of this question—namely, what happened to living tissue when it was exposed to extreme cold? The society had seen extreme cold primarily as a threat to human life, but Hunter saw icy media as a technical means for performing an "analysis" of life, one that allowed him to distinguish between two different modes of vitality. As Hunter noted in his posthumously published lectures on surgery, one normally becomes aware of living beings through their actions, whether these activities are associated with particular organs or systems (e.g., respiration and blood circulation) or with more obviously intentional movements (e.g., locomotion or communication). We are, in other words, most familiar with "practical life," life as it is manifested in actions. Yet Hunter employed extreme cold in order to suspend all organic actions without producing death. This implied that practical life was grounded in a more fundamental mode of vitality Hunter called "simple life."[15] Simple life denoted life's capacity of "preventing . . . matter from falling into dissolution."[16] In this state of suspended animation, simple life continued to maintain temperature differences between the body and its surrounding medium, by, for example, opposing cold through the generation of heat. (Implicit in such a claim is the premise that the creation of heat is not an action.) As the cold became more intense, to the point that it "threaten[ed] destruction" of the animal, simple life, in a last-ditch effort to maintain differences between the living body and its medium, "rous[ed] the animal powers to action for self-preservation."[17] Thus, as Hunter noted in the 1786 version of his essay on the resuscitation of persons apparently drowned, a state of suspended animation should be understood as a *self*-suspension of activity, and thus as a sign of life: "[a] state of relaxation should therefore (in cases where an universal violence has not been committed) be considered as a criterion of life; and . . . should be for some time admitted as a probable reason for supposing life still to exist."[18] The moment of death, on the other hand, was generally indexed not by stasis but rather by a violent spasm, as living tissue desperately sought to save the organism: "When the stimulus of death takes place, the whole animal is thrown into action, in which contracted state, absolute death is produced."[19] For Hunter, both of these capacities—the capacity to suspend actions and the spasm as a last-ditch attempt to employ action to prevent death—were consequences of the capacity of living matter itself to maintain differences between internal and external heat.

On the one hand, the concept of simple life seemed to explain the phenomenon of suspended animation, for the "suspension of the actions of life" that Hunter produced artificially was equivalent to that state from which the Royal Humane Society sought to revive patients. On the other hand, though, even as Hunter's concept of simple life helped to explain the state of suspended animation, it also called into question the traditional Aristotelian linkage between life and movement. While many aspects of the Aristotelian philosophical legacy had been rejected by the late eighteenth century, the link Aristotle had established between life and movement (*kinesis*) was rarely questioned, and for most eighteenth-century theorists, to live—to be animate—meant engaging in movement, whether in the sense of actual locomotion or in the sense of organ activity (e.g., respiration). Yet Hunter's category of simple life suggested that vitality should not be defined by movement, since in the state of suspended animation, a being remained alive though all its vital movements had ceased; that is, the link between vitality and animation was temporarily bracketed. Suspended animation thus constituted a kind of ontological *epoché* of animation, one that revealed a more fundamental mode of vitality (simple life). Hunter was not the only Romantic-era author to disconnect life from movement: Alexander Humboldt, for example, proposed that the life-force (*Lebenskraft*) of living beings could suspend the chemical actions at play in living bodies, while F. W. J. Schelling claimed that nature and life emerged only when "an infinite . . . productive activity" was "inhibited, retarded."[20] However, Hunter's authority as an experimentalist lent his claims about suspended animation significant weight among physicians, surgeons, and interested laypeople and highlighted the extent to which suspended animation confused distinctions between physiology and psychology, since this state suggested that sensation remained even in the absence of bodily animation.

While the Royal Humane Society was interested in simple life only insofar as it supported the position that individuals could be revived from their states of suspended animation, Hunter was intrigued by the possibility that, in addition to its "conservative" dimension, simple life might also serve as the foundation for the emergence of *new* forms of existence. In his *Lectures on the Principles of Surgery*, for example, he noted that a cryonic dream had inspired his work with ice. Hunter wrote that he had hoped that extreme cold might enable him to extend a single human life over many generations; that is, that

> it might be possible to prolong life to any period by freezing a person in the frigid zone, as I thought all action and waste would cease until the body was thawed. I thought that if a man would give up the last ten years of his life to this kind of alternate oblivion and action, it might be prolonged to a thousand years: and by getting himself thawed every hundred years, he might learn what had happened during his frozen condition. (76)

While Hunter abandoned this cryonic dream following his failure to revive frozen carp, the image of using suspended animation to produce quasi-immortality emphasizes his approach to simple life as a source of innovation. Rather than simply maintaining life as it was, simple life promised for Hunter *new* forms of social organization, such as an individual who fades in and then out of social existence every hundred years (and who thus could, in principle, outlive his children, grandchildren, great-grandchildren, great-great children, etc., which presumably would in turn require the development of new forms of inheritance).

Yet for Hunter, the new forms of social organization that simple life promised were themselves functions of the new kinds of *ontological* possibilities that were implicit in this mode of vitality. Hunter's understanding of simple life as the precondition for the emergence of new forms of biological organization came to the fore in his discussion of the "non-organic" life of a fertilized poultry egg. The poultry egg functioned as a sort of eighteenth- and nineteenth-century "model organism" for embryological investigations, and Hunter experimented with the eggs of different fowl.[21] For Hunter, the fertilized poultry egg exemplified living matter that was non-organic—that is, "devoid of apparent organization"—for though the egg was made up of many different elements, such as yolk, white, membranes, and cicatricula, none of these could be described as organs.[22] The egg was also devoid of vital action, for it initially lacked all of the processes, such as digestion, that constituted the actions of life.[23] Nevertheless, the non-organic body of the newly fertilized egg *lived*, a fact that Hunter established by demonstrating that the fertilized egg possessed the same powers of resistance and opposition to cold as organic living bodies. Thus, when a fertilized chicken egg was placed in an ice bath, "life"—just as in the case of organic living bodies—"allowed the heat to be diminished 2° or 3° degrees below the freezing point, and then resisted all further decrease," though when "the powers of life were expended by this exertion . . . the parts froze like any other dead animal matter."[24]

The simple life of the fertilized egg was of interest to Hunter because it did not conserve an existing organization, but rather served as the necessary precondition for the development of organic form. Hunter emphasized that the simple life of the egg should not be understood as a passive medium for an otherwise autonomous process of organic unfolding. The whole of the fertilized egg, and not simply the "cicatricula," or nucleus, evidenced processes of transformation:

> When heat is applied to an impregnated egg, the living parts are put into motion, and an expansion of what is called the cicatricula takes place. This very probably begins at the chick as a centre; but it would appear that the whole [of the egg] did not derive its expansion immediately from the chick . . . and the further from the chick these powers [of the rest of the egg] are at an early period, the strongest is this expansion of parts; for we find changes taking place in this

circle near to the circumference, sooner than to the chick, which afterwards become distinct vessels.[25]

Within the egg, organic form did not propagate from a central point outward across a neutral medium but was rather the result of what we might describe as the creativity of matter, a creativity that enabled organs to come into being only when a germ within the egg (what Hunter calls "the chick") became linked to quasi-autonomous chemical and physical processes of expansion and contraction in the rest of the egg; these quasi-autonomous processes of expansion and contraction were themselves dependent upon pressure and heat differences between the inside and outside of the egg. Hunter's example of the egg suggested that simple life did not merely maintain existing organic form in cases of extreme deprivation (e.g., freezing) but also referred to those non-organic processes, such as chemical and physical processes of expansion, that enabled organs and organic form to emerge in the first place.[26]

Though the fertilized egg initially seems to be a very particular and effectively unrepeatable instance of the capacity of simple life to serve as catalyst for the emergence of organic form, Hunter's reflections on the therapeutic potential of suspended animation suggested that simple life did not necessarily exhaust its capacities for innovation after processes of embryological development had concluded. He proposed several technologies for reviving people who had fallen into states of suspended animation, and these technologies depended upon the ability of the suspended living body to create or accept new "organs" that would link it in new ways with its environment. A pair of double bellows, for example, could be transformed into a new lung that would enable "artificial breathing," while a tube of eel skin wrapped around a spiral wire and inserted into the stomach formed a new hybrid digestive tract that extended into spaces that could be manipulated by the physician.[27] These constellations of living tissue, dead animal skins, and metal testified in their own way to the creativity of matter that became possible when active life was bracketed and simple life came to the fore. They emphasize as well Hunter's understanding of suspended animation as more than simply a state in which existing powers were preserved or conserved but one on the basis of which new human powers and social relations could be constructed.

SUSPENDED ANIMATION AND POETIC SUSPENSION

The impetus of Hunter's experiments in suspended animation—namely, his desire to extend a single human life across many centuries—bears a generic resemblance to the medieval and Renaissance search for the "philosopher's stone," which was purported to enable immortality. Yet the specifically cryonic nature of Hunter's proposal highlights the extent to which his version of this fantasy is fundamentally modern in nature. Being periodically frozen and unfrozen over the course of several centuries presumably would have had little appeal in earlier European social contexts, in which the judgments of one's peers were taken as the ultimate horizon of meaning or in which social life was understood as either constant or moving toward a religious apocalypse. The only context in which Hunter's proposal makes sense is one in which continuous and progressive social change is perceived as the norm. Only when social change is understood as endemic, swift, and neverending does it make sense to dream of periodically being frozen and revived in order to "learn what had happened during [one's] frozen condition." The rapid changes in surgical and experimental techniques that Hunter himself experienced during the course of the eighteenth century no doubt established the phenomenological basis of his fantasy of coordinating the temporality of a living human body with that of incessant social change.[28] Yet insofar as interest in suspended animation was sparked by canal accidents, the fantasies attached to this new concept were also bound up with the rapid and massive increase in transportation structures that enabled ever more expansive movements of people and goods across Britain and the Continent.[29] Through the figure of the canal, Hunter's uneasy fascination at the prospect of being left behind by processes of social change flows into other sites of transformation, such as the rapid urbanization of Britain, the development of colonial commercial and scientific networks, and "innovations" in political structures, all of which contributed to a sense of modernity at the end of the eighteenth century.

Suspended animation emphasized the modernity of these institutions by calling attention to a more fundamental fracturing, or heterogeneity, of time itself. It is common to understand the eighteenth and early nineteenth centuries as the period in which a modern sense of history—history understood not simply as chronicles of past events but rather as an all-encompassing narrative of human "development"—first emerged, and there is little doubt that the stadial histories proposed by authors as varied as Adam Smith and G. W. F. Hegel are premised on an understanding of human time as having only one real directionality, even if small eddies of "savages" had not yet been caught up in the rush of progress.[30] But the phenomenon of suspended animation pointed to a disturbing ontological split in time, for it suggested that different parts of the natural world imposed asynchronous temporalities on living beings. Hunter's experiments with ice are fascinating precisely because they imply that as one living being—a carp, for example—begins to freeze in the ice-filled tub, it ceases to occupy the same temporality as that of the experimentalist who stands directly next to it; from the perspective of the experimentalist, the organic functions of the fish begin to slow down, perhaps even stop entirely. As historian of science Alexandre Koyré has noted, the thought experiment of two stones simultaneously dropped, one from the mast of a ship and another from a bridge under which the ship passes, had served early modern philosophers such as Giordano Bruno and Galileo Galilei as a means for abandoning the Aristotelian conception of space as "a closed and finite world" in favor of a notion of an "open infinite Universe."[31] Yet the image of two bodies contiguous in space but each occupying its own time constituted a much more fundamental dislocation, for it suggested that time—at least time as it bore upon living bodies—differed depending on local conditions and there was no overall time or invariant measure that encompassed all of these local times; a man frozen in the past might nevertheless return and become part of the future.[32] The phenomenon of suspended animation suggested that this temporal variation could in principle be harnessed by means of technologies, but it also implied, as the converse of this hope, that temporal differences and asynchronicities might be *imposed* on living beings by technologies.

Romantic literary authors were especially attentive to the ways suspended animation emphasized the dislocations between an individual lifespan and the temporalities of modern social change. The strange role of sensation in Hunter's experiments, its persistence even in the absence of bodily actions, highlighted the extent to which suspended animation was neither simply physiological nor simply psychological (or "spiritual") but somehow both physiological and psychological at the same time, and literary authors were quick to exploit this liminal status of the concept. Suspended animation fascinated authors such as Coleridge and the Shelleys in part because of its promise to alter the nature of lived time, enabling states that merged life and death, dreaming and waking, past and future. Moreover, suspended animation seemed to demand the invention of new concepts and modes of thought. Hunter himself had contended that "a new bend of the mind" was necessary to understand his concept of simple life, while in A *Dissertation on the Disorder of Death* (1819), lay author Walter Whiter argued that "a new language" was required to understand how suspended animation might make it possible to overcome death entirely.[33] Literary authors drew on the concept of suspended animation precisely for this inventive potential, for it encouraged authors and readers to imagine new forms of connection between the past and the future, death and life.

Romantic-era poets and novelists engaged the inventive potential of suspended animation in part by representing links between suspension and different modern institutions or experiences.[34] The primary narrative of Mary Shelley's account of the "modern Prometheus" unfolds, for example, after Robert Walton restores animation to a "nearly frozen" Victor Frankenstein, but Walton is in a position to perform this modern act of resuscitation only because his voyage to the North Pole is itself part of the modern project of expanding global trade (Walton is in search of the Northwest Passage) and the scientific rationalization of nautical navigation (Walton hopes to discover the northern magnetic pole so that navigators can "regulate" nautical compasses).[35] And the queer vitality of the "stubble plains" and "soft-dying day" of Keats's "To Autumn" (1820)—a vitality exemplified by the simple life of the "store" of granary seed that persists in a form of suspended animation during the winter—is marked by the very fact of the poem as something worthy of notice; for the poem's primarily urban readership, an awareness of the strange vitality upon which agriculture depends has been dissociated from its traditional locus in the experience of farming and instead positioned as a means for cultivating a sense of repose.[36] To take another example from Keats, his 1820 "Ode on a Grecian Urn" also links suspended animation with institutions of modernity, despite the apparent classicism of this "cold pastoral" on frozen time. The poem describes figures of suspended animation pictured on the urn (e.g., the boughs of the trees that never shed their leaves), yet insofar as this is an ancient *Grecian* urn, these suspensions are linked to that newly emergent museum culture that made it possible for members of the lower middling classes, such as Keats, to see ancient art newly "recovered" from countries such as Greece.[37]

Romantic-era authors did not limit themselves to representing states of suspended animation within the content of their poetry and prose; they also developed literary techniques for introducing forms of suspension into the experience of literature itself, treating literature as a technology for creating "altered states." Keats's "Ode on a Grecian Urn," for example, suspends the forward movement that early-nineteenth-century readers had come to associate with that genre. Stuart Curran has noted that by the late eighteenth century, the ode was often understood as a form that emphasized dialectical movement; the passage from strophe to antistrophe was to reveal "a progression by polar opposites" that led to the final synthesis or "Stand" of the epode.[38] Yet if Keats's title thus solicited expectations that his poem would move forward toward a resolution and "message," the verse itself stages a cyclical movement of repeated return. The poem begins with the speaker's questions about the referents of the events represented on the urn: "What leaf-fringed legend haunts about thy shape . . . What men and gods are these?" (lines 5, 8). The second and third stanzas initially exploit the expectation of forward movement by transforming a limitation into a virtue—the silent melodies to which the urn alludes are "sweeter" than melodies actually heard (line 11)—and the poem then seems to synthesize these positions, suggesting that suspended movements and voices enable a kind of immortality. Yet the third stanza ends by undercutting this expectation of synthesis, describing intense sensations of burning and parching produced in the narrator by the movement of the preceding two stanzas. The fourth stanza returns to the questions of the first stanza, asking again after the referents of the events represented on the urn—"Who are these coming to the sacrifice?" (line 31). And the final stanza, rather than resolving this rhythmic movement of the ode, emphasizes that the urn is an object that "dost tease us out of thought" (line 44). The poem situates this experience within the context of "deep time"—"When old age shall this generation waste, / Thou shalt remain, in midst of other woe / Than ours" (lines 46–48)—and concludes with the cryptic contention that the urn's "message" is that "Beauty is truth, truth beauty,—that is all / Ye know on earth, and all ye need to know" (lines 49–50). Rather than movement forward toward resolution, Keats's ode employs movement in the service of stasis and suspension, returning the narrator and reader to questions already asked and found unanswerable.

Though earlier criticism has by no means ignored this combination of suspension and animation in Keats's ode, a shared, albeit implicit, critical commitment to the value of movement has often obscured Keats's commitment to the *suspension* of animation. Critics intrigued by the ekphrastic dimension of the poem tend to assign movement and suspension to different registers of the poem: Grant F. Scott, for example, distinguishes between the stasis of the poem's referent (the Grecian urn) and the movement of the words of the poem, while for W. J. T. Mitchell, the poem privileges a "paralyzing eternity" in its content in order to ward off the inevitable "dialectical" movement of ekphrasis.[39] Critics interested in exposing the ideological investments and functions of Keats's poetry also tend to separate, and privilege, what moves from what remains still. Marjorie Levinson, for example, argues that while much of Keats's earlier poetry made his desire for upward class mobility manifest in its content and form, the "Ode on a Grecian Urn" masks Keats's class aspirations by appealing to a pure enlightenment that the poem purports to deliver. Moreover, because the "working contradictions" that generated this poem are "arrested" at "the level of form and idea," Keats's ode could then become "the ideal commodity" for a middling class seeking to secure its own cultural capital.[40] From this perspective, we might also see the stasis of the poem as responsible for its canonical status, for though

Keats's ode ends after fifty lines, the suspension produced by its form encourages practices of close reading designed to return its auditors and readers again and again to the poem.[41]

These otherwise quite different interpretive approaches all assume that a poem *should* produce movement, whether by overcoming the stasis of sculpture through a flow of words or by encouraging a movement forward through an economic contradiction. Yet reading Keats's ode in the context of Romantic-era interest in suspended animation suggests that this effort to parse suspension and animation—to align vitality with what moves and death with what remains still—forecloses the possibility of *vital* suspensions of animation; that is, suspensions that revivify, rather than paralyze. Deconstructive readings of Romantic-era verse have been much more attentive to this possibility of thinking suspension and vitality together, often positioning what Murray Krieger called the "specially frozen sort of aesthetic time" of poetic discourse as a consequence of poetry's paradoxical commitment to *both* movement and stasis.[42] In the context of Keats's poetry more specifically, Tilottama Rajan has emphasized the suspensions of "work" at which Keats's verse aimed, a perspective that allows us to distinguish a passivity of "indolence" (which seeks to accumulate sensations as though they were capital) from the "wise passiveness" of suspension proper.[43] Yet insofar as deconstructive readings understand stasis and movement primarily in terms of epistemology and discourse, they often miss the *sensory* suspensions that Romantic-era poetry aims to engender. Though Keats's ode may employ images of stasis to dramatize epistemological paradox, its emphasis on coldness signals that the poem is also a technology for producing "trances" in readers and auditors. The suspensions produced by Keats's verse were, of course, quite different from those Hunter's experiments produced, but for both Keats and for Hunter, "to become cold" meant bringing to the surface a form of vitality in which sensation was liberated from its usual subordination to action.

The desire to treat language as a technology for creating altered states of sensation is especially evident in the irregular rhyme of P. B. Shelley's great hymn to suspended animation, "Mont Blanc; Lines Written in the Vale of Chamouni." Originally published as an appendix to Mary Godwin's anonymous travelogue *History of a Six Weeks' Tour* (1817), Shelley's poem followed several of his own letters in which he described Mont Blanc as a "vast animal," through the veins of which "frozen blood for ever circulated."[44] The poem depicts a narrator for whom Mont Blanc appears suspended "far, far above" him, while "Its subject mountains their unearthly forms / Pile around it, ice and rock; broad vales between / Of frozen floods, unfathomable deeps" (lines 60–64). The suspended animation of the mountain in turn reduces the narrator to a state of passivity, a "trance sublime and strange" (line 35). Yet as in the case of Keats's ode, suspended animation denotes not a lack but a surfeit of sensation, for it is from within this state that the narrator senses the tensions that link living beings to their environments. The frozen power of the mountain suspends the activities of plants and animals, prolonging that otherwise seasonal "torpor of the year when feeble dreams / Visit the hidden buds, or dreamless sleep / Holds every future leaf and flower," yet the narrator is able to sense that "bound / With which from that detested trance they [i.e., plants and animals] leap" (lines 88–91). The poem positions "torpor" as a state within which it is possible to *sense* the vital difference between "simple" and practical life.[45] What appears from outside as simply the paralysis of plants and animals is in fact the site of a differential—a productive difference—between the living being and its environment, one that enables the movement of living things (their "leap") when environmental conditions change.[46]

Through its peculiar use of rhyme, Shelley's poem seeks to produce a similar state of tense suspension in its readers or auditors. As William Keach has noted, "Mont Blanc" is not, as it is sometimes taken to be, a poem in blank verse but is rather poetry suspended between blank verse and rhyme. Shelley employs rhyme consistently and throughout the poem: as Keach notes, "of the 144 lines in *Mont Blanc*, only three end in words which have no rhyme elsewhere in the poem."[47] However, Shelley's use of irregular forms of rhyme—imperfect rhyme, internal rhyme, homonymic rhyme, and large distances between rhyming words—makes the poem feel like one written in blank verse, especially as one gets further into the poem. "Mont Blanc" begins with tightly bound rhyme:

> The everlasting universe of things
> Flows through the mind, and rolls its rapid waves,
> Now dark—now glittering—now, reflecting gloom
> Now lending splendor, where from secret springs
> The source of human thought its tribute brings
> Of waters—with a sound but half its own,
> Such as a feeble brook will oft assume
> In the wild woods, among the mountains lone,
> Where waterfalls around it leap forever,
> Where woods and winds contend, and a vast river
> Over its rocks ceaselessly bursts and raves. (lines 1–11)

In this first stanza, each of the first five end words has at least one rhyming complement, and where the first instance of a rhyming sound occurs in a noun, the second instance generally occurs in a verb. This initial regularity of rhyme is aligned with an abstraction of content: though the latter half of this stanza emphasizes apparently concrete particulars—a feeble brook; a vast river—these are in the service of a complicated analogy intended to explain the embeddedness of the mind within the natural world. Yet if this first stanza establishes with the reader a rhyme-based variant of what John Hollander calls the "metrical contract"—an implicit agreement about what kind of poem this is to be—it is a contract the limits of which are soon tested.[48] As the poem moves from the philosophical contemplation of the first stanza to descriptions of trance, suspended animation, and vital tension in subsequent stanzas, line-end rhyme becomes more and more unpredictable. Rhyme begins to connect the ends and middles of lines, and extended rhythmic units are created, in which line-end words receive rhyming complements only much later in the poem. "Things," the word with which Shelley ends the first line of the poem, loops, for example, in a long rhythm through the poem—appearing also in lines 40, 46, 85, 94, and 139, sometimes at the end of lines and sometimes in the middle—and serves as a rhyming pair for words such as "influencings," "wings," and "brings" (lines 38, 41, 68). Moreover, as Keach notes, three line-end words—"forms," "spread," and "sun" (lines 62, 65, 133)—are fully suspended, having no line-end rhymes elsewhere in the poem, frustrating any expectation for a final, harmonious rhythmic organicism.[49]

The irregular rhyme of Shelley's poem points toward a tension between rhyme and meaning that is far "wilder" than that valorized by New Critical and structuralist readings of rhyme. In his classic account of the relation between verse rhyme and reason, W. K. Wimsatt suggested that verse instantiates a virtuous form of "wildness" in cases in which the semantic role of rhyming words remains in tension with the repetition inherent in rhyme. Such is the case, Wimsatt argues, when two rhyming words play different grammatical roles—in "Mont Blanc," for example, when the noun "springs" is rhymed with the verb "brings"—or, conversely, when rhyme plays the role of a "binding force" that connects a semantic "incongruity or unlikelihood."[50] Yet for Wimsatt, as well as for later structuralists such as Roman Jakobson, the tension of wild rhyme is understood as dependent upon readerly judgments, whether conscious or unconscious, about specific and well-defined verse elements (for example, a judgment about the relationship between the grammatical roles of two rhyming words).[51] The irregular rhyme of Shelley's poem, by contrast, keeps readerly judgment in a state of constant tension and incompletion. Because the rhymes of Shelley's poem occur across such large and irregular stretches of verse, auditors or even readers of the poem are hard-pressed to keep track of the relative semantics of rhyming words (tracking that would be necessary if one is to adjudicate relationships of particular rhymes to the meaning of sentences or phrases).

Neither so regular as to produce simple patterns of repetition nor so irregular as to enable the fact of rhyme to be lost, Shelley's looping rhymes enable an ever-shifting modulation—what Immanuel Kant would have called a nonlegislative free play—between sensation and consciousness. The auditor or reader of the verse remains suspended, constantly in search of the ever-changing measures of the poem. In place of Wimsatt's adjudicating consciousness, which calmly compares the data of sensation with the concepts of the understanding and marks the difference between the two, Shelley's poem privileges readers, speakers, and listeners who give themselves over to a kind of trance in which the relationship between the sonic medium and the meaning of the poem continually shifts and modulates.[52] Grasping the extended and irregular rhymes of Shelley's "Mont Blanc" requires a suspension of judgment, rather than judgment itself. Rather than seeking or establishing an alternate foundation for proper measure—along the lines, say, of Ezra Pound's later Imagist injunction to poets to "compose in the sequence of the musical phrase, not in sequence of a metronome," or Charles Olson's efforts to move the basis of measure to the living breath so that the forces of what Olson called a "field" could be properly registered—Shelley's poem ensures that we must continually seek, without ever being sure that we have found, any proper measure for its rhymes.[53] Sensation thus plays for the reader or auditor of the poem a role not unlike that of the living body in Hunter's experiments with cold: rather than handing up data to consciousness so that judgments can be made, sensation instead retains a kind of autonomy as the rhymes of the poem encourage continuous modulation or search for measure. Shelley, in short, does not employ rhyme in "Mont Blanc" as a medium in the sense of a stable *channel* that communicates information and thereby enables an observer to make judgments; rather, the medial function of rhyme functions more in the sense of a shifting *milieu* to which a living auditor or reader feels bound, and to which that reader or auditor continually seeks to bind him- or herself.[54]

THE POLITICS OF SUSPENSION

Romantic-era techniques of poetic suspension were not, of course, unprecedented. Milton's poetry—a precedent that loomed large for many Romantic poets—is marked by frequent instances of grammatical suspension, as Milton deferred for multiple lines the completion of a sentence or the referent of a pronoun.[55] Yet however much Romantic-era techniques of poetic suspension may have drawn formally from such precedents, poets such as Keats and Shelley as-

sumed a largely secular framework that distinguished the *telos* of their verse practices from that of their predecessors. Milton's techniques of suspense were an
aspect of his effort to encourage "patience under adversity" until the arrival of the
moment (*kairos*) within which God's plan would be revealed; as Milton put it in
his sonnet on his blindness, "they also serve who only stand and wait."[56] Keats's
and Shelley's techniques of poetic suspension, by contrast, did not assume this
validating background of religious faith. Rather than understanding suspension
as an interval between the present and a future kairos, Keats and Shelley presented the suspension of animation as an end in itself; if anything like redemption
was to be had, it occurred within, rather than beyond, suspension. For Keats and
Shelley, poetic suspension was neither a mimesis of nor a technique for encouraging political patience; rather, it served as a means for enabling relationships
between sensation and consciousness that could emerge only when animation
and movement were halted.

The contrast with Milton highlights the political stakes of this Romantic-era
belief that something productive could emerge from states of torpor and passivity. If Keats's and Shelley's interest in stasis and torpor are not to be understood
as poetic analogues of a political practice of patience, what then are the politics
of their techniques of poetic suspension? Is their emphasis on suspended animation simply an attempt to replace political praxis and movement with quietism,
encouraging individuals to take pleasure in sensory tension and an apolitical "life
of sensations," to appropriate Keats's famous phrase?[57]

Coleridge developed one answer to this question, positioning suspended animation as an object of critique. As I noted above, Coleridge was suspicious of any
form of suspension that affected the will, and he vilified "modern" mass media
institutions that he felt encouraged trances. He contended, for example, that the
relatively recent institution of the circulating library promoted an indiscriminate
approach to reading, thus encouraging

> a sort of beggarly Day-dreaming, in which . . . the mind furnishes for itself only
> laziness and a little mawkish sensibility, while the whole *Stuff* and Furniture of
> the Doze is supplied *ab extra* by a sort of spiritual Camera Obscura, which (*pro
> tempore*) fixes, reflects, & transmits the moving phantasms of one man's De-
> lirium so as to people the barrenness of a hundred other trains [of associations]
> . . . under the same morbid Trance, or '*suspended Animation*,' of Common
> Sense, and all definite Purpose.[58]

What differentiated the "suspension of disbelief"—a state that Coleridge valorized—from the "morbid Trance, or '*suspended Animation*,' of Common Sense"

was the relationship of will and sensation in each. Coleridge happily endorsed the "bracketing" of belief and disbelief so long as this suspension occurred under the legislation of the will—so long, in other words, as it was a state of *willing* suspension of disbelief—but he was critical of states in which the will itself seemed suspended. Coleridge was comfortable with suspension when it remained contained within an epistemological register, and thus remained subject to the legislative authority of the will, but suspicious of suspensions that led to a non-legislative free play among knowledge, acts of willing, and sensation.[59]

Coleridge's striking allusion to the camera obscura highlights the resonances between his critique and that of later critics of mass media, particularly those associated with the Frankfurt School. His fear that media institutions suspend the critical faculties of audiences by "fix[ing], reflect[ing], & transmit[ting] . . . moving phantasms" is an early articulation of the logic that underwrites Susan Buck-Morss's more recent history of "anaesthetic" uses of media technologies, a history that begins in the nineteenth century and eventuates in twentieth-century mass media such as cinema and television. Drawing on the work of Walter Benjamin, Buck-Morss argues that the early-nineteenth-century history of this lineage revolved around both substances that temporarily numbed bodily sensation, such as surgical ether, and technologies that "flood[ed] the senses," such as coffee, cocaine, and phantasmagoria (e.g., magic lanterns). Buck-Morss contends that both the numbing and overstimulation of the senses involved a "conscious, intentional manipulation" of what she calls the "synaesthetic system" (that "circuit from sense-perception to motor-response [that] begins and ends in the world"). This technological manipulation produces an erasure of "experience"—that is, a failure to integrate individual experiences into collective and more temporally extended meaningful wholes—and a consequent atrophying of political awareness, will, and action.[60]

Yet Hunter's physiological experiments with simple life and Keats's and Shelley's experiments in the poetics of trance suggest that this critical perspective depends upon an incomplete analysis of the links between animation and automaticity in modern society. For Coleridge, as well as for Buck-Morss, the problem with modern techniques of sensory manipulation is that they *introduce* automaticity into the "synaesthetic system"—and thus, to the extent that Romantic poetry and modern mass media technologies produce trances, they risk automatizing motor responses along lines dictated by consumerist or authoritarian political structures. For Keats and Shelley, by contrast, automaticity is always already a fact of human life; it does not depend in any essential way upon new media technologies, nor can we look back to a past state within which relationships between sensation, consciousness, and motor-response were in some way more free. For Keats and Shelley, then, the problem was not the suspension of animation but rather its ceaselessness. The pressing task was thus locating means for *suspending* this automatic animation of the world, for it was only by "bracketing" habitual links between sensation and motor-response that new habits became possible.

This emergence of suspension as a valorized "negative capability" is especially evident in Shelley's different representations of animation in Queen Mab, an early poem, and "Mont Blanc." Drawing on the materialist philosophy he had adopted from David Hume, Baron d'Holbach, and William Godwin, Shelley suggested in Queen Mab that humans were particularly complicated bundles of habits impelled by the same natural forces of animation that kept the rest of the natural world in constant motion. However, he also suggested that mind and thought were causally linked to the physical interactions of the material world, with the consequence that both mind and matter were slowly "re-forming" along utopian lines. He depicted this progressive materialism in the image of one "wide-diffused . . . / spirit of activity and life, / That knows no term, cessation, or decay."[61] Thus, in this early work, Shelley employed "animation" and "activity" as linked concepts that allowed his readers to think systemically: by imagining the universe as ceaseless "activity and life," readers would be able to sense, and orient themselves toward, global and systemic changes already under way, changes of which they themselves were parts. From this perspective, anything that hindered animation—for example, the earth's polar ice caps—was problematic, and the poem includes a vision of a future in which "those wastes of frozen billows that were hurled / By everlasting snow-storms round the poles" are "unloosed," thereby destroying the frigid zone with which Hunter had once hoped to extend human life spans (102; lines 59–63).

"Mont Blanc," however, represented a significant revision of the progressive materialism that underwrote Shelley's earlier poem. Rather than valorizing images of ceaseless animation, "Mont Blanc" suggested that humans facilitate the progressive movement of matter only *after* they place themselves into altered states of suspension in which they can sense, and thus bind themselves to, the differentials that traverse both the natural world and human bodies. Precisely because "the everlasting universe of things / Flows through the mind" (lines 1–2), the differentials these "waves" produce can then be linked to the smaller-scale differentials that the sensing human body establishes within itself. Frozen glaciers are understood no longer as a threat to ceaseless animation but rather as a point of vital pause within the movement of matter. "Mont Blanc" suggests that it is from within the altered state produced by submitting oneself to the rhythms of this poem that new modes of synthesis with the "measures" of the world become possible.

Speaking more generally, we can say that where Coleridge understood suspension in terms of a powerful exterior force that controlled a susceptible interiority (thereby preventing otherwise autonomous acts of interior willing), Keats and Shelley understood suspension as that which enabled and established the distinction between a psychological inside and a nonpsychological outside—it was suspension itself, in other words, that first made possible that very capacity for willing that Coleridge sought to protect. For Coleridge, suspension was a being-held-by-something; for Keats and Shelley, by contrast, suspension denoted a holding-against that served as the basis for the subsequent development of particular forms of willing. For Shelley especially, this understanding of suspension had direct political application, for he insisted that the inequities and violence of the oppressive British state could be overcome only through strategies of "passive resistance." Passive resistance was neither a deferral of action nor a form of patience but rather a non-action—a suspension of action—able to suspend the automatism of violent action and reaction.[62]

SUSPENDED ANIMATION AND THE BEING OF SENSATION

These differing assessments of suspended animation were due in large part to a basic disagreement about how best to understand the nature of sensation. Coleridge and more recent commentators such as Buck-Morss are part of a long tradition in which sensation is approached solely from within an epistemological framework—that is, in terms of its contribution to the formation of knowledge.[63] Within this epistemological framework, sensation denotes the passive reception of data by the senses, or, as Coleridge put it, sensations are "experimental notices of the senses" conveyed to the faculty of understanding.[64] Seen as a faculty or capacity for generating "notices," sensation then ought to remain in the service of the understanding and the will.[65] As I noted above, suspended animation disturbed him because he feared that if it was indeed a state in which neither the will nor reason legislated, then sensation must have grabbed the reins. Keats and Shelley, by contrast, did not approach sensation from within the frame of epistemology but instead approached it ontologically. Keats and Shelley were interested in sensation as a capacity—the limits of which were not necessarily known—of parts of living bodies, rather than simply as a faculty that played a role in the construction of knowledge about the world. From an ontological perspective, the question of how sensation contributed to the formation of knowledge was secondary; the more fundamental question was how, through sensation, a living body linked itself with other forms and states of matter and motion.[66] This onto-logical perspective did not contest that sensation could function as a middleman that handed up "notices" to consciousness, but it emphasized that such a hand off was possible only because, prior to such acts of transfer, sensation functioned as a corporeal power that bound the body into the world, and which had the capacity to establish and preserve differences between the inside and outside of the body.

The ability to think sensation ontologically, rather than epistemologically, in the Romantic era was no doubt in part a consequence of the debate among anatomists in the late eighteenth and early nineteenth centuries about the dis-tinction between "irritability" and "sensibility." In the mid-eighteenth century, the Swiss physician Albrecht von Haller employed the distinction between ir-ritability and sensibility to focus attention on the specific powers of different kinds of living tissues: rather than seeing the body as a series of tubes that were con-nected hydraulically to, and controlled by, a central agency in the brain, Haller instead contended that the body was made up of different kinds of fibers that had their own immanent powers. For Haller, irritable tissues (e.g., muscles) had an autonomous power to contract when touched, while sensible tissues (e.g., nerves) had an autonomous power to create mental impressions.[67] Debates sub-sequently emerged around this distinction both because authors disagreed about the precise powers to attribute to irritability and sensibility and because it was often difficult in practice to assign only one of these powers to specific tissues.[68] However, the significance of this debate lies less in the specific claims made by commentators than in the impetus that it provided for empirical investigation of the powers of tissues. The distinction between sensibility and irritability encour-aged researchers to investigate empirically a variety of capacities and thresholds: the specific stimuli to which different tissues were able to respond; the refractory times following stimuli during which irritable tissues were unable to respond; the quantitative or qualitative minima and maxima within which sensible tissues were able engage the world outside the tissue. Irritability and sensibility—and, by extension, sensation—had to be investigated ontologically, as capacities (in some cases quantifiable) of tissues, rather than simply as functions within an epis-temological problematic. In some cases, investigation could occur in vivo on the basis of self-reports (e.g., Blagden's and Fordyce's experiments in heated rooms), while in other cases, tissues had to be separated from the body and investigated in vitro (e.g., Hunter's experiments with ice). In either case, though, discussions of sensibility and irritability emphasized ontology, rather than epistemology, and in this sense established a more general problematic on the basis of which late-eighteenth- and early-nineteenth-century authors could begin to understand sensation as a synthetic capacity of living matter, rather than simply the transmission of information from the body to the mind.

In medical circles, this alternative approach to sensation emerged primarily in discussions and analyses of anomalous phenomena, rather than in textbook accounts of the normal, habitual capacities of the body. In his reports on the effects of inhaling nitrous oxide, for example, the chemist Humphrey Davy used the term "sensation" to describe an absolutely new experience, one that could not be explained simply in terms of the conceptual ordering of data received by the senses. Despairing of his ability to describe his experience in language, Davy wrote, "I have sometimes experienced from nitrous oxide, sensations similar to no others, and they have consequently been indescribable."[69] Other subject-witnesses who participated in Davy's experiments used similar language, describing their experiences of strong sensations that could not be described in terms of content but only compared with other sensations in terms of intensity. J. W. Tobin, for example, noted that for him it was "not easy to describe [the] sensations" he experienced after inhaling nitrous oxide and could only assess these as "superior to anything [he] ever before experienced" (16). George Burnet provided a particularly intriguing report, suggesting that his sensations were experienced even before the stimulus reached the brain: "There is now a general swell of sensations, vivid, strong, and inconceivably pleasurable. They still become more vigorous till they are communicated to the brain, when an ardent flush overspreads the face" (16). Coleridge, himself one of the participants in these experiments, suggested that inhaling nitrous oxide created an atmosphere of sensation that then altered the "impressions" one received from external objects: "The first time I inspired the nitrous oxide, I felt a highly pleasurable sensation of warmth over my whole frame . . . On removing the mouth-piece, the whole sensation went off [i.e., ended] almost immediately. The second time I felt the same pleasurable sensation of warmth, but not, I think, in quite so great a degree. I wished to know what effect it would have on my impressions; I fixed my eye on some trees in the distance, but I did not find any other effect except that they became dimmer and dimmer, and looked at last as if I had seen them through tears" (17). In describing his altered state, Coleridge made a distinction between impressions, which here referred to the "notices" sent by the senses to consciousness, and sensation, which established something more like a foundational mood or atmosphere within which the referents of sensory impressions were situated.

Hunter's experimental work with states of suspended animation illuminated what was at stake in the ontological approach to sensation. For Hunter, suspended animation was striking precisely because even when the actions that depended upon the will and understanding were suspended, sensation nevertheless remained, regulating the temperature of the living being in relationship to its environment. This fact emphasized that sensation was not solely a capacity that translated aspects of an environment into notices sent to consciousness (if simply because in cases of suspended animation, there was no consciousness to which such notices could be sent). Rather, sensation was more fundamentally the capacity of living tissue—and by extension, a living being—to maintain differences between itself and its environment (in this case, differences of temperature). Sensation enabled a living being to suspend itself within the world by establishing a border that maintained differences between the living being and the world. In the absence of this capacity of sensation (i.e., after a living being had died), the body became simply a conduit through which the forces of the world were directly transmitted.

For Hunter—as well as for Keats and Shelley—the "being of sensation" thus could not be grasped solely as a matter of physiological notices or mental representations but rather had to be understood as what Gilles Deleuze and Félix Guattari describe as a "compound of nonhuman forces of the cosmos."[70] Sensation is a compound in the sense that it holds together otherwise heterogeneous forces or vectors. When a living being is placed in a cold external environment, for example, sensation binds together the cold external environment and the warm interior of the living being, enabling these two temperatures to exist side by side, rather than allowing the cold environment to extend its reach into the interior of the living being. Living bodies can indeed derive data about the external world from this ontological capacity of sensation, but sensation is not *fundamentally* a mirroring of the world by the organism; it is, instead, the ontological capacity of the living body to bind itself into the measures of the world.

Poets such as Keats and Shelley sought by means of their techniques of poetic suspension to solicit this capacity of sensation to compound differences. As Davy, Coleridge, and their fellow experimentalists had discovered, one could briefly illuminate the ontological side of sensation by means of altered states produced by drugs such as nitrous oxide. However, such states were fleeting, inconstant, and incommunicable. Poetry, by contrast, offered Romantic authors a means for preserving and amplifying the ontological dimension of sensation. In a general sense, poetry—or, at any rate, eighteenth- and nineteenth-century poetry—"couples" two different levels or orders: a semantic order of meaning and a rhythmic order of meter and rhyme.[71] Eighteenth-century poets often sought to subordinate one of these orders to the other: to use rhythm, for example, to render the propositions of the semantic order more "forceful," as in Alexander Pope's formulation, and in this way make the order of meaning appear to be more primary than the order of rhythm.[72] Keats and Shelley, by contrast, aimed to produce what Hans Gumbrecht calls a "constitutive oscillation" between the semantic and rhythmic orders.[73] Such an oscillation preserves and amplifies the compound between the orders of meaning and rhythm rather than privileging one over the other, and sensation names the ability of auditors or readers of the poem to suspend themselves within this compound.

The sensation produced by this constitutive oscillation between rhythm and meaning is strikingly impersonal; it is less something that a reader or auditor "has" or "possesses" than an experience that is briefly occupied or endured. Thus, though both "Ode on a Grecian Urn" and "Mont Blanc" describe the experiences of a first-person narrator, these lyric subjectivities do not provide points of identification that allow readers or auditors to mirror the lyric voice of the poem, either to accept or contest the conclusions reached by that voice. Instead, the lyric voice functions as a sensitive line, like that of a thermometer, that allows auditors or readers to register the impersonal tensions, both semantic and rhythmic, that traverse the poem.

As Keats's ode "To Autumn" demonstrates, Romantic poets could establish a sensitive line even in the absence of a subjective lyric voice; rather than registering the deformations of a subjective voice, "To Autumn" employs the figure of apostrophe to create an impersonal line within a field of tensions. "To Autumn" has long been understood as a poem of both sensation and suspended animation, its "gorgeous sensuousness" and "sensuous firmness" leading readers to a sense of "repose" (or, less generously, "stagnation").[74] Yet as Geoffrey Hartman has noted, "To Autumn" is also a poem of "true impersonality," one that lacks the lyric "I" of either "Ode on Grecian Urn" or "Mont Blanc."[75] "To Autumn" enables this feeling of impersonal repose by employing apostrophe to compound reference and self-sufficiency, narrative and atemporality, and organic and non-organic life. As Jonathan Culler notes in his brilliant essay on apostrophe, the blatant impossibility of the apostrophic address—the fact that apostrophes address absent or inhuman entities, such as "autumn," that cannot respond to human speakers—serves to suspend the referential function of language.[76] The "thee" and "thy" of line 12 of "To Autumn," for example—"Who hath not seen thee oft amid thy store?"—do not refer us to an entity outside the poem that could answer such questions, nor do they refer to a "feeling" Keats had about autumn, nor even to an "image of voice" that would allow a poet like Keats to bring a poem into being by addressing inhuman entities as though these latter could respond. Instead, the apostrophic address functions by suspending the referential use of language, which in turn allows the poem to suspend narrative—that is, those temporal sequences that arrange referents within the frame of past, present, and future. Culler suggests that the goal of this double suspension is the emergence of a peculiar poetic temporality, one that "might be called the timeless present but is better seen as the temporality of writing," that is, that "set of all moments at which writing can say 'now.'" By suspending reference and narrative, apostrophe allows us as readers and auditors to bracket "our empirical lives . . . forgetting the temporality which supports them" in favor of the "now" of the poem (149).

"To Autumn" reminds us, though, that apostrophe creates its suspended present not by severing language from referentiality, but rather by suspending particular references. The specific experience of the suspended present at which "To Autumn" aims is that of "autumning," the title of the poem functioning as an address *to* autumn but also as the infinitive form of a new verb: "to autumn." The apostrophic address of the poem enables this experience of repose by referring us to a "referent" that is itself characterized by suspension, for autumn is a season of transition in which the practical life of organic development begins to be suspended in favor of the simple life of seeds and hibernation. The structure of the poem adopts this rhythm of suspension, moving the reader or auditor from the first stanza's images of organic unity—Autumn and the sun conspiring to "swell the gourd" (line 7) and "set budding more,/And still more, later flowers" (lines 8–9)—to the final stanza's image of a non-organic mode of life, in which it is now the "barred clouds," rather than organic beings, that "bloom" the "soft-dying day,/And touch the stubble fields with rosy hue" (lines 25–26). This impersonal field of non-organic life represented in the final stanza of the poem is populated by hosts of organic entities: wailing gnats, bleating lambs, singing hedge-crickets, a whistling red-breast, twittering swallows. Yet the adjectives that modify these organic entities (wailing, bleating, singing, whistling, twittering) draw them back into the diffuse field of non-organic life that blooms around them by emphasizing the non-semantic rhythmic forces that traverse the human language within which the poem itself is written. With the arguable exception of "singing," all these verbs denote rhythmic sounds, or refrains, that occur outside of the register of human meaning. "Twitter," for example, *means* something—that is, the word denotes for the auditor or reader of the poem a particular kind of sound—but it *means* an instance of sound that does not operate in the register of human meaning; to say that a bird (or a person) twitters means that we, as listeners, can in principle grasp only the form, not the content, of this rhythmic speech. The adjective "twittering"—along with "wailing," "bleating," and "whistling"—employ denotative meaning to point to the suspension of meaning. The "repose" of the poem is thus produced not because apostrophe severs reference and temporal narrative but rather because apostrophe creates a field of tensions that holds together reference and self-sufficiency, narrative and an impersonal present, and organic unities and the impersonal refrains that traverse them.[77]

Understanding sensation as an impersonal ontological compound rather than a representational mirroring of the world shifts as well our sense of what it means for Romantic poetry to aim at what Keats called "a Life of Sensations rather than Thoughts."[78] Keats's preference for a life of sensations was *not* oriented toward the "accumulation" of sensory stimuli in the sense of adding to the storehouse of personal experiences that would reconfirm the primacy of the poet's or reader's subjectivity. Rather, Keats's life of sensations referred to that sense of rending, tearing, and disorienting vitality that occurs when a living being binds itself, or is bound, to the inhuman forces of its environment in new ways. Sensation is fundamentally synthetic in nature, continually binding and rebinding conscious-ness to the inhuman measures of the world, and while this synthetic capacity of sensation normally operates in the service of existing habits, states of suspension produce forms of bracketing that enable new modes of synthesis with these world-measures. Keats's "Life of Sensation" thus names those moments when poetry functions as a technology for opening organic life to the inhuman forces of non-organic life.

There remains, of course, the possibility, feared by both Coleridge and Buck-Morss, and desired by the authors of the first epigraph to this chapter, that this synthetic capacity of sensation—the capacity to reconfigure the relationship be-tween a living being and its environment—could be exploited to produce "sug-gestibility." Yet producing this kind of automatism would demand a near-absolute regulation of an individual's environment, and even in such cases, suggestibility would still depend upon that more primal capacity for synthesis that suspension enables. Thus, though suspension may indeed produce a drugged-like state, this must be understood as an addiction-to-synthesis or a craving for openness (with "openness" understood as the desire for ever-new modes of synthesis).[79] A crite-rion of recursive expansion or feedback guides Keats's and Shelley's poetics of trance: suspensions are good insofar as they enable, rather than foreclose, future suspensions. It could thus only be in extraordinarily limited environments—total institutions like secret jails, for instance, in which the environment is managed so completely that it is drained of the potential for engagement—that this addiction-to-synthesis might be channeled into suggestibility. Both Coleridge and Frankfurt School critics feared that mass media would be able to create such managed environments for the populace at large. Yet this fear seems to overestimate the power of mass media to create environments while at the same time it depends (as I discuss in chapter 4) upon a limited and problematic concept of "media."

SUSPENSION, ANIMATION, AND THE SUSPENDED HISTORY OF ROMANTIC EXPERIMENTALISM

By allowing us to link late-eighteenth-century vitalist experiments on animal heat with an early-nineteenth-century poetics of cold pastoral, the concept of suspended animation expands our understanding of the ways in which Romantic-era literature was experimental. As I noted in chapter 1, Wordsworth and Coleridge understood their experiments in *Lyrical Ballads* under the rubric of "fitting." In the 1798 "Advertisement" to *Lyrical Ballads*, the reader's task was described as that of "ascertain[ing] how far the language of conversation is adapted to the purposes of poetic pleasure," but in the 1800 preface, these judgments of "adaptation" were more explicitly described as a matter of determining a "fit"; that is, the reader was to "ascertain, how far, by fitting to metrical arrangement a selection of the real language of men in a state of vivid sensation, that sort of pleasure and that quantity of pleasure may be imparted, which a poet may rationally endeavour to impart."[80] From this perspective, a work is experimental when it re-"fits" relationships between literary form, content, and a certain kind of readerly sensation. Keats's and Shelley's experimental practices of suspension, by contrast, did not strive to fit poetic form and content to one another in order to produce a particular species of sensation (e.g., pleasure) but rather sought to engender trances. Their mode of poetic experimentalism thus aimed at a non-legislative free play among sensation, consciousness, and the milieu of those nonhuman forces within which an individual existed. Such free play might indeed produce a sense of exhilaration akin to pleasure. However, this was not a feeling that operated within the limits of an organic pleasure principle but was instead an affect produced by that distension of faculties—the overcoming of their former limits—that Shelley described in *A Defence of Poetry*.[81] In place of the activity of "fitting," which always relies on some given measure, the experimentalism of suspension aimed to enable an experience of the emergence of measure itself.

The interest of Romantic poets in suspended animation also has implications for our understanding of the historicity of these poems and their poetics. As I noted in chapter 1, Clifford Siskin has argued that the experimental nature of Romantic lyrics should be understood as an integral part of a larger disciplinary project that helped to create and valorize the category of Literature. Romantic literary experimentalism helped to consolidate the concept of Literature, Siskin argues, by "coupling . . . the lyric to personal, subjective feeling—the *I* expressing itself"; this coupling "naturalized" the "'experimental' role [of the lyric]: lyrics functioned as 'data' in innovative efforts to know the present by linking it developmentally to the past."[82] Yet, as I note above, Shelley's and Keats's poetics of cold pastoral did not aim to link subjective expression with developmental narrative. Their interest in suspended animation was instead oriented toward an experience of impersonality, and rather than plotting a narrative chain from the past through the present into the future, suspended animation confused temporality, emphasizing that there was not *one* time—the premise of the schema of "development"—but rather multiple times. In place of perceptions of temporal and historical development, suspended animation enabled sensations of straining and intensity, as the trance of suspended animation slowed down development and history to the point that both could be reconfigured.

The post-Romantic history of the concept of suspended animation emphasizes this open understanding of temporality and historicity, for rather than revealing a narrative of development, the history of the concept of suspended animation is instead a narrative of multiple suspensions. Within the fields of medicine and psychology, interest in suspended animation waned in the 1820s, only to be revived in the 1840s in the context of debates among surgeons about the respective benefits of ether and mesmerism. Though Hunter's understanding of suspended animation as a possible therapeutic technology had not been shared by all of his medical peers, the introduction of surgical anesthesia in the 1840s linked suspended animation to the image of docile and easily managed patients and allowed surgeons who promoted the use of anesthetic drugs to explain (and justify) the state into which such gases placed patients.[83] However, following the ascension of surgical anesthesia, the concept of suspended animation again faded from view until the late nineteenth century, when it was then revived in the wake of the emergence of technologies of refrigeration and freezing.[84] In the twentieth century, interest in suspended animation has been suspended and revived every couple of decades, with recent interest in personal cryogenics (i.e., freezing one's almost-dead body in the hopes of future revival) simply the most recent variational revival of Hunter's original cryogenic dream.

Within what we might loosely call the humanities, interest in suspended animation has followed a similar, yet at the same time more complicated, pattern. Where Coleridge's suspicions of the suspended animation produced by mass media found echoes in both modernist and Frankfurt School–inspired critiques of the narcotic effects of mass media, Keats's and Shelley's interest in the productive power of suspended animation has remained alive in twentieth-century reflections on the productive potentials of suspension. A sense of suspension as the condition of possibility of organic life itself, for example, was taken up at the start of the twentieth century both in the work of Henri Bergson, who claimed that living beings are able to produce a "delay" within otherwise automatic relations of action and reaction, and in early-twentieth-century phenomenological efforts to "bracket," or suspend, habitual ways of engaging with the world.[85] More recently, elements of these Bergsonian and phenomenological traditions of interest in suspension were synthesized by Gilles Deleuze and Félix Guattari in their work on sensation, "bodies without organs," and "non-organic life." For Deleuze, our subjective feelings and perceptions are best understood as "contractions," or syntheses, which can provide "notices" about aspects of the world only because they

> refer back to organic syntheses which are like the sensibility of the senses; they refer back to a primary sensibility that we *are*. We are made of contracted water, earth, light and air—not merely prior to the recognition or representation of these, but prior to their being sensed. Every organism, in its receptive and perceptual elements, but also in its viscera, is a sum of contractions, of retentions and expectations.[86]

Deleuze contends that organic life—life structured into systems of organs and oriented toward consciousness and self-consciousness—is possible only because of another mode of vitality, a "non-organic" life that contracts and holds elements together. Like Hunter's simple life, Deleuze's non-organic life is difficult to isolate from organic life, though Deleuze, like Hunter, points to the fertilized egg as an example of the operations of non-organic life.[87] Moreover, for Deleuze and Guattari, as for Keats and Shelley, art can make non-organic life intuitable, and it does so by establishing compounds of impersonal "percepts" and "affects."[88] It is thus no coincidence that Deleuze found himself solicited by the scene of Riderwood's suspended animation and subsequent reanimation in Charles Dickens's *Our Mutual Friend* and that he argued that Dickens employed the trope of suspended animation to reveal an "immanent life carrying with it the events of singularities that are merely actualized in subjects and objects."[89]

Traces of the Romantic experimental approach to suspended animation also persist in the late-twentieth-century popular fascination with the confusion of temporalities that characterize various forms of "half-life." These traces are most evident in horror and science-fiction novels and film, genres in which suspended animation often serves as a narrative device that allows authors to carry a traumatic past into the present (horror) or reconstellate possible futures through potentials contained in the present (science fiction). In both cases, suspended animation produces experiences of intensity—delightful horror, for example, or the feelings of quasi-sublimity associated with science fiction—by allowing the perceptions and feelings of one time period to bear on, and alter, another quite distant time period. Yet even when employed simply as a narrative device, the trope of suspended animation revives elements of Keats's and Shelley's practice of cold pastoral, forcing readers beyond a purportedly "realistic" sense of time and enabling new rhythms of temporal connection.

Given this history of suspensions, the specificity of Hunter's, Keats's, and Shelley's Romantic-era approaches to the logic of suspension remain especially vital, for they collectively suggest that the "critical" approach to the paralyzing tendencies of modern society and mass media focus attention on a secondary level and thus risk missing the extent to which suspension makes possible both resistance and the possibility of creating something new. Keats's and Shelley's work suggests that if art facilitates the construction of a better world, it does so not solely on account of its content but in addition because it encourages a state of suspension that frees sensation, enabling new forms of linkage between elements of the open system of the world. Keats's and Shelley's practice of cold pastoral thus suggests that the *telos* of a politically engaged aesthetics cannot be solely to awaken a population frozen in automaticity but must also seek to produce suspensions in those who are already far too animated. Suspension, in this latter sense, involves empowering the differential capacities of sensation, which in turn makes possible new forms and objects of willing. In place of Hunter's literal dream of traveling to the future, Keats's and Shelley's poetics seek to produce the future by attuning readers to the rhythms of slow time.

CHAPTER 3

Life, Orientation, and Abandoned Experiments

I am already far north of London, and as I walk in the streets of Petersburgh, I feel a cold northern breeze play upon my cheeks, which braces my nerves and fills me with delight. Do you understand this feeling? This breeze, which has travelled from the regions towards which I am advancing, gives me a foretaste of those icy climes. Inspirited by this wind of promise, my daydreams become more fervent and vivid. I try in vain to be persuaded that the pole is the seat of frost and desolation; it ever presents itself to my imagination as the region of beauty and delight.

Mary Wollstonecraft Shelley, Frankenstein, 49

We must reach a secret point where the anecdote of life and the aphorism of thought amount to one and the same thing. It is like sense which, on one of its sides, is attributed to states of life and, on the other, inheres in propositions of thought. There are dimensions here, times and places, glacial or torrid zones never moderated, the entire exotic geography which characterizes a mode of thought as well as a style of life.

Gilles Deleuze, The Logic of Sense, 128

Mary Shelley's *Frankenstein* is, perhaps above all else, an analysis of the linked problems of orientation and disorientation, of finding one's path and losing one's way. As the first epigraph reminds us, the novel begins with a representation of certainty and clear orientation: the narrator of the book, would-be polar explorer Robert Walton, seems sure both of his way and of what he will find at the North Pole. Inspirited by a wind of promise, unable to doubt himself, Walton has assumed the position of a man capable of leading and inspiring conviction in others. Yet even in this apparently sure beginning, one can already sense a wavering that will eventually come to throw Walton off track entirely—for why, we wonder, would someone so sure of himself ever think to mention this, to write about it to others? Wouldn't certainty of the sort Walton describes be more of the order of an invisible medium within which one acts and exists, rather than a conscious fact of which one could become aware; more on the order of still and silent air, in other words, than a bracing, inspiring breeze?

One can sense Walton's anxiety about orientation, moreover, in the very nature of his voyage. His is, quite simply, a voyage intended to put an end to nautical disorientation, for he seeks the North pole in part to "discover the wondrous power which attracts the needle," a discovery that would then "regulate a thousand celestial observations that require only this voyage to render their seeming eccentricities consistent forever" (50). Walton's search for the pole thus makes sense only on the basis of a more fundamental anxiety about errancy and eccentricity; a worry that, in the absence of a regulating mechanism, one can always go—and perhaps already has gone—astray. Given these initial waverings and anxieties, it is not entirely surprising that, by the novel's end, Walton's ship has become frozen in the ice, his crew has begun to mutter about mutiny, and the pneuma of promise and certainty that earlier inspirited Walton has guttered and flagged. And though the events depicted in the novel have their narrative origin in Victor Frankenstein's famous act of abandoning his animated experiment—the living creature he created—the *novel* can really only end when the external impediments to Walton's quest are allowed to trigger an internal crisis, a state of disorientation and doubt that catalyzes his decision to abandon his own experimental quest to discover the polar source of orientation.

Taking inspiration from Shelley's novel, this chapter reflects on the relationships among orientation, life, and what I will call "abandoned experiments." As I noted in the introduction and chapter 1, scholars of literature and the arts have not shied away from describing different kinds of literature as experimental, but the concept of the abandoned experiment is far less familiar territory in literary studies (as well, for that matter, in science studies). As I use the term, an abandoned experiment is not the same as an unpublished draft, for it does not denote simply the fact that a text remained unfinished or unpublished during the author's lifetime. Nor is it equivalent to a textual "fragment," whether published as such (e.g., Coleridge's "Kubla Khan") or rendered as such by unanticipated events (e.g., P. B. Shelley's *The Triumph of Life*, which remained unfinished because Shelley drowned during a sailing trip undertaken during the time he was revising the poem).[1] Rather, the concept of an abandoned experiment denotes both that an author chose not to publish a work and an intentional act or stance on the part of the author linking that unpublished work with a biographical life crisis. To return to my opening example of Robert Walton, an abandoned experiment is not simply a matter of failing to reach a goal—it isn't just becoming frozen or stuck—but it is also an internalization of that failure, via either an overwhelming sense or an explicit decision, that an earlier line of orientation has been lost and a new one must be taken up. Abandoned experiments are located at the intersection of two borders: on the one hand, the border between literature and science (though an abandoned experiment retains an orientation toward the sciences, it does not enable "progress" *within* either a scientific or literary field) and, on the other hand, the border between text and biography (the abandoned experiment can be a reflection, instigator, or resolution of an extratextual life crisis).

What follows is an attempt to map some of the poles and dimensions of Romantic-era abandoned experiments. Drawing on work from the field of science studies on the economy of experiments—work, for example, that considers how modes of gift exchange enable collaboration on experiments but at the same time introduce dynamics of rivalry—will provide us with some guidance in thinking about the abandoned experiment. I will also draw on research by scholars of science such as Peter Galison on the question of how experiments "end" (though, as we will see, the abandoned experiment is one that cannot truly end, precisely because it is abandoned). Yet the confusion of literature, science, text, and biography that constitutes the abandoned experiment is sufficiently disorienting that Romantic literature and biography will turn out to be our best guides to its dynamics. And though I will continue to draw on Shelley's novel as a resource for thinking about abandoned experiments, my primary case study will be another, less well-known, text, Samuel Taylor Coleridge's reflections on "the theory of Life." Begun (like Shelley's novel) in 1816, this essay has a peculiar biographical context, for Coleridge started writing it just after moving into the house of his new physician, James Gillman, as part of a last-ditch attempt to help moderate his (Coleridge's) debilitating opium dependency. Coleridge's essay was to be a return gift of sorts, for he intended to give what he had written to Gillman so that Gillman could add to it and then submit the complete essay, under his own name, to a medical essay competition. Unfortunately, Coleridge's text quickly exceeded the borders of the scientific essay, taking up the much more general question of how one orients a life, and the two men were forced to abandon their experiment in joint authorship. Coleridge's opium dependency, combined with the relentlessly philosophical lens through which he viewed the world and his problems, has always made it difficult to determine where Coleridge's "life" ends and where his "texts" begin, and the example of his abandoned essay on life will help us to think more closely about the relationship among literature, experiments, and biographical orientation.

While much of this chapter is focused on the economy of experiment and abandonment specific to this particular text and moment in Coleridge's and Gillman's intertwined biographies, I also use Coleridge's abandoned essay on life as a proxy for what has seemed to many literary critics as his more significant abandonment, at about this same point in his life, of poetry in favor of philosophical and theological handbooks. Though Coleridge did not in fact cease writing poetry after 1816—perhaps unfortunately, he continued to write it until the end of his life—the vital authorial impetus that formerly inspired his verse seems to have been shuttled into other genres of literary endeavor. Literary critics have mourned Coleridge's de facto abandonment of poetry, but I focus here less on loss than on the relationship between that act of abandonment and Coleridge's attempt to create a new "vital" genre, the life-manual or life-handbook. Reflecting on the emergence of the life-manual in Coleridge's corpus helps us to think further about the interweaving of life and work that is fundamental to the abandoned experiment, for the relationship between the abandoned experiment and life-manual forces us to move beyond our default critical understandings of the relationship of texts to lives—for example, texts as entirely separate from real life, texts as expressions or reflections of real life, or "real life" as itself something that can only ever be known through texts—and instead understand that texts serve as ways of ordering, or creating lines of orientation between, biographical states of affairs that are extrinsic to the text itself.

OPIUM, SPIRALS, AND SUSPENSION

When Coleridge began composing a text on the theory of life in the last months of 1816, he was forty-four years old and at a crossroads in both his personal life and his career as an author. He had come to the conclusion earlier in the spring that he was, at least on his own, entirely incapable of controlling his habit, now in its second decade, of nearly daily opium use. Though Coleridge seems to have begun using opium guided by the belief that it was a relatively risk-free medication, he had, by at least 1808, recognized that whatever its medical utility, it had also become for him an end in itself.[2] Coleridge was frank in letters to friends and acquaintances about the despair he felt over his inability to stop using laudanum.

Coleridge wrote to his friend J. J. Morgan on May 14, 1814, that he felt as though he had been "crucified, dead, and buried, and descended into *Hell*," a state in which

> tho' there was no prospect, no gleam of Light before, an indefinite indescrib-able Terror as with a scourge of ever restless, ever coiling and uncoiling Ser-pents, drove me on from behind.—The worst was, that in *exact proportion* to the *importance* and *urgency* of any Duty was it, as of a fatal necessity, sure to be neglected: because it added to the Terror above described.[3]

And though Coleridge frequently solicited the help of both physicians and friends to help him stop, or at least cut down on, his use of laudanum, he had not had any long-lasting success; as he noted in a letter written in April 1816, for a long time "no sixty hours *have yet passed* without my having taking [*sic*] Laudanum" (4:630). This frequency of use had—in Coleridge's own estimation, as well as in the estimation of others—nearly paralyzed his critical and authorial capacities.

What makes Coleridge's dependency particularly compelling and tragic was his tendency to parse it through exceedingly nuanced philosophical and theologi-cal paradigms, with the result that his opium use called into question his most fundamental beliefs about the nature of the human will. Coleridge's despair was not simply a consequence of the fact that his habit of opium use made it difficult for him to write and thus to earn a living, but also that he habitually interpreted these incapacities as evidences of more fundamental human capacities and in-capacities. His personal dependency took on an eschatological dimension, for it called into question the freedom of the human will. Since at least the late 1790s, Coleridge had repeatedly claimed to be convinced, and had sought to find or develop a philosophy that would prove, that the will was fundamentally inde-pendent of bodily inclinations.[4] Yet his persisting habit of opium use seemed to constitute a living disproof of this purported autonomy of the moral will, and it thus forced Coleridge to make finer and finer distinctions if he wanted both to ac-knowledge his incapacity while holding on to the position that the will remained free. He wrote in 1814 to J. J. Morgan, "By the long Habit of the accursed Poison my Volition (by which I mean the faculty instrumental to the Will, and by which alone the Will can realize itself—it's [*sic*] Hands, Legs, & Feet, as it were) was completely deranged, at times frenzied, dissevered itself from the Will, & became an independent faculty" (3:489). Such a claim indeed left the will free, since it distinguished between the will and volition, that by which the will acted in the world. Yet this distinction also meant that acts of willing, no matter how free in principle, might nevertheless remain impotent and unrealized should the organ of volition not be sufficiently disciplined.

Whatever philosophical consolation the distinction between will and volition may have provided Coleridge, it clearly did little if anything to allay his sense that his own life was drifting beyond the possibility of recovery. What especially oppressed him was his sense that he had, through lack of control of his volition, wasted—or at least left to lie fallow—those gifts of intellect and love of truth with which he had been blessed. As he wrote in April 1814 to his publisher, Joseph Cottle,

> For years the anguish of my spirit has been indescribable, the sense of my danger *staring*, but the conscience of my GUILT worse, far, far worse than all!—I have prayed with drops of agony on my Brow, trembling not only before the Justice of my Maker, but even before the mercy of my Redeemer. 'I gave thee so many talents. What has thou done with them'? . . . my Case is a species of madness, only that it is a derangement, an utter impotence of the *Volition*, & not of the intellectual Faculties. (3:476–77)

It is not clear from this letter, or from Coleridge's notebook entries written around the same time, whether Coleridge understood his impotence of volition as itself the consequence of a moral failing or rather of some other kind of incapacity (for example, a physiological weakness).[5] However, the result was a sense of oppressive indebtedness, a feeling of having received gifts that were to be used by passing the fruits of these gifts on to others but which instead had remained locked inside Coleridge, unexpressed and—like his "moral Will"—suspended.[6]

Coleridge's unsuccessful struggle to master his habits and volition bore not only upon his sense of indebtedness but also upon his own self-definition as a lover of truth. Though he was willing to admit many personal failings, Coleridge habitually presented himself in his correspondence and in print as someone who cleaved to truth to such an extent that he had even cultivated in himself an incapacity to lie consciously to others (though, as he admitted, he sometimes produced the same result as lying by simply omitting to say certain things). In an 1816 letter to Dr. James Gillman, Coleridge assured the doctor—whom he hoped to convince to help break his habit of opium use—that he had developed "prior habits" that in fact "render[ed] it out of my power to tell an untruth."[7] Yet this claim, no doubt intended to assure Gillman that caring for Coleridge would remain within reasonable limits, was itself a lie, for Coleridge had already admitted to another correspondent that the habit of opium use had perverted these "prior habits," transforming what had been a virtuous incapacity (his inability to lie) into a capacity for vice (an active capacity to lie). Thus, Coleridge wrote in 1814—two years before his claim to Gillman that he could never tell a lie—that his opium use had repeatedly forced him to cross this border as well: "I have in this one dirty business of Laudanum an hundred times deceived, tricked, nay, actually & consciously LIED."[8] Coleridge's letter to Gillman is thus an instance of a desperate attempt to force a bad habit to serve the cause of good. Having apparently lost faith that truth-telling could on its own break the bad habit of lying—a habit he had picked up as a means for perpetuating his opium dependency—Coleridge instead sought to use this latter habit against itself by lying his way into a situation that might end his dependency.

Plotting such a path was at best a gamble, for Coleridge had already recognized the extent to which his opium dependency itself depended upon such a recursive logic, that is, a path by means of which he repeatedly sought to end his pain by making recourse to precisely that which caused it. So, for example, in a letter dated 26 June 1814, he asked Josiah Wade to "conceive a poor miserable wretch, who for many years has been attempting to beat off pain, by a constant recurrence to the vice that reproduces it" (3:511). Given Coleridge's attentiveness to this logic of spiraling suspension, one can imagine that he was aware of the risk of employing an uncultivated habit of lying as a means of finding a path back to his "prior habits" of truth-telling. Yet by the spring of 1816, Coleridge clearly could see no other way out of his state of suspended animation, and so lie he did—this time, he hoped, in the cause of truth.

WASTED GIFTS, STOLEN GIFTS

As it turned out, Coleridge's gamble paid off: Gillman, though just beginning his practice a few miles north of London, agreed to lodge Coleridge without cost while attempting to wean him off his habit. Gillman was essentially agreeing to serve as a temporary surrogate for Coleridge's volition, isolating the author from those disorienting and disabling forces—druggists, friends, and acquaintances willing to supply the author with opium—with which the poet had found himself unable to contend. By providing Coleridge with ersatz volition, Gillman would allow Coleridge to renew a trajectory he had initiated a year earlier—one that had seemed, at least temporarily, to promise a release from the long "suspended animation" of his literary powers that had gripped him for the preceding decade and a half. In the spring of 1815, Coleridge had found a publisher for a volume, to be entitled *Sibylline Leaves*, which would bring together some of his previously uncollected poems and include a preface outlining his theory of poetry. During the

summer and fall of 1815, the preface evolved into a sprawling literary biography—
it would, in fact, eventually become his *Biographia Literaria* (1817)—but the publisher still seemed willing to accept the project, and Coleridge had been able to
reduce his consumption of laudanum. However, in the winter of 1815, following
an illness, Coleridge had fallen again into his old habits, and—as if to underscore
his problem—had taken up residence above an apothecary.[9] Thus, the recupera-
tive residency with Dr. Gillman, begun in the spring of 1816, was a way of trying
to reconstellate his biography and this new volume of poetry, positioning them
as the starting points of a path that would allow Coleridge to regain his former
reputation as a practicing and vital author. And by the fall of 1816, it was clear the
arrangement was working: Gillman had been able to moderate Coleridge's use of
opium, and Coleridge himself was again able to write.

From this perspective, Coleridge's offer, in the winter of 1816, to collaborate
with Gillman—his physician, his source of free housing, and by that date, also his
friend—by jointly composing a Jacksonian Prize entry takes on the sense of a gift,
a return of thanks. Gillman, still at the start of his medical practice, was certainly
interested in consolidating his own reputation, and though he had already won
the 1810 Jacksonian Prize, awarded annually by the Royal College of Surgeons,
for an essay on hypochondria, an additional prize would increase his renown. It
seems likely that Coleridge and Gillman settled upon the project of writing a
joint essay during a lengthy and relaxing fall trip to the coast, a trip that—at least
in Coleridge's estimation—brought further healing, allowing him to "pas[s] for
three or four hours every day through a current of changed air."[10] The themes for
the 1816 Jacksonian Prize had been announced—submissions were to focus on
either scrofula or syphilis—and Gillman and Coleridge decided to write on the
former topic, though the entry would be submitted under Gillman's name alone.
Coleridge's part of the essay would thus constitute a reciprocal gift of force and
direction: in exchange for Gillman's help in setting into movement once again a
life and a career that had become paralyzed midway through its course, Coleridge
would provide Gillman, now only at the start of his career, with impetus and
orientation.

One guesses that Coleridge also hoped that simply *giving* Gillman a few sections of an essay might avoid a repetition of those problems with joint authorship
that had emerged as a result of the most famous of Romantic-era literary experiments, the 1798 collection *Lyrical Ballads* that Coleridge had written together
with William Wordsworth. When Coleridge and Wordsworth first published their
collaborative volume, they advertised its poems as "experiments," designed "to
ascertain how far the language of conversation in the middle and lower classes

of society is adapted to the purposes of poetic pleasure."[11] Though these poetic experiments brought Coleridge and Wordsworth relative fame, it turned out that the volume had also, and unwittingly, involved another experiment: namely, to determine how far co-authorship could persist without producing rivalry between the two authors. Unfortunately, the answer to this latter trial was not long. In 1800, as the two began work on a second edition of *Lyrical Ballads*, Wordsworth's dissatisfaction with Coleridge's contributions became evident. He reordered the poems in ways that downplayed Coleridge's contribution: where the 1798 version began with Coleridge's "The Rime of the Ancyent Marinere," that poem was retitled "The Ancient Mariner" and also shifted to the less important position of penultimate poem in the first of the two volumes of the 1800 edition.[12] Wordsworth also subtly redescribed what, precisely, was experimental about *Lyrical Ballads*: rather than being a collection composed of multiple experiments, the volume as a whole was now "an" experiment, and one, moreover, that was mapped within Wordsworth's authorial biography.[13] In the next edition in 1802, Wordsworth removed Coleridge's poems entirely, appropriating sole authorship of *Lyrical Ballads*. And in 1815—a year before Coleridge had decided to seek Gillman's help—Wordsworth added insult to injury by issuing a new edition with an "Essay, Supplementary to the Preface," in which he seemed (to Coleridge, at least) to appropriate many of Coleridge's ideas about poetry and philosophy.

Wordsworth's perceived theft had been enough to rouse Coleridge a bit from that opium-encouraged suspended animation of his own literary powers, giving him the energy to compose his own experiment in biography, *Biographia Literaria*. As Raimonda Modiano has noted, the "publication of Wordsworth's [1815] Preface turned the disillusion and sense of inferiority that Coleridge had so often felt toward Wordsworth into vehement anger and self-assertive energy," and *Biographia Literaria* was Coleridge's attempt to outflank his former friend, for he suggested in his literary biography that Wordsworth had appropriated as his own gifts of insight and ideas that had in fact originally come from Coleridge (245). This fury toward Wordsworth suggests that Coleridge's sense of the cause of his wasted gifts may have shifted: in 1814, it was his own aberrant volition that had left these gifts suspended, while in 1815, it was Wordsworth who had effectively stolen them (or had, at the very least, taken advantage of the pause in Coleridge's life produced by opium dependency to appropriate aspects of Coleridge's philosophy). *Biographia Literaria* was thus to function as a textual time machine: by narrating the sources and chronology of his own philosophical education, Coleridge could reappropriate those gifts, first wasted then stolen, and thereby recover that earlier literary trajectory that had been suspended by opium use.

GIFTS AND THE ECONOMIES OF EXPERIMENT:
LYRICAL BALLADS

Though this bitter rivalry between Wordsworth and Coleridge unfolded as a consequence of individual decisions and idiosyncrasies, Modiano has emphasized that these biographical events cannot be divorced from a more general logic of gift exchange that the two authors could neither control nor abandon. Submission to the logic of gift exchange is no doubt always a possibility when authors collaborate, but in retrospect we can say that Wordsworth's and Coleridge's appropriation of the term "experiment" for their collaboration on *Lyrical Ballads* was likely to lead to problems. Their attempt to use this "scientific" term to reconfigure the audience for and history of poetry forced them to reconfigure the relationship of gifts to experiments and authorial attribution that was characteristic of the eighteenth-century "economy of experiment."[14] To experiment with the term "experiment" in the field of literature was, in other words, also to experiment with the function of the gift in experiments, and Wordsworth and Coleridge ended up hybridizing, in ways neither could have anticipated or controlled, aspects of sixteenth-century patronage science with eighteenth-century institutional science.

For historians of science, gift dynamics are most closely associated with sixteenth-century patronage science, of the sort practiced by individuals such as Galileo Galilei. Mario Biagioli notes, for example, that in sixteenth-century Italy, "gifts and other economically non-quantifiable services and privileges were the medium through which patronage relationships were articulated and maintained," and thus for an author such as Galileo, the advance of *scientia* was inextricably linked to gift-giving: the more discoveries and instruments Galileo could dedicate or give to his patrons (or would-be patrons), the more likely he was to receive gifts such as salaries (and even food), and thus the more likely he would be able to continue to make discoveries and create inventions.[15] However, the more prestigious his patron, the more careful an author such as Galileo had to be in positioning his gifts:

> Paradoxically, one had to donate exceptional gifts to an absolute prince in order to be accepted as a distinguished client, and yet one could not present his exceptional gifts as really coming from him because that gesture might be read by the prince as an unacceptable "challenge," thereby jeopardizing the client's access to patronage, legitimation, and credibility. (53)

The patronage structure had direct implications for questions of authorship and attribution, for, as Biagioli notes, "within court patronage, one could gain legitimation as a scientific author only by effacing one's individual authorial voice" (53).

By the late seventeenth century, patronage cultures had been largely displaced by institutional cultures, in which collective groups such as the Royal Society, rather than individual patrons, validated scientific discoveries and inventions. Biagioli has suggested that the rise of the experiment and collective witnessing of "matters of fact" enabled science to escape from what he describes as the "deadlock of the patronage system," which encouraged competition between clients but discouraged the patron from "stooping" to the level of deciding between competing scientific claims made by clients (356–57). The shift to institutional science also meant a reconfiguration of the relationship between authors and attribution: natural philosophers no longer needed to efface their contributions, and—as Steven Shapin has emphasized—the site of self-effacement shifted to technicians, who created and maintained the experimental apparatus but were rendered anonymous by the natural philosopher who authored the print description of the experiment.[16]

Yet this shift from patronage to institution did not so much mean the eclipse of the gift from the economy of science but rather its realignment within the emerging economy of experiment. A modern scientific institution such as the Royal Society, for example, was itself founded through the form of the royal gift, and members of the society were expected to give (rather than, for example, sell) copies of their publications and inventions to the society.[17] Contemporary "historians" and advocates of the Royal Society, moreover, stressed the importance of gift-giving for this new organization of knowledge production. Thomas Sprat, the first historian-advocate of the Royal Society, contended that the success of the institution was in part due to the fact that its founders were willing to donate their own money to fund its activities, since the *"publick faith of experimental philosophy, was not then strong enough, to move Men and Women of all conditions, to bring in their Bracelets and Jewels, towards the carrying of it on . . . It was therefore well ordained, that the first Benevolence should come from the Experimenters themselves"* (77). For Royal Society member Robert Boyle, the new figure of the natural philosopher was to be understood in terms of a gift exchange between God and mankind, with "knowledge being a gift of God, intrusted to us to glorify the giver with it," and science serving as a kind of return gift to the deity.[18] The site of the gift thus shifted from the presentation of the results of science, as had been the case in sixteenth-century patronage science, to its divine and secular framings: to do institutional science was to give thanks to God, and the king's gift of a royal patent to a "society," as well as the mutual gift-giving of the society's members, enabled disputes to be adjudicated without involving the royal figure himself in these debates.[19]

When Wordsworth and Coleridge appropriated the term "experiment" for *Lyrical Ballads* and poetry at the end of the eighteenth century, it was not clear what role gifts would play in their unprecedented mode of experimentation. Since Samuel Johnson's famous 1755 public letter in which he rejected a patronage relationship with Lord Chesterfield, literary authorship itself was understood as having shifted from a patronage system to more purely market-based relationships.[20] While this opened the possibility that Wordsworth and Coleridge's new mode of experimentation might simply divorce itself completely from the function of the gift, the situation was complicated by the fact that both Wordsworth and Coleridge were beneficiaries of various forms of patronage in the 1790s; moreover, both authors were interested in establishing modes of mutual "benevolence" not unlike those of the early Royal Society members.[21] Literary authorship thus provided no clear guidance in how the gift ought to function in this new mode of experimentation, while at the same time Wordsworth and Coleridge's experimentation with experiments also meant that they could not rely on traditional validating gift frames of either the patronage or the institutional models of science. It was thus not clear what kind of dynamics were to hold among would-be literary experimentalists: were these to be the dynamics of prestige, rivalry, and attribution specific to the patronage model, the institutional model, or some other, unprecedented model of collaboration?

Seeking to distance themselves from agonistic models entirely, Coleridge and Wordsworth shifted the site of the gift from an enabling frame to a private relationship between two authors. Each author would contribute his own poems to a common pool. This private exchange of gifts between two authors seemed, in turn, to promise the possibility of attribution-free experiments: rather than marking out to whom each poem belonged, the collection would instead be published anonymously. If, that is, *Lyrical Ballads* was to provide its readers with the means for making their own experiments, what matter which "technicians" actually constructed these apparatuses?

Except that attribution and the parsing of gifts did subsequently turn out to matter to both Wordsworth and to Coleridge. Rather than enabling that hierarchy-free and ever-expanding sphere of experimenting readers for which Coleridge and Wordsworth seemed to have hoped, pooling gifts in the construction of *Lyrical Ballads* enabled instead a situation of immanent rivalry—one in which it would always, and necessarily, remain difficult to tell precisely where mine and thine began and ended. Both poets found themselves again and again attempting to escape from that authorless and attributionless vortex of anonymity within which they had originally sought to contain their private exchange of gifts. In *Bio-*

graphia Literaria, Coleridge sought to sever his connection to this unprecedented mode of literary experimentation by attributing the mode of experiment itself to Wordsworth: it was Wordsworth alone, wrote Coleridge, who had presented the *Lyrical Ballads* as an experiment, and it was thus Wordsworth's experiment that we must "suppose to have failed."[22]

SCROFULA, LIFE, AND SUSPENSION: THE *THEORY OF LIFE*

If *Biographia Literaria*—begun a year before Coleridge met Gillman—was Coleridge's attempt to achieve orientation by clarifying those distinctions between mine and thine that *Lyrical Ballads* had confused, Coleridge's collaborative essay with Gillman set off in another direction entirely. Rather than seeking, as he had in 1815, to negate the experiments of *Lyrical Ballads*, Coleridge instead sought to recombine and reconfigure many of the driving forces of that experimentation with experiments that he and Wordsworth had initiated in 1798. Thus, his collaboration with Gillman was also to orient itself toward, while departing from, contemporary science; it was also to require a form of gift-giving; it was also to involve a reconception of collective authorship; and it was also to involve a form of witnessing. Yet each of these elements would be reconfigured in ways that seemed to promise an experimental collaboration free of rivalry. For example, rather than proposing that he and Gillman *pool* their individual gifts to create a joint experimental work, Coleridge would instead return to Gillman a sort of equivalent of what he had already received from his physician: a gift of orienting text in return for Gillman's gift of surrogate volition. Open to the healing airs of the coast and inspirited by his own wind of promise, Coleridge began working on those sections of the essay he would give to Gillman. Coleridge's task was to outline the history of opinions about scrofula and the implications of that history for present treatment of the disease; these two sections would then lead into Gillman's medically more specific discussions of the pathology of and cure for the illness.[23]

Scrofula was a topic that seemed to play to both men's strengths. By the early nineteenth century, "scrofula" was used to denote hard swellings, often (though not always) located around the neck, and Gillman had claimed significant success in treating this condition. Scrofula was also understood as the modern name for what had for many centuries been termed "the King's Evil," so called because it was purported that the touch of a king could heal the disease. This peculiar, quasi-mythological history of the illness would provide Coleridge with an opportunity to employ his skills of historical research and exegesis. Finally, scrofula had an intensely personal dimension for Coleridge, for he had earlier (self-) diagnosed his "indolence" as due to scrofula.[24] By 1816 he had instead come to see opium as the source of his failures of volition, and an essay on scrofula thus promised to secure theoretical mastery over his past misdiagnosis.

Coleridge quickly wrote up a concise but sophisticated history of scrofula, developing an interpretation of the disease that was "cultural" in two different senses. Noting that the incidence of "the King's Evil," and its supposed cure by the royal touch, had increased massively in the wake of the Reformation, he began by arguing that references to the disease had been used as a tactical device by rival religious authorities as means for gaining converts. At the same time, though, he maintained that incidence of the disease *had* increased significantly since the time of the Reformation but that this was due to an increase in both "depressive causes," such as lack of adequate food among the poor, and an "excess of mental and bodily stimulants" among the middle and upper class (460, 471). The fact that incidence of scrofula could increase as a result of both "depressive" and "excessive" factors thus suggested that scrofula had to be understood as a "constitutional disease," a disease that depended upon a "derangement of some one or all of the primary powers, in the harmony or balance of which the health of the human being consists" (478).

With this conclusion, Coleridge had in principle reached the end of his contribution, leaving for Gillman the discussion of the pathology of, and treatments for, scrofula. Yet in the final paragraph of his section, Coleridge abruptly reoriented the goals and scope of the essay as a whole. For if, he argued, scrofula was a "constitutional disease," and if this meant a "derangement of some one or all of the primary powers, in the harmony or balance of which the health of the human being consists," it suggested that an adequate understanding of scrofula would require "a distinct conception of life itself, or, as it has been the fashion of late years to name it, of the living principle" (478). This in turn required that, before Gillman could append his two sections on pathology and therapies, Coleridge would first have to write *another* section, outlining—of all things—the nature of life itself.

Perhaps not surprisingly, this reorientation was to be the undoing of the project, for though Coleridge apparently did finish this new section before the December essay contest deadline, he did not do so in time for Gillman to be able to make sense of the addition. Coleridge himself was clearly aware that his decision to locate the solution to the problem of scrofula at the far end of a discussion of the nature of life itself might seem like a daydream or delusion, for at the end of his section on scrofula, he wrote that he "anticipate[d] that the investigation of this question [of the principle of life] . . . nay, even the mere reference to it as a problem . . . will be conceived by many as visionary & seducing us beyond the bounds of clear conceptions" (478). Yet, he continued, comprehension of the principle of life was *not* beyond "rational thinking"; moreover, its "solution" was "regarded of such importance by the greatest thinkers of our profession [that is, Gillman's profession of medicine], from Harvey & from Haller to John Hunter" (479). Nevertheless, the necessity of solving the problem of scrofula by means of an explanation of the nature of life remained unclear to Gillman. Coleridge wrote a long letter to Gillman in mid-November in which he tried to allay the latter's "anxiety" and "lively perception of the inappropriateness of what I have hitherto written," but the two were apparently not able to come to an understanding in time to submit a completed essay.[25]

Coleridge's decision to launch into a discussion of life midway through an essay ostensibly devoted to scrofula was clearly a misstep. Yet it was not a completely unwarranted stumble. He was seeking to heighten the importance of Gillman's essay on scrofula by recontextualizing it within what was becoming an increasingly heated debate, carried out by members of the Royal College of Surgeons, and focused on John Hunter's understanding of the principle of life. Though the illustrious surgeon had died in 1793, Hunter's heirs—assisted by surgeons eager to appropriate Hunter's renown for their profession as a whole—had established a lecture series on surgery and anatomy.[26] This lecture series began in 1806, and in 1813, the trustees of Hunter's estate established the yearly Hunterian Oration, designed to provide surgeons with an opportunity to pay "tribute to the memory of those practitioners of surgery who have contributed during their lives to the advancement" of the discipline (146). The first orations tended to focus on Hunter himself, and praise for the deceased surgeon was lavish. However, several of the early orators also sought to explicate for their audiences Hunter's "theory of life." This was certainly a reasonable task, for though Hunter was believed to have held a more or less vitalist theory of life, his precise understanding of the relationships among life, matter, and organization was not entirely clear, at least in the essays and books he had published during his lifetime. From this perspective, Coleridge's decision to include a lengthy section on the principle of life in Gillman's prize essay submission should be seen as attempt to use the essay to create a sort of unified field, or synthesis, within the various institutions—orations, lecture series, publications, and so on—sponsored by the Royal College of Surgeons. If successful, such an endeavor could well position Gillman as a regulating pole of modern medicine, one whose therapeutic successes could then be seen as a consequence of his deep understanding of life itself.

Coleridge had his work cut out for him, for this debate between Hunter's followers concerning the principle of life had become increasingly polarized. Many of Hunter's followers had "gone transcendent," interpreting Hunter's distinction between organization and simple life as equivalent to the claim that life was an immaterial principle that existed separate from living matter. Richard Saumarez was one of the first to promote this interpretation of Hunter's concept of simple life: six years after Hunter's death, Saumarez argued in his massive two-volume *New System of Physiology* (1799) that the recently deceased surgeon had been "the dignified chieftain of a small band" who "proclaimed the existence of a principle of life, which was the cause (not the effect) of organization and action, and to which it had prior existence" (1:vii–viii). Saumarez's book was intended for nonspecialists, but the image of Hunter as a theorist committed to a transcendent interpretation of simple life was advanced within the Hunterian Oration series as well, particularly by Hunter's former student, John Abernethy.[27] In his 1814 oration, Abernethy depicted Hunter as "a man of genius," who had proven that life could not be understood in terms of "the matter of animals and vegetables" alone but required that one acknowledge the "superaddition of some subtle and mobile substance."[28] For Hunter, Abernethy claimed, "life must be something independent of [material] organization."[29]

On the other side of the debate were those who argued that Hunter had *not* intended life to be understood as a transcendent principle that was separate from matter. Just a few months before Coleridge and Gillman began working on the scrofula essay, one of Abernethy's former students, William Lawrence, argued in his own 1816 Hunterian Oration that "life" should not be understood as something superadded to matter but rather was simply that word that we used to denote particular organizations of matter.[30] This claim provoked Abernethy, who responded that Lawrence's claim was not only false to Hunter's philosophy, but also "detrimental to society," for the belief that "life" denoted nothing more than particular organizations of matter smacked of the "materialist" philosophy promoted by *philosophes* living in Britain's recent military rival, France.[31]

Where Abernethy and Lawrence had established two poles within this debate about life—transcendent vitalism versus materialism (or, arguably, materialist vitalism)—Coleridge sought, in his essay section on life, to reorient, or repolarize, the debate. Following a few paragraphs of perfunctory praise of John Hunter ("that the true idea of Life existed in the mind of John Hunter I do not entertain the least doubt"), Coleridge quickly arrived at the central premise of his essay: that all previous "attempts to explain the nature of Life . . . presuppose the arbitrary division of all that surrounds us into things with life, and things without life—a division grounded on a mere assumption."[32] Coleridge acknowledged that such a division may seem phenomenologically self-evident, in the sense that the world appears to present itself to us as divided into nonliving entities (rocks, water, and air) and living entities (plants and animals). He acknowledged as well that such a division was certainly functional, in the sense that it was sufficient for the purposes of "ordinary discrimination" (488). Yet he contended that simply to assume this distinction between the living and the nonliving prevented the attainment of a truly *philosophical* conception of life, a conception that goes beyond mere description and observation and actually *explains* life (488). To explain life meant,

> the reduction of the idea of Life to its simplest and most comprehensive form or mode of action; that is, to some characteristic *instinct* or *tendency*, evident in all its manifestations, and involved in the idea itself. This assumed as existing in *kind*, it will be required to present an ascending series of corresponding phenomena as involved *in*, proceeding *from*, and so far therefore explained *by*, the supposition of its progressive intensity and of the gradual enlargement of its sphere. (504–5)

From Coleridge's perspective, all theories of life premised upon a distinction between living and nonliving things were fundamentally flawed, for they "confin[e] the idea of Life to those degrees or concentrations of it, which manifest themselves in organized beings, or rather in those the organization of which is apparent to us" (494).

What Coleridge thus demanded of his readers was a fundamental suspension of the given as regards life, a suspension of the apparent self-evidence of vitality in the case of living things and deadness in the case of nonliving things. This was not to be a partial suspension of the given, through which, for example, a prior sense of life was simply extended to cover objects that one formerly took as nonliving (e.g., rocks), for such an expansion of the domain of life would still be based on the assumption that life could be understood by investigating the qualities of particular things. Rather, to explain life, one had to abandon the premise that life was fundamentally a "property" or something that could be thought within the schema of things; anticipating Heidegger's distinction between Being and beings, Coleridge proposed that to understand Life, one had to suspend the link between *life* and *things*.[33] From this perspective, the opposition between Abernethy's transcendent vitalism and Lawrence's materialism was in fact only superficial, for it hid a more fundamental agreement: both thought of life as a property of things, though they disagreed about the precise nature of that property. In this sense, Coleridge's claim to be championing Hunter's approach to life was arguably accurate, for (as I noted in chapter 2), Hunter too had claimed that life could be understood only by facilitating a "new bend of the mind."

Look for tendencies, orientations, and intensities that *connect* elements of the world: these were the imperatives of Coleridge's approach to life. To see the world intensively—in terms of manifestation, and manifestation in terms of dynamism and intensity—was then to understand life as a power that intensified itself by generating new forms of unity. Life unifies, linking that which has not yet been linked; it is the power of binding. Life thus begins, Coleridge contended, as identical with the "elementary powers of mechanism," producing, in this phase, the metallic elements; those unified, basic forms of metals, such as gold (514). Gold was not itself "living," but gold and the other elements were the result of life's process of self-development: life brings into being unities, and elements such as gold constitute the simplest form of unity. Each element constituted a "special union of absolute and of relative gravity, ductility, and hardness" that "may be endlessly modified, but can never be decomposed" (514). Life had then further developed itself by producing crystals. These latter are like metallic elements, in that each crystal is formed of a single substance, but are more complex than the elements because in crystals, single substances are grouped into specific forms. As elements had formed the basis for crystals, so crystals had formed the basis for even more complex forms, such as plants and animals: "Nature, by the tranquil deposition of crystals, prepared, as it were, the fulcrum of her after-efforts," such as the "increasing multitude of [geologic] strata, and in the relics of the lowest orders, first of vegetables and then of animal life."[34] Plants, and then animals, instantiated even more complex forms of unity, with human culture situated as the most complex form of animal life.

Coleridge contended that human culture evidenced this same drive toward increasing complexity. The most complex form of culture was that in which individuals willingly and consciously coordinated liberty and independence with submission and resignation to both the political state and to God. "In Man," Coleridge wrote, "the centripetal and individualizing tendency of all Nature is itself concentered and individualized," but the acme of sociality and politics involved a polity capable of balancing polar forces:

> The intensities must be at once opposite and equal. As the liberty, so must be the reverence for law. As the independence, so must be the service and the submission to the Supreme Will! As the ideal genius and the originality, in the same proportion must be the resignation to the real world, the sympathy and the inter-communion with Nature. (551)

"Individuation" was the name by means of which Coleridge grouped these imperatives (look for tendencies, look for orientation, look for intensity). Life, that is, was not a quality that distinguished some things from others; rather, it was *"the principle of individuation,* or the power which unites a given *all* into a *whole* that is presupposed by all its parts" (510). "Individuation" was admittedly a curious term for Coleridge to have chosen, for the essay's would-be readers would have been most likely to connect it with medieval scholastic philosophy. Yet Coleridge had in mind not the medieval sense of this term—that is, individuation as the principle that *distinguished* multiple particulars in a class from the universal to which they all belonged—but rather was modifying a term he had found in the German *Naturphilosophie* of F. W. J. Schelling and Heinrich Steffens.[35] Thus, to be on the lookout for individuation meant to look beyond, or behind, those differences between beings and toward the processes of expansion and intensification that produced *new* kinds of "individuals": individuation as a power of emergence, rather than of distinction.

To become oriented toward individuation meant orienting oneself toward orientation itself, for the most general tendency under which life acts was that of polarity, and even the polar nature of the earth was a consequence of life's essential polarity.[36] In place of that starry-eyed pantheism of his youth, which had encouraged a diffuse but shallow sense of immanence ("Everything is alive and full of spirit!"), Coleridge was now able to formulate protocols for much more focused and deeper ways of getting high on life ("Attune oneself to the poles, and be on the lookout for individuation"). And in place of that suspended animation of will opium dependency had produced, he was able—with the support of Gillman's ersatz volition—to produce a positive mode of suspension: the suspension of limits rather than of the will.

THE LIFE-MANUAL

Seeking to end the sterile contest between Abernethy's and Lawrence's different theories of life, Coleridge emphasized individuation in the scrofula essay in order to encourage a form of subjective experimentalism, one that would allow readers to overcome existing disciplinary assumptions about the nature of life. And insofar as the essay in which these protocols were described was to be attributed to Gillman, the doctor would then himself become an orienting pole of modern medicine, pushing surgeons and anatomists beyond their respective professional borders, and even beyond biology, understood as a regional mode of knowledge. Yet as a consequence of this eccentric itinerary of Coleridge's essay sections, what had begun as a medical analysis of scrofula had transformed into an all-purpose treatise that offered orientation in a much more fundamental sense. The implicit premise and promise of Coleridge's sections of the essay remained the same: namely, that a proper orientation toward life would facilitate a medical orthopaedeia, pointing out for physicians how best to care for scrofula. Yet Coleridge's contributions also promised to place at hand a more general therapy, with a proper orientation toward life now also serving as a means for regulating one's place within society more generally.

The deictic nature of Coleridge's essay sections—their attempt to point, rather than to argue—highlight the extent to which this purported "essay" on scrofula had in fact abandoned that genre for a new genre, one that we might best describe as the "life-manual." The genre of the life-manual was philosophical in nature but aimed neither at the deductive conclusions of a philosophical system nor at the desultory links of an essay. It instead sought to provide readers with a set of protocols that would allow for life experiences, which meant both an experience of the disorientating nature of life, and (as a consequence) an experience that altered one's life. The life-manual was thus something one kept at hand, rather than reading once and discarding; the therapeutic disorientation that the life-manual provided was not a one-time achievement but a perpetual struggle, something one had to experience again and again.[37] A life-manual was a text, in short, by means of which an individual allowed Life to transform a life; a means by which Life intensified itself precisely by shifting the orientation of a living being.

This collaborative life-manual Coleridge sought to write with Gillman was thus also to be, like *Lyrical Ballads*, an "experimental" text, but one that relied on a quite different sense of experiment. Wordsworth and Coleridge had sought to reconfigure poetry by developing a new mode of experimentalism that engaged, but also transformed, key elements of the eighteenth-century scientific economy of experiment, such as the relationships among gifts, witnessing, and attribution. They nevertheless retained from eighteenth-century sciences an understanding of experiments as *technical means* that enable a kind of *measuring*. Thus, each reader of *Lyrical Ballads* was to treat his or her responses to the poems as experimental data for determining "how far" the language of conversation was in fact "adapted to the purposes of poetic pleasure," while the final measure of the literary experiment would be the collective response to the volume (a response that would be measured at least in part by the demand for further editions and the willingness of other poets to participate in this "school" of experiment). The life-manual Coleridge hoped to write with Gillman was oriented not toward the reconfiguration of poetry but rather toward a complete transformation of the sciences themselves, and it was premised on an understanding of experiment as a vital experience rather than a technical means for measuring. Where *Lyrical Ballads* sought to reorient taste by enabling individual and collective measurements,

Coleridge and Gillman's life-manual was to produce its effects in a register prior to taste and measure, encouraging reorientations that were the result of the measureless activity of individuation. And where the experiments of *Lyrical Ballads* could still be understood as a kind of "trial" that depended upon an analytic separation of witnesses and experimental results, the experiments of the life-manual sought to collapse this distinction by transforming lives such that medicine would no longer be simply a technical means for curing illnesses but part of a scientific-philosophical pursuit of truth.

The form of gift-giving that was to enable this unprecedented form of experimentation was not that of the reciprocal contribution of texts to a common pool, as had been the case in *Lyrical Ballads*, but rather asymmetrical transits between lives and texts. The scrofula essay was possible only because Coleridge had received from Gillman a life-gift of surrogate volition and biographical orientation. It was in one sense impossible for Coleridge to reciprocate this gift, for the very premise of Gillman's help was that the physician was *not* lacking in either volition or biographical orientation. Gifts that cannot be reciprocated often place the recipient in a difficult, even intolerable, position, for a feeling of irremediable indebtedness can develop into hatred and violence, especially when the gift seems to bear on life itself. *Frankenstein* is, among other things, an analysis of this potential for violence that is always harbored in life-gifts: aware that he has been abandoned, the creature's search for his maker is in part an attempt to reestablish an economy within which gifts—or their violent converse, namely, blows—can be reciprocated.[38] Though Gillman's life-gift to Coleridge was not literally a spark of life, it nevertheless harbored this same potential for violence, for Coleridge was also unable to return the gift of orientation, at least in the same register in which he had received it. Coleridge thus gave a gift in another register, a gift of orienting text rather than of surrogate volition—or, more accurately, a gift of theoretical orientation that would ground Gillman's practical successes with scrofula by allowing readers to conclude that the doctor's prior therapeutic successes had been a consequence of his superior gift of vision into the nature and functioning of life, which would in turn enable the physician to reorient the practice of medicine (and, eventually, all of the sciences).[39]

Though Coleridge's transmission of text to Gillman *seems* like a return gift, potentially discharging Coleridge's debt to Gillman, it was in fact a more complicated gesture, since Coleridge does not seem to have understood Gillman's failure to use his writings as a rejection of this gift. *Had* Coleridge understood Gillman as simply rejecting his textual contribution, the debt that Coleridge owed Gillman would have remained in place. Yet what Coleridge seems to have intuited is that the gift of orientation is quite unlike other forms of the gift, for it is not oriented toward a stable community but rather toward the emergence of a new community in the indefinite future. In cases of mutual gift-giving that aim at enabling a fusion of identities into a school or community—the path taken by both the early members of the Royal Society and by Coleridge and Wordsworth in the case of *Lyrical Ballads*—there is always the possibility of the gift going wrong. Someone can refuse a gift (and hence implicitly refuse the community), or someone can fail to return a gift (and hence implicitly identify the community with a single individual rather than the anonymous collective). In these instances, reciprocity depends upon, and is intended to reinforce, a preexisting collective unity. The gift of orientation, by contrast, named precisely the capacity to create new, unprecedented forms of collective endeavor. Because this is the gift of the *capacity* to create a new collective in the future, there was no need for Gillman to "prove" that he had accepted the gift (by, for example, using it as the basis for the Jacksonian prize essay submission that was the initial impetus of the collaboration). Insofar as Gillman oriented himself in any way toward this gift of orientation—whether by incorporating it into "his" essay on scrofula or by abandoning it among his papers—it was a gift that had thereby been received.

As it turned out, the life-manual was more than Gillman expected or wanted at that point in his life. The dilemma it presented Gillman was, in part, that it seemed to confuse ends and means. Gillman and Coleridge had set out with the intention of producing an essay on a specific topic, and from this perspective, each of the sections should be a means leading to this end. In Coleridge's gift, though, the telos of the essay became confused, for insofar as Coleridge was successful in providing Gillman with a regulative life-manual, scrofula would cease to be the real subject of the text, functioning instead as simply one example of this regulating power at work.[40] As a consequence, to use Coleridge's text contributions would also be implicitly to question precisely that professional trajectory that the essay sections initially seemed designed supposed to support: instead of underscoring those professional limits within which Gillman could be seen as excelling, the life-manual was intended to disorient its professional readers, asking them to think anew the relationship between life and medicine. Arguably a noble goal, it had raised the stakes considerably for Gillman, for he either had to incorporate this gift of theoretical orientation into a project that promised to lead both him and his readers into uncharted waters or he had to abandon Coleridge's gift. Playing Walton to Coleridge's Victor, Gillman chose the latter option: no longer inspirited by those coastal winds of promise that had inspired the project in the fall, Gillman turned away from Life and the regulative pole for safer and more certain climes, leaving the unfinished essay to drift amid his notes and papers.[41]

Gillman's decision to remain on the path of professional progress emphasizes the peculiar temporality of abandoned experiments and the eccentric terrain that they inhabit. Like the famous *experimentum crucis* (crucial experiment), first described by late-seventeenth-century natural philosopher Robert Hooke as an experiment that serves "as a Guide or Land-mark, by which to direct our course," the abandoned experiment enables orientation.[42] However, unlike the crucial experiment—or, for that matter, the failed experiment—the abandoned experiment does not provide orientation from *within* a given field, whether that of a science or an art, but rather enables one to find one's path only by means of a solicitation from *beyond* that given field. In 1816, Gillman chose his path by choosing *not* to give in to the solicitation to go beyond his field; he chose, that is, to seek success within the already established terms of his profession. Yet Gillman's decision not to respond to the solicitation of Coleridge's text did not thereby make the abandoned experiment go away: Coleridge's text remained suspended in Gillman's papers and could have served the physician, at any point in the next two decades, as the catalyst for an attempt to renew and reorient the sciences. In this sense, the abandoned experiment does not possess that key characteristic of every properly scientific experiment, namely, an "ending." Peter Galison has stressed that we should not understand a scientific experiment as having any kind of "natural" end: since "the experimentalist can never, even in principle, exhaustively demonstrate that no disturbing effects are present," there is "no *strictly logical* termination point inherent in the [scientific] experimental process," and thus individuals and groups must develop, often by means of antagonistic debates and implicit skills, mechanisms that allow the individuals involved to agree that the experiment has ended.[43] Yet even if the end of a scientific experiment is the result of work and debate, rather than something that occurs automatically, every scientific experiment must come to an end. The abandoned experiment, by contrast, does not end but remains suspended—a point that remains outside of narratives of progress, but to which one can nevertheless continue to return for disorientations and reorientations.

DISORIENTATION, THE HELIOTROPIC IMPERATIVE, AND EXPERIMENTATIVE FAITH

Though Coleridge's essay sections remained frozen in Gillman's papers, they nevertheless represented the start of both a genre and a biographical trajectory that Coleridge would pursue for the rest of his life. By engaging the condition of possibility for change and orientation, Coleridge was able to redirect his own life, using both Gillman and the thought of polarity as the pivots that allowed him to transform circling and drifting into a new line of navigation. Where *Biographia Literaria*, begun a year before Coleridge met Gillman, had represented Coleridge's effort to recover his bearings by reappropriating those gifts that he thought Wordsworth had taken from him in the wake of their literary experiments, Gillman's help allowed Coleridge to invent the orienting life-manual. And this, as it turned out, was an approach he would pursue again and again in both his subsequent published and unpublished projects, texts that included *The Statesman's Manual* (1816), *A Lay Sermon* (1817), his revision of *The Friend* (1818), *Aids to Reflection* (1825), *The Constitution of Church and State* (1829), and his unfinished *Logic*.

That these latter texts are indeed life-manuals is perhaps obscured by the fact that Coleridge clearly hoped in these cases to steer the vital reorientation at which the life-manual aimed toward a Christian port. This goal had been largely submerged in the scrofula essay sections but came much more clearly to the fore in *The Statesman's Manual*, which Coleridge wrote at the same time, in the fall of 1816. This text, as well as a companion text—*A Lay Sermon*, published in March 1817—and a projected (but never written) third volume had been solicited from Coleridge by the publisher Rest Fenner, intended to provide readers with guidance and direction during the period of domestic economic hardship occasioned by the end of the Napoleonic Wars.[44] Addressed to the "higher classes of society," *The Statesman's Manual* encouraged in its readers a "sober and meditative accommodation to [our] times" and thus opposed those "self-realizing alarms" caused by improper habits of thinking, such as "restless craving for the wonders of the day" and a tendency to think in abstractions.[45] Coleridge suggested that the "antidote" to these faulty habits of thinking was, in principle, study of the Bible and, more specifically, an ability to parse current political events through scripture. However, the premise of *The Statesman's Manual* was precisely that its readers, though presumably devout, needed a guide that would teach them *how* to be guided by the Bible in this way. This meant learning to attune oneself to the "enthusiasm" of reason, cultivating a habit for an "oblivion and swallowing-up of self in an object dearer than self, or in an idea more vivid" (23). To read the Bible without the capacity for such enthusiasm was to read it without dexterity and, as a consequence, to miss its capacity to unify a life. What *The Statesman's Manual* aimed to provide, then, were protocols by means of which readers could learn to give themselves over to these enthusiasms of reason.

That same science of Life that Coleridge had outlined in his scrofula essay sections turned out to be one of those means for encouraging such self-oblivion. *The Statesman's Manual*, while in principle underlining the centrality of the Bible, consistently emphasized the importance of *disturbing* one's habitual relationship to scripture, for the "main hindrance to the use of the Scriptures, as your Manual lies in the notion that you are already acquainted with its contents" (25). In what may seem a paradox, Coleridge aimed to produce such disorientation by emphasizing the importance of balance, stressing that readers had to learn to acknowledge equally the claims of reason, religion, and the will. This balance could be sought and engendered, he suggested, by focusing on that distinction between Life and living things that he had outlined in his sections for the essay on scrofula, for such a distinction focused attention on a vital process that unified reason, religion, and the will. In place of the limited view of nature that sought to separate the living from the inanimate as though they were fundamentally different, Coleridge urged instead a more polar thought, one that could discern the "one attraction" that bound all living things together, and that recognized the function of the human mind as a "compass" (77–78).

Where the scrofula essay sections tended to emphasize the ascent of Life *away* from the elements and plants, *The Statesman's Manual* by contrast presented its readers with a heliotropic imperative, encouraging them to aspire to an "earlier" stage of life, that of plantness. Coleridge suggested that one could reach a properly balanced relationship among reason, religion, and the will by orienting oneself by means of scientific understandings of the vegetable, for these latter allowed one to see how the plant symbolized the capacity of reason to unify:

> I feel an awe [before the plants], as if there were before my eyes the same Power, as that of the REASON . . . with the rising sun [the plant] commences its outward life and enters into open communion with all the elements, at once assimilating them to itself and to each other. At the same moment it strikes its roots and unfolds its leaves, absorbs and respires, steams forth its cooling vapour and finer fragrance, and breathes a repairing spirit, at once the food and tone of the atmosphere, into the atmosphere that feeds it. LO! at the touch of light how it returns an air akin to light, and yet with the same pulse effectuates its own secret growth, still contracting to fix [chemically] what expanding it had refined. (72)

As this passage suggests, it was not by means of a general "awe" before the complexity of plants that one developed a proper feeling for Life but rather through a focused attention on those concrete particulars of vegetable existence that had been revealed by the botanical sciences (for example, the plant's capacities of chemical "fixing" and its relationship to its "atmosphere"). By means of such attention, one could cultivate one's own therapeutic heliotropism—feet planted firm, head turned toward the sky and following the orienting beam, an expressive exhalation answering the gift of life:

> O!—if as the plant to the orient beam, we would but open out our minds to that holier light . . . [then] ungenial, alien, and adverse to our very nature would appear the boastful wisdom which, beginning in France, gradually tampered with the taste and literature of all the most civilized nations of christendom, seducing the understanding from its natural allegiance. (73–74)

This feeling for the light would, in turn, allow readers to locate that middle passage between liberty and independence, on the one hand, and submission and resignation, on the other, to which Coleridge had referred in his sections of the scrofula essay. (As we will see in chapter 6, Coleridge's interest in vegetable atmospheres participated in a more general interest on the part of Romantic-era authors in soliciting the strange vitality of plants.)

While it is tempting to interpret Coleridge's dream of becoming-plant in terms of allegory—to see in it a barely disguised imperative that one remain passive and obedient to religious (and presumably also political) authority—such an interpretation misses the function of *The Statesman's Manual* as a life-manual. A life-manual is a text less interested in asserting a series of truths than in providing readers with protocols for disorientating themselves, for changing their lives. The practice of plant-ing is a practice of re-attuning, of checking and hindering existing habits in favor of the cultivation of new relationships, of slowing down so that a trajectory produced by a "mechanical impulse of some power of which [one is] unconscious" can be altered.[46] In this sense, both "God" and "Life" function primarily as regulative poles—concepts that help one push oneself beyond existing limits, each serving to complexify the other—rather than as stable concepts with any specific content tied to any specific imperatives. "Become other!": this is what both the thought of Life and the thought of God meant for Coleridge, and to produce protocols for such change, such disorientation and self-abandonment, remained Coleridge's mission for the rest of his life.

Coleridge emphasized in *Aids to Reflection*—arguably the most successful of his life-manuals—that this capacity to become-other required of the reader a mode of "*experimentative* faith in the Writer."[47] It was by means of experimentative faith that readers could orient and reorient themselves toward an author such

as Coleridge, who continually experimented with new styles and who continually tuned his authorial harp in public, to use a metaphor Coleridge himself had
employed a few years earlier in his 1818 revision of *The Friend*:

> The musician may tune his instrument in private, ere his audience have yet
> assembled; the architect conceals the foundation of his building beneath the
> superstructure. But an author's harp must be tuned in the hearing of those,
> who are to understand its after harmonies; the foundation stones of his edifice
> must lie open to common view, or his friends will hesitate to trust themselves
> beneath the roof.[48]

However, the real purpose of experimentative faith was not simply to regulate
the relationship between writer and reader but rather to make possible a readerly
mode of "continued attention."[49] Such continued attention was the precondition
for the disorientation that would in turn allow readers to feel, and to align themselves with, the orient(ing) beam of Life. Coleridge's own route to the concept of
experimentative faith exemplified the method itself, for it was only by abandoning his experiment in collaboration with Gillman that he was able to develop the
genre of the life-manual and its experimental method of disorientation.

POETRY AND LIFE

I do not want to end this discussion of experiments and abandonment without
acknowledging that Coleridge's discovery of the genre of the life-manual and experimentative faith also seems to have damped the vital inspiration that had made
possible his earlier poetry. As Paul Youngquist notes, it "is a commonplace of
Coleridge criticism that he turns from poetry to philosophy and morals about the
time he tries finally to kick his 'accursed Habit'"—or, perhaps more accurately,
that his attempts at poetry after 1816 are lacking in comparison with the verse
he composed before that date.[50] Though many of Coleridge's critics—beginning
with contemporaries such as William Hazlitt and Robert Southey—have regret-
ted his loss of poetic voice, Youngquist arguably takes us furthest in thinking about
how to understand this fact, for he connects Coleridge's shift in favored genre to
what we might call Coleridge's "sense of life." In what is only an apparent anachronism, Youngquist parses both Coleridge's opium habit and early verse through
the lens of Friedrich Nietzsche's analysis of the "glowing life of the Dionysian
revele[r]," that is, that ability to link intense pleasure and artistic achievement
with the loss of subjective control that Nietzsche both analyzed and in some sense
championed in *The Birth of Tragedy*.[51] Coleridge's opium addiction and his early
poetry were thus, from this perspective, simply two expressions of the same self-

nullifying life drive, one that did indeed separate will from volition but was also capable of memorializing this loss of control in the form of art.

While Youngquist's analysis of the Dionysian drive behind both Coleridge's habit and early poetry seems to me fundamentally correct, I find problematic the ultimate conclusion that he draws. For inasmuch as Youngquist wants to "rehabilitate" the earlier opium-indulging Coleridge, he also wants to chastise him for not remaining a Dionysian poet, arguing that Coleridge's sober turn, around 1816, to

> philosophy is a turn away from—and submergence of—the Dionysian, the very force which so invigorated his earlier poetry. His philosophical alternative institutes assimilation to moral discourse as the condition of health and happiness—as indeed it was for Coleridge himself in his desperate efforts to abandon opium: "In the one crime of OPIUM, what crime have I not made myself guilty of!"[52]

Youngquist acknowledges that Coleridge's residence with Gillman did *not* in fact cure the former poet of his opium habit—Gillman never seems to have promised more than a reduction of opium use, and Coleridge occasionally found ways around even this—but he argues that the *form* of Coleridge's attempted cure, the search for a "disciplinary" environment that could provide him with ersatz volition, echoes the moral stringency of his new philosophy.[53] For Youngquist, in short, what changed after 1816 was that Coleridge lost sight of Life, that dizzying, ecstatic, nauseating force that nullifies the self.

I have suggested that it was precisely to this sense of life that Coleridge continued to cleave, even in and after 1816, but what changed were the generic means by which he sought to solicit its forces of disorientation. Though the life-manual is no doubt a different means than lyric verse of soliciting the disorientation of Life, I am not certain that it is actually a more sober or "disciplinary" genre. The life-manual is far more explicitly committed than lyric poetry to *changing* a life, though its protocols for producing such transformation demand a different mode of pleasure than that at which the lyric was generally understood by Romantic poets to aim.

None of this is to deny, of course, that I too would have been happy for a Coleridge who had been able to write more poetry like "Kubla Khan," "The Rime of the Ancient Mariner," and "Frost at Midnight," and like many others, I suspect that Coleridge's inability to become what we feel he should have been, as a poet, was a direct consequence of his problems with opium. Yet I feel uneasy in favoring the Coleridge who-had-been over the Coleridge who-came-to-be, or seeing only lack after 1816—for in fact there is still plenitude after 1816, even if not of the sort for which we would have hoped, had we had any control over such things. This regret over the Coleridge who-came-to-be—that is, the Coleridge who turned away from poetry—seems to me to imply an understanding of poetry that keeps us from understanding how deeply literature can be part of a life. For what would poetry be if we could indeed compel into being a Dionysian poet who kept on writing for the entirety of his life? Wouldn't this suggest that poetry, however much we might like to think of it as an embodied expression of Dionysian ecstasy— or, alternatively, a working-through of crises—nevertheless keeps its distance from any ecstatic experience or crisis sufficiently deep that it could question poetry itself as genre and life activity? Doesn't criticism of this loss of the Dionysian Coleridge, in other words, imply that poetry is nothing more than an expressive tool in the service of a continuous trajectory of self-development, rather than an expression of those rending, disorienting forces that can truly call a life into question?

Coleridge is an extraordinarily useful figure in thinking through these kinds of questions, for his life, despite all his efforts, never demonstrated a sinuous curve of self-development; rather, it marked out an erratic trajectory characterized by peri-odic disorientation and perpetual return. When we read Coleridge's life in terms of errancy and disorientation, his loss of poetic potency no longer appears as the loss of a potential but rather as a consequence of his active attempt to locate a form of writing that would provide him with orientation. This is not to suggest that poetry is intrinsically unable to provide such orientation; it is simply to say that it could not do so for Coleridge after 1816. Our commitment, in short, ought to be to follow— and thus to understand—the *changing* relationships of life and letters in those who have written great poetry. This means seeing poetry as one medium among others by means of which authors seek to produce events—that is, points of inflection that cast meaning over what has come before and what is to come—in their lives.

The concept of the abandoned experiment gives us, at least in Coleridge's case, purchase on these intersections of life and text, for the concept forces us to go beyond any single corpus, whether that of a particular genre (e.g., "Coleridge's poetry"), a productive authorial totality ("Coleridge's writings"), or a biographical unity ("Coleridge's life"). As I noted above, Coleridge's collaboration with Gill-man should be understood as an attempt to take up again the elements that had enabled the experiments of *Lyrical Ballads* (an orientation toward science; gift-giving; collective authorship) but to rework these elements in a way that would not lead to rivalry. In this sense, Coleridge's attempt to write an essay with Gillman was not an event that could be contained in the intersection of Coleridge's and Gillman's individual biographies, but it was also an event that bore upon the per-sonal relationship between Coleridge and Wordsworth, and—perhaps even more important—upon their earlier verse experiments. Gillman found, of course, that he could not fit Coleridge's gift of text and orientation into an essay that was to go out under his own name, and he and Coleridge had to abandon their experiment in collaboration. Yet precisely because this new attempt at collaboration involved reconfiguring the components of Coleridge's original collaboration with Wordsworth, Gillman and Coleridge could abandon their experiment without the danger of rivalry and conflict.

The abandoned experiment should thus not be understood as a misstep or a false path that, by revealing a wrong direction, points the way toward stable accomplishments. The abandoned experiment is, rather, an experience of disorientation that initially calls into question, and can continue to question, what it means to live a life and what part texts can or ought to play in such a life. The abandoned experiment is, in this sense, external to both science and to art (or an art, such as poetry): it is not a mode of experimentation by which either science or art advances but rather an experience that calls into question the very nature and worth of such pursuits.

To conclude this reflection on lives and texts by returning to the comparison between Coleridge and his dark literary doubles, Robert Walton and Victor Frankenstein, we can say that Coleridge's experiments in abandonment allowed him to escape both the violence and disappointment that Mary Shelley mapped out in her novel. In 1816, Coleridge was, like Victor, still bound to dynamics of pursuit and rivalry that were the consequence of an earlier experiment. In Shelley's novel, Victor is unable to free himself from or abandon these dynamics, and he finds a "solution" only in death. Though Walton is arguably able to learn from Victor's failure, the cost of such pedagogy is disappointment, for it is only with reluctance and regret that Walton abandons his polar inspiration and aspirations. Coleridge, by contrast, links abandonment and experiments in such a way that he is able to accept surrogate volition and thereby begin to chart a path away from the trajectories marked out by Victor and Walton: rather than remaining bound to his past experiment or giving up on his polar aspirations, Coleridge instead creates a new genre, the life-manual, that pursues the poles from another angle. Coleridge's approach makes one wonder, in fact, whether Victor's problem was that he didn't continue to write: perhaps a handbook—a life-manual—for his creature could have ended their violent double-bind. The life-manual was, in any case, the means by which Coleridge was able to link abandonment and experiment and thereby create points of inflection—that is, points at which a text, while remaining distinct from a life, nevertheless can change the direction of the latter, while at the same time a life, even as it retains its distance from a text, gives to the latter a new meaning and mission.

CHAPTER 4

Nausea, Digestion, and the Collapsurgence of System

That nauseous, rank and heaving matter, frightful to look upon, a ferment of life, teeming with worms, grubs and eggs, is at the bottom of
the decisive reactions we call nausea, disgust or repugnance.

Georges Bataille, Erotism, 56–57

Disorders of the stomach are become, in a manner universal . . .

J. Hill, Centaury, the Great Stomachic, 3

The end of the world begins with a stomachache.

Or it does so, at any rate, for the paranoiac. According to medical accounts that
range from early-nineteenth-century physiology to contemporary clinical diag-
nostics, paranoia is associated with feelings of disgust and nausea. As literary critic
David Trotter notes, disgust and nausea are understood to play a functional role
in the production of paranoid fantasies: paranoia "mobilizes disgust at moments
of crisis, moments when the 'still persisting central core of the personality' . . . is
felt to be under threat" (67). In the normal course of things, of course, disgust and
nausea bring to conscious awareness a sense that something has gone awry, that
something is out of place. Nausea, for example, is a premonition of an imminent
digestive reversal—a felt sign that the stomach may expel something upward and
outward—while disgust has often been described as a culturally coded echo of
this desire to distance oneself from something unclean or unhealthy. For the
paranoiac, though, the corporeal semiosis of disgust and nausea are bound to
an epistemological rupture: disgust and nausea occur as one understanding of
the world is expelled and replaced with another, generally much more private
and idiomatic, understanding of the system that governs the world. Thus, in the
context of paranoia, disgust and nausea signal not simply the imminent expulsion
of something, but also, and simultaneously, the incorporation of something else

("Now I see that they were *all* part of the plot!"). We might call this phenomenon *collapsurgence* in order to capture those dynamics, both affective and epistemological, of the process by which one system for understanding the world collapses, and another, purportedly deeper, understanding of order surges forth. Nausea is part of a paranoiac's project of making sense of the world in terms of systems—or, as Trotter puts it, "Paranoia's commitment to system has as its obverse a commitment to nausea" (66).

From this perspective, what should we make of the fact that a significant strand of modern literary practice seems specifically to aim at *inducing* nausea in its audiences as a means for making these latter aware of a controlling system? I am thinking here less of novels such as Jean-Paul Sartre's *Nausea*, which describes the existential sickness of the novel's protagonist but does not seek to produce nausea in readers, and more of the graphic depictions of sperm, blood, urine, and shit in the novels of Georges Bataille; the references to ejaculation, repeated ad nauseum, in William S. Burroughs's cut-up novels; or the lengthy descriptions of scatological practices and even voyages through shit in Thomas Pynchon's *Gravity's Rainbow*.[1] Nor is this endeavor to produce nausea by framing bodily excretions limited to literature: consider sculpture and installation and performance art that present human effluvia, such as Piero Manzoni's *Artist's Shit* (1961); raw meat and blood, as in Hermann Nitsch's *Orgies-Mysteries-Theater* (1962–1998); or living tissue, for example, the "frog steak" artificially grown and then consumed by the Tissue Culture and Art Project in *Disembodied Cuisine* (2000–2003).[2] There is little point, of course, in approaching such works from the perspective of a clinical understanding of paranoia. Nevertheless, we can see these pieces as employing the connection between paranoia and nausea in a *tactical* sense. Such works suggest that in a world in which living human bodies come under the sway of systems, or a System, one can counter this control only by engendering corporeal shocks and affective counterflows. The mission of art, from this perspective, is to commit itself to sticky, slimy, oozing forms of transgression that will enable a controlled form of collapsurgence, and in this way allow audiences to achieve the extra-artistic goal of recognizing and freeing themselves from an otherwise hidden system.

Though this goal of nausea-production emerged most clearly in twentieth-century art, the origins of this strategy are to be found in Romantic-era literary texts. Depictions of "low" bodily functions have a long history in Western literature; such representations were central to Juvenalian satire, for example, as well as to the "grotesque realism" of the Renaissance.[3] However, it was only in the Romantic era that the two premises necessary for the practice of collapsurgence emerged. First, the Romantics approached their world in terms of systems and their obverse, paranoia. Romantic-era authors understood their social world as dominated by invasive and imperializing social systems, such as the inhuman system of slavery or the dehumanizing system of British law and politics.[4] The ever-expanding but perpetually uncertain extent of systems (could one ever actually escape "the system"?) encouraged a hermeneutics of suspicion, the dynamics of which are well captured by Thomas Pfau's description of the "axiomatic premise" of Romantic paranoia: "the experience of the real hinges on a constant preparedness to distrust experience, and to treat experience as ideological frames conspiring against our genuine access to the real."[5] Second, and equally important, the Romantics found plausible a materialist understanding of reality that suggested that systems took hold of, and in fact animated, individuals at the level of their bodies. Systems extended into the interior of the individual, which meant that the life of the individual expressed the system and, conversely, that the system expressed itself in the life of the individual. This intimacy of life and system was indexed in the link, proposed by many authors, between thoughts and digestion: thoughts, like food, were understood to have a certain tangibility and thus were, in a certain sense, swallowed and assimilated by the body. Given these premises, an attempt to produce nausea by means of texts could emerge as a coherent strategy for freeing readers from their living participation in systems, for reading a text could mean reversing the flow of "mental digestion." To induce nausea by means of texts was a first step that would allow an individual to grasp that seemingly autonomous choices were in fact animated by a system, and thus also a first step in enabling readers to retreat or escape from a system.

Yet Romantic-era efforts to induce collapsurgence were not all of a piece, and this chapter outlines three different literary experiments by means of which authors in the Romantic era sought to use literature to free the living and thinking body from its entrapment by systems. I begin with some background, outlining the emergence in the eighteenth century of what I call "chylopoietic discourse," by means of which both medical and nonmedical authors debated the physiological and social implications of foreign foods for individual and national bodies. Insofar as chylopoietic discourse sought to encourage critical—and often panicky—reflection on the systems of trade that brought new foods into the bodies of Britons, it also tended to encourage paranoia about the reach and importance of these new systems. My first case study, Thomas Trotter's A View of the Nervous Temperament (1807), reveals what happened when authors too quickly assimilated the materialist position that systems took hold of individuals by forcing them to "digest" thoughts. Trotter's account, which purported to describe a national epidemic of nervous illness brought on, in part, by foreign foods, in the end suggested that even attempts to point out this epidemic were in fact likely simply to perpetuate it. The next case study, William Godwin's novel *Caleb Williams* (1794), focuses on an instance of paranoid chylopoietic discourse more attuned to the question of language. A literary exploration of the materialist political philosophy that Godwin had developed in *Of Political Justice* (1793), Godwin's novel explored the consequences of un-digesting systems and suggested, in good moderate fashion, that a disabling form of nausea results when systems are expelled too quickly. Yet this lesson was complicated by Godwin's choice to employ a first-person narration, for such a form could provide no explanation of *how* one could escape all-embracing system. The final variant of paranoid chylopoietic discourse that I consider, exemplified by Samuel Taylor Coleridge's antislavery lectures from the 1790s, represents a solution to the problems encountered by Trotter and Godwin. Coleridge recognized that the key to collapsurgence lay in making a sharp distinction between the material causality of systems and the capacity of language to express and link events. The peculiar rhetorical function of the key trope that Coleridge employed—he suggested to his listeners and readers that they had ingested the blood of slaves—was premised on the principle that though the expression of such an event possessed no literal referent or causal force in the usual sense, it could nevertheless alter relationships between elements of a social system.

As I document in this chapter, paranoid chylopoietic discourse bears a peculiar, aporetic relationship to both the sciences and the arts, and it is, as a consequence, difficult to locate methodological models in either science studies or literary criticism capable of illuminating its dynamics. Chylopoietic discourse relied both on the particular results of eighteenth-century sciences of digestion and on the founding premise of eighteenth-century science more generally: namely, that all of reality, including the human body, formed a coherent, unified system that in principle could be exhaustively explained by means of the tools and methods of science. Yet paranoid chylopoietic discourse was neither a popularization of the sciences, in the senses discussed by scholars such as Jan Golisnki, Simon Schaffer, Larry R. Stewart, and Michael Lynn, nor an attempt to create a new (quasi-)science, as, for example, outlined by Robert Darnton in his account of mesmerism in the era of the French Revolution.[6] Nor was paranoid chylopoietic discourse an artistic exploitation of scientific uncertainty or ambiguity of the sort described by Dennis Todd in his study of eighteenth-century literary interest in medical debates about the origins of embryological monsters.[7] Rather than seeking to extend the legitimacy of the sciences, create a new science, or exploit points of uncertainty in the sciences, paranoid chylopoietic discourse instead aimed at something like a *sur*-science: a sort of mist that would surround the sciences, drawing from their results and leaving unquestioned their methods, while at the same determining the *affective tenor* of the aspirations to systematicity of the sciences. The sciences, in other words, would be left just as they were, but the knowledge that they produced would be parsed through a totalizing "mood" in order to produce transformative events within the life of the individual.

The relationship of paranoid chylopoietic discourse to the arts and aesthetics is equally vexed, for the affective experience sought by the experimental genres I consider here—namely, the experience of nausea—is not properly an aesthetic experience but rather a site of aporia within aesthetic discourse. As I note in the penultimate section of this chapter, both the concepts of nausea and disgust tend to confuse oppositions between aesthetic judgments and nonaesthetic sensations: nausea and disgust both *delimit* aesthetics, in the sense that they are experiences that signal something outside aesthetic experience, yet both can also result from *too much* aesthetic experience. Despite their similarities, though, nausea is in fact more difficult to think than disgust, for it cuts across the distinction between nature and culture that still underlies most discussions of the phenomenon of disgust. When pursued as a goal of discourse, nausea is not, fundamentally, a matter of judgment and taste (the *gustus* in disgust) but rather a matter of setting out on a voyage into the unknown, with all the implications of transformation that that entails.[8] As a consequence, understanding nausea and collapsurgence requires that we go beyond traditional Romantic aesthetic categories (the picturesque, the beautiful, and the sublime), as well as beyond the dichotomy of pain and pleasure that underlies them, in order to consider other ways in which texts provide audiences with opportunities to experience transformative events.

Neither the sur-science nor the sub-aesthetics of paranoid chylopoietic discourse can be well understood on the basis of the topographic assumptions about social space that dominate histories of eighteenth-century science: that is, understandings of science as a discrete "realm," or set of interests, that can be extended into, contested by, or imitated by other social realms or interests. I have thus drawn much of my methodological inspiration from recent discussions in science studies that consider—or in some cases even seek to create—modes of contemporary discourse that also channel the sciences through affective, ecstatic *Stimmungen*.[9] Yet insofar as these contemporary discourses lack the paranoid dimension I consider here, the challenge of this chapter is to understand how Romantic authors could bring science and affect together to create a sense of a totalizing system yet also position an awareness of that oppressive totality as the necessary, almost ritualistic, passage-point that would enable self-transformation and freedom.

FOREIGN FOODS AND THE EMERGENCE
OF CHYLOPOIETIC DISCOURSE

Historians and literary critics have noted that the massive expansion of European, and especially British, trade and colonization during the course of the eighteenth century made an unprecedented number of new commodities available to British consumers, even as an increase in wages and a drop in commodity prices meant that these goods were available to an ever-increasing percentage of the national population.[10] Even at the start of the eighteenth century, Britons were fascinated with the apparent folding of time and space that international trade engendered, as materials and products gathered in distant lands became available to, and part of, British goods and bodies. In the *Spectator*, Joseph Addison provided an early figure for this fascination, noting that the "single dress of a Woman of Quality is often the product of an hundred Climates."[11] Addison saw in the layers and folds of the clothing of the nobility a symbol of global interconnection, one that made colonial and trade links between different parts of the world visible. The power of Addison's image—its phenomenological efficacy—depends upon its capacity to link two different scales of experience, mediating between the temporally and spatially distributed scale of international trade, which enables the elements of expensive dress to be derived from "the different ends of the Earth" and the scale of embodied experience; insofar as clothing functions as a sort of second skin, his image allowed British readers to map international trade onto their own bodies.

Yet if expensive clothing provided an early-eighteenth-century image for the temporal and spatial foldings produced by the increasingly global nature of British trade, mid- and late-eighteenth-century commentators tended to focus on food in their efforts to understand the increasing complexity of the modern world. Interest in food grew as more and more Britons encountered traces of foreign locales through the intermediary of imported foodstuffs.[12] Sugar, for example, was available only to the very wealthy in the late seventeenth century but had become a staple of all classes by the end of the eighteenth century, and previously exotic foods such as tea, chocolate, coffee, spices, and tropical fruits were an increasingly common part of the British diet by the early nineteenth century.[13] In books and pamphlets such as Simon Mason's *The Good and Bad Effects of Tea Consider'd* (1745), Simon Pauli's *A Treatise on Tobacco, Tea, Coffee and Chocolate* (1747), and Joseph Hanway's "Essay on Tea" (1757); novels such as Tobias Smollett's *The Adventures of Peregrine Pickle* (1751) and *The Expedition of Humphrey*

Clinker (1771); prints such as William Hogarth's *Gin Lane* (1751); and antislavery poems that linked the blood of slaves with the sugar these laborers produced for British consumers, British authors considered the effects of foreign foods on British physiology and morals.[14] Though British writers continued to discuss the transformation of global space and time by means of commodities such as clothing, furniture, and dishware, food became especially important in these discussions of the causes, effects, and consequences of global trade.

Food was of especial interest in large part because it bore a more complex relationship to the embodied experiences of consumers than did other commodities. Though a product such as clothing might serve as a kind of second skin, food was literally assimilated into consumers' bodies. The obvious complexity of the processes involved in digestion and assimilation encouraged modes of genre-folding, as authors interested in the effects of global trade upon consumers incorporated bits of contemporary medical discourses about the body and digestion into their accounts. The result was what I call a "chylopoietic discourse" that sought to understand the new importance of foreign foods in terms of the body's capacity to produce (poiesis) nutriment (chyle).[15] Chylopoietic discourse was not a medical discourse per se—that is, it was not a discourse solely carried out by and among physicians or surgeons—but rather represented a medicalization of discussions of the causes and impacts of global trade. While some of the authors who wrote about the effects of foreign foodstuffs on British bodies were physicians or surgeons, many more were not, and both physicians and non-physicians sought to encourage in their readers a dietetic awareness of the physiological effects of domestic and foreign foods.[16] Authors considered whether, for example, Chinese tea in fact contributed to "the Health of our *English* Constitution" by "invigora[ting] the minutest Fibre, brac[ing] the relaxed, inelastic Tubes" of our physiology; whether coffee "pricks and stimulates the Fibres of [the] Stomach . . . thereby oppos[ing] . . . noxious Matter"; and whether foreign "spices and peppers" wore out "the excitability, or vital power of the stomach."[17]

The development of chylopoietic discourse depended upon the significant and widespread experimental interest in digestion in the late eighteenth and early nineteenth centuries. Because of its location deep inside the body, the mechanisms of the stomach and digestive tract had been notoriously difficult to determine for classical and early modern medical theorists. Where the classical Galenic tradition had been willing to ascribe digestion to vaguely understood modes of "concoction" (*pepsis*) of food into vital substances, seventeenth- and eighteenth-century theorists sought to isolate and theorize more precisely the mechanism that transformed food into chyle.[18] Many early-eighteenth-century theorists likened the stomach to a mill, arguing that the primary activity of digestion was a mechanical grinding that reduced food to fine particles, which could in turn be "strain'd thro' the narrow Orifices of the lacteal Veins."[19] Other medical theorists argued that digestion required a physical transformation of food, whether through processes akin to "fermentation" or to "putrefaction."[20] By the late eighteenth century, experiments by Lazzaro Spallanzani and John Hunter suggested that digestion was primarily a consequence of chemical reactions, with mechanical forces playing a minimal role in most animals.[21] Nevertheless, the precise mechanisms of digestion remained murky. It was not clear, for example, what distinguished the nutritive from the nonnutritive parts of food, nor was it clear how the stomach and digestive tract were able to distinguish between these different elements of food, absorbing what the body could use and eliminating what it could not.

The evident discrepancy between the complexity of digestion and the explanatory capacity of available physiological models encouraged two trends in chylopoietic discourse. First, this discourse focused increasingly on a normalized human body. Even as Spallanzani's experiments demonstrated the diversity of animal digestive systems, eighteenth-century authors argued for the homogeneity of the digestive tract across humans. Where early eighteenth-century physicians had often distinguished the physiologies of different classes of human bodies — describing, for example, the peculiarities of the aristocratic body or the "Chinese" body — late-eighteenth- and early-nineteenth-century writers interested in the effects of global trade on British consumers suggested that the bodies of the upper and middling classes were, for all practical purposes, identical, even if their differing patterns of consumption might necessitate different prescriptions for health.[22] Romantic-era physician Thomas Beddoes's three-volume *Hygëia*, for example, focused on those sources of ill health *shared* by the "*Middling and Affluent Classes*," while Thomas Trotter suggested that though individuals laboring in different occupations might suffer different health problems, this was due to differences in the occupational environment rather than to any intrinsic corporeal differences.[23]

Second, chylopoietic discourse increasingly sought to explain digestion in terms of a transmutation and transformation of food, rather than simply a change in scale. The shift in the late eighteenth century from mechanical to chemical models of digestion was a shift in the implicit paradigm that allowed researchers, physicians, and the lay public to understand how food was incorporated into the body. The earlier stomach-as-mill model was premised upon the belief that the stomach simply changed the scale of food, without changing its elements; that is, by the Concoction that is performd [*sic*] in the Stomach, the Food is divided into integral Parts, not differing from what they were before, but in obtaining lesser Bulk; in the same Manner altogether as Coral is ground upon a Marble with Water, and reduced into an impalpable Powder, whose Parts are only small Pieces of Coral, and not any Principles in to which Coral is resolv'd.[24]

The deep premise of this earlier model was that of an *existing* continuity between the body and food; from this perspective, the function of the digestive tract was simply to change the size of aliment, grinding food into smaller and smaller chunks until these particles could slip into the porous parts of the body. The later chemical model of digestion, by contrast, implied that the chylopoietic organs produced both quantitative and qualitative transformations; the digestive tract not only reduced food from the macro- to the micro- scale but also translated, or transformed, food into a form (or forms) that could be engaged by the body. Rather than assuming an existing continuity between food and the body, the chemical paradigm suggested that the digestive tract changed the very structure of the food material it engaged. It was the obvious complexity of such processes, in fact, that encouraged late-eighteenth-century physician authors and lay authors to stress proper digestion to their readers: precisely *because* digestion involved so many factors, some of which were especially difficult to understand, consumers could not afford to be lax in their attention to matters of diet.

CHYLOPOIESIS AND PARANOIA IN TROTTER'S
A VIEW OF THE NERVOUS TEMPERAMENT

These two trends in chylopoietic discourse—a turn toward "typical bodies" and an interest in qualitative transformations—encouraged authors to engage in generic experimentation. Designed to intervene in widespread patterns of food consumption, chylopoietic discourse was almost always oriented toward lay audiences, but the proper literary form for such discourse was by no means obvious. Individual authors labeled their contributions with a variety of formal descriptions: "essays," "remarks," "treatises," "views," and "dissertations." However, these formal descriptions were in general untrustworthy guides, for the accounts of food, digestion, and consumption contained within these works drew from multiple genres, as authors folded medical claims into satire, for example, or emplotted their medical accounts within narratives of suspense.

The paranoid version of chylopoietic discourse that appears in Thomas Trotter's *A View of the Nervous Temperament* (1807) provides a particularly interesting illumination of generic folding. Trotter was well positioned to reflect on the systemic aspects of global trade, for he gained most of his medical experience as a physician in the British navy, within which he eventually rose to the position of Physician to the Fleet. His personal interest in systematic accounts of what we would now call "public health" was evidenced in his three-volume treatise on British nautical health, *Medicina Nautica* (1797–1803), as well as *An Essay Medical, Philosophical, and Chemical on Drunkenness and Its Effects on the Human Body* (1804), which documented the debilitating effects of alcoholism on individuals and the social body.[25] In A *View of the Nervous Temperament*, Trotter remained interested in the effects of drink and food on individual and social health. However, he shifted his attention to the more protean category of nervous illnesses and developed an image of the digestive tract as the organ through which the pressures of international trade and recent political history exerted their effects and became phenomenologically available.[26]

A *View of the Nervous Temperament* was oriented toward a lay audience, and Trotter's primary goal in writing it was to convince his readers that an epidemic increase in "nervous illnesses" threatened both the health and national security of Great Britain. Trotter begins the book with an attempt to induce a statistical panic, claiming that Britain had witnessed, over the course of the eighteenth century, a marked increase of nervous diseases: where earlier in the eighteenth century, nervous illness accounted for only one-third of illnesses in Britain, "at the beginning of the nineteenth century," nervous diseases were accounting for two-thirds of all disease in Britain.[27] Trotter acknowledged that other physicians had failed to notice this epidemic but attributed this oversight to the peculiar capacity of nervous illness to mimic nonnervous diseases. The "protean shape and multiform appearance" of nervous diseases meant that physicians often misdiagnosed their patients' ailments, which both masked the increasing frequency of nervous diseases and facilitated their spread (xv). The urgency of Trotter's account was premised upon the invisible and protean nature of this epidemic, a nature that also made it difficult, if not impossible, to provide a "generic definition" of nervous illness, though Trotter suggested that one might at least look for characteristics such as an individual's "inaptitude to muscular action . . . an irksomeness, or dislike to attend to business and the common affairs of life; a selfish desire of engrossing the sympathy and attention of others to the narration of their own sufferings."[28] In any case, Britain needed to recognize, and hinder, this epidemic: left unchecked, the spread of nervous illness would "inevitably sap our physical strength of constitution; make us an easy conquest to our invaders; and ultimately convert us into a nation of slaves and idiots" (xi).

Urbanization, Trotter reasoned, was responsible for the increase in nervous illness in Britain. "The remote causes of nervous diseases," he wrote, "are chiefly to be sought in populous towns; and increase in proportion to the deviation from simplicity of living," and "the state of civilized mankind is thus exposed to causes of bad health, which have no power among rude nations" (54). He argued that the structure of the human body was at odds with the requirements of urban occupations and living, and the second chapter of A *View of the Nervous Temperament* analyzed how the demands of specific professions injured human physiology. "Many species of manufacture are unhealthful" as a result of the "noxious metallic fumes to which the workmen are exposed," while the owners and managers of the means of production were injured by "the posture of leaning over a desk, [which] contracts the motion of the lungs, and impedes the functions of the stomach."[29] In the case of "literary men," the process of writing and thinking required a "posture of the body [that] . . . is unfavourable to health," as "the lungs are seldom expanded to full inspiration" and "all the secretions, and their excretories, fall into inaction from want of muscular motion" (38). Even the aristocracy—the class of the "idle and dissipated"—could not escape the evils of the urban environment, for their excessive eating and drinking wore out the vitality and capacities of their organs (43–46).

Though each urban occupation injured the human body in its own way, all of these corporeal insults tended to disturb the structure of the nervous system as a whole. The fumes inhaled by factory laborers, the cramped lungs of managers, the physical inactivity of authors, and excessive stimulation of aristocratic bowels all encouraged the formation of a "nervous temperament." Once damaged, the nervous system found itself locked into a pathological feedback loop, "storing" that which had disturbed it in the first place. Those who became afflicted with a nervous temperament could not expel their sensory impressions in the normal, healthy way: the nervous body, like a miser, "hoard[es], . . . retains, or records as it may be termed, all the effects of vicious indulgence," and these "stored" impressions then cause further nervous symptoms (210–11). Yet because a nervous temperament was a systemic problem, symptoms became difficult to interpret, for the origin of a symptom was not the afflicted body part but rather the nervous system itself: "The migratory power which these affections possess, of traversing every part of the body is the inscrutable *idiosyncracy* of the NERVOUS TEMPERAMENT" (216).

Though Trotter was interested in all the ways the physical environment could injure the nervous system, the stomach played a particularly important role in his account, and he returned repeatedly to the problems that urban occupations posed for the "chylopoietic organs." He claimed, for example, that the extreme mental focus demanded of literary men disturbed their chylopoietic organs, and "the powers of digestion, with all the viscera subservient to them, partake in a particular manner of this derangement, and grow unequal to their office." Moreover, "the debility and inactivity which take place in the chylopoietic organs, react on the nervous part of the frame; and the faculties of intellect, as sympathizing in a great degree, with all these highly sensible bowels, are influenced by the general disorder" (39). The sympathetic capacity of the digestive tract was important for Trotter because, as he explained in a later chapter, the "chylopoietic organs" have an "exalted" station in the "animal economy," a result of "the pre-eminence they occupy in preparing the nourishment of the whole system."[30] Because of its vital role in nourishing the body, the "human stomach is an organ endued by nature, with the most complex properties of any in the body; and forming a *centre* of sympathy between our corporeal and mental parts, of more exquisite qualifications than even the brain itself" (203).

The central role of the stomach in the production of nervous temperaments and illnesses authorized Trotter to consider carefully the dietary regimen of the British, for, he suggested, what the British ate directly affected the health of the nation. In the sections of his book devoted to possible cures of British nervous illness, Trotter forbade all "high seasoned food" (70), including "all the spices and peppers" (75), "soy, catchup, and all the warm sauces" (76), and mustard, claiming that through the use of these sorts of condiments, "the excitability, or vital power of the stomach, is worn out" (70). Trotter recommended beef and mutton over veal and lamb, and roasted meat over boiled; he argued against the use of "*legumina*, such as peas and beans"; and he urged abstinence "from every species of spirit or fermented liquor" (276, 279). Trotter was critical of overeating, contending that "the organs of digestion are thus oppressed and weakened by the surplus; the liver in particular is enlarged in its volume: the blood is charged with a richer quality, and every vessel and gland stimulated to excess" (68–69).

This image of the stomach being "oppressed and weakened" by surplus is the master image that dominates Trotter's analysis of the causes of increased British nervousness. Yet the stomach played this role for Trotter because he understood it as *both* an individual and a social organ. The stomach was not simply part of the individual body but also a nexus that linked individual desires to social dynamics. In his discussion of the nervous damage induced by tea and coffee consumption, for example, Trotter pointed not just to what he believed to be the immediate corporeal effects of these substances—namely, the destruction of the dietary tract—but also to the transformations of social relations that tea and coffee drinking encouraged. So, for example, as the consumption of these beverages spread from the upper and middle classes to the lower class, their use became even more harmful, for they were seen by members of the lower class as a necessity—and worse yet, a necessity that took precedence over more nutritious foods. "A man or woman who has to go through much toil or hardship has need of substantial nourishment," yet if "the humble returns of their industry are expended on this leaf [tea], what remains for the purchase of food better adapted to labour?" (72–73). Trotter also contended that tea functioned as what we would now call a "gateway drug," for it "pav[ed] the way" to liquor (73), at least among the poor, and, in the context of coffeehouses, to the unsupervised use of medicines by the middle classes (318–19).

As beverages that originated outside of Britain and flowed back along colonial trade routes, coffee and tea showed the extent to which Britain had tied its very existence to urban commercial centers. Trotter warned that, as a general principle, "a commercial people merely, can never be an independent nation," for "they owe to foreigners the consumption of their manufactures" (150). The British were also consuming the raw goods of foreigners, causing Trotter to despair: "Commercial Britain, enriched by manufacture and colonial wealth, when compared to her barbarous state, does not exhibit a nobler spirit of independence, or show more fortitude in opposing French invasion, than what was done nineteen hundred years ago" (146–47). The purported fruits of global commerce, in short, poisoned their own roots, in part by debilitating the nervous systems of British laborers and consumers, in part by encouraging the consumption of other damaging substances, and in part by making Britain dependent upon other countries and lands. The stomach thus served Trotter as a figure by means of which one could understand the capacity of commerce and trade to produce systemic sociophysiological changes over time.

Trotter's emphasis on the stomach was motivated in part by the intense experimental interest in digestion that I described in the last section, but his model of digestion pushed even further an understanding of the stomach and digestive tract as organs of extraordinary complexity. The digestive tract served Trotter as a useful figure for understanding historical dynamics and change because he understood the stomach not simply as an organ of selection and filtering but, more fundamentally, as an organ of translation, a series of corporeal folds that forced the body to translate the distributed and far-flung scales of urbanization and global trade into the much more local scales of individual human experience. Where eighteenth-century physiologists often sought to understand digestion as an isolated and ahistorical bodily system, Trotter viewed the digestive system as a sort of sticky surface that mapped and retained the intersections of a series of intertwined social and physiological systems. The digestive system connected the rest of the organs to the individual body; this individual body was folded into the system of an urban community, which was itself folded into a nation, and this latter was folded into a global network of trade.

Two consequences followed from Trotter's understanding of the digestive system. First, the stomach, and the body to which it was linked, were granted a historical dimension. For Trotter, as food passed through the body, it left bad or good corporeal traces, and these traces in turn motivated future individual actions and possibilities. Because the nervous body acted like a miser ("hoard[ing] . . . all the effects of vicious indulgence"), and because it was capable of maintaining a corporeal history, it found itself forced into recursive ("vicious") feedback cycles, which impelled the individual to repeat "vicious" forms of indulgence.[31] Nor was this corporeal memory limited to the individual; Trotter argued that it could be passed on hereditarily, from parent to child, and thus from generation to generation.[32]

Second, Trotter's account also suggested that the body's historical nature—its capacity for an individual and a collective memory—had to be understood at least in part in terms of force and shock. The stomach transmitted to the rest of the body more than simply alimentary elements. In the case of tea consumption in the working classes, for example, the purchase of this beverage at the expense of food acts "with double force" (73), since consumption of tea caused its own ill effects, but in order to purchase tea in the first place, the working classes had to forego more nourishing food, a form of abstinence that also sickened the body. As a consequence, the role of the digestive tract in producing nervous illness could not be understood solely from the point of view of its immediate relationship to particular kinds of aliment. In addition, one also had to consider the ways in which urban environments established fields of tension—such as economic and gustatory "pressure" to drink tea—that could overwhelm the organs and capacities of individual bodies.

Yet the fact that the digestive tract registers force also leads us to the central aporia of Trotter's account, for though he is quite clear on the means by which nervous illness is produced, it is not at all obvious from his account how—or even if—nervous illness can be combated. While the stomach and digestive tract are able to select only the alimentary elements of food, they could not in similar fashion filter out the shock of the medium within which these foods were produced and distributed. The physician thus had to combat shock with shock, "forc[ing] new ideas on the patient" and "forc[ing] a new train of actions upon the temperament" (156) in order to counter the effects of nervous illness (215). Yet in what way did this therapeutic shock differ from the shock produced by the environment itself: what differentiated "good" from "bad" shock? Trotter's difficulty in theorizing shock and force seems to account for his indecision about whether to attribute the evils of spices, coffee, and other foods to their material specificity (that is, their direct action on the body) or to the more general social dynamics of trade that made such commodities available.

Trotter's failure to provide an account of the distinction between good and bad shock also begs the question of what, precisely, he understood to be the point of *publishing* this paranoid account of an epidemic that was creeping, mostly silently and unseen, throughout the body politic, turning Britain into a nation of slaves and idiots. Trotter clearly intended his book to be shocking, and presumably it was in this sense intended to serve as a good shock that could counter the bad shocks of the urban environment. Yet to what could fearful Britons turn, should they become convinced by his claims? At some points in A *View of the Nervous Temperament,* Trotter seems to suggest that doctors possess the expertise necessary to continue the shock therapy that his book initiates. Yet the form of the book—an address to lay audiences rather than physicians—reminds us that he had already disqualified the expertise of most physicians in his opening pages; it was precisely because physicians had failed to recognize this epidemic of nervous disease that the book makes an end run around medical professionals to address the public directly. Consumer self-regulation, Trotter's account implied, was the most viable solution to this epidemic of nervous illnesses. Yet aside from a diet of good British beef and potatoes, it is not at all clear what ought to be the elements of and domain for such self-regulation, for it is difficult to imagine that one could actually live in a city and remain free from nervous illness. Trotter's description of metallic fumes provides a figure for this problem of medium that he both sought to expose and which at the same time confuses his account: just as the cloud of metallic fumes determines the health of the factory worker who labors within its miasma, so too does the urban environment function as the determining medium for its inhabitants. There was, in a sense, no escape from this factory cloud of the nineteenth century, nor, given the momentum of the dynamics of urbanization and global trade, was it likely that individuals could choose a "pantisocratic option" by retiring in small groups to a life of countryside farming.

In short, despite Trotter's intense interest in the dynamics of feedback and memory, he does not consider the relationship between those dynamics and his own medium, namely, language and books. Though he attends to the corporeal effects of textual excess, he does not account for how his own account can have a beneficial effect. The problem is not that of expertise, for he explains at length the capacities a physician must have in order to recognize nervous illness. The problem, rather, is that of application, for he does not explain how language and books could counter the contagion that he intuits everywhere around him.[33] Since everything—occupations, food, and even the activity of reading—has become a medium, or vector, of contagion, it is by no means clear why we ought to exempt his book, despite its good intentions, from this power to facilitate nervous illness. Though Trotter's early-nineteenth-century readers might have responded with appropriate shock and nausea to his account of a hidden epidemic of nervous illness, A View of the Nervous Temperament also seems to close off every possible exit and escape.

A View of the Nervous Temperament thus places the readerly experience of collapsurgence within a narrative structure that is formally tragic—knowledge of the danger facing the community comes too late for action—while at the same time positioning the "audience" of the book *within* this tragic structure, such that no cathartic act of purging is possible. Insofar as Trotter viewed the human digestive tract as capable of translating both the elements and the force of its surrounding medium into the body itself, he could not simply *describe* to readers the dangers of nervous illness but also had to counter the force of the surrounding medium through a rhetorical production of nausea and panic. Yet in seeking to induce nausea by overwhelming readers with multiple examples of the vectors through which the body absorbs its medium, Trotter also denied the body the capacity to purge itself of its unhealthy modes of consumption; flows move only from environment through the body, rather than the reverse. Readers will always learn about this nervous epidemic too late for any kind of real resistance, and Trotter himself was forced to rely on one of the very media he claimed further perpetuated the social systems he hoped to change. The more that Trotter presented the body and mind as materially dependent upon its surroundings, the greater the rhetorical shock of his account; by the same token, though, the more that the body and mind were materially dependent upon their surroundings, the less plausible it was that this shock could have any curative value.

DIGESTING THOUGHTS AND THE NAUSEA OF SYSTEM:
GODWIN'S *CALEB WILLIAMS*

Because of its explicit emphasis on the physiology of digestion, Trotter's A View of the Nervous Temperament provides us with a way of understanding the link between nausea and paranoia in less physiologically oriented examples of Romantic-era texts that also focused on the hold of systems over individuals. Many of the canonical political texts that we now associate with this period—Edmund Burke's

Reflections on the Revolution in France, Mary Wollstonecraft's vindications of the rights of men and woman, William Godwin's *Enquiry Concerning Political Justice*, and William Cobbett's writings on paper money and the national debt, to name just a few examples—were characterized by the twin obsessions of exposing and combating the hold of social systems over individuals, and many of these authors also struggled, like Trotter, to locate a literary form that could reveal, and at the same time combat, the power of system. However, as in the case of Trotter's *A View of the Nervous Temperament*, the results were often ambiguous, as authors found that they could heighten the rhetorical force of their accounts only at the cost of positioning subjects as incapable of breaking the hold of systems.

William Godwin's novel *Things as They Are; or, the Adventures of Caleb Williams* (1794) occupies a particularly important place in this history of Romantic-era paranoid approaches to systems. Often depicted as a sort of "paranoid gothic" novel, *Caleb Williams* presents a narrative that documents the mechanisms by means of which social systems and oppression are internalized, as the eponymous protagonist "comes to feel part of the very things that oppress him."[34] The novel is Caleb's first-person account of his discovery that his employer, the apparently virtuous aristocrat Lord Falkland, is a murderer who has successfully covered up his crime for many years. After Falkland intuits that Caleb has discovered his secret, he dismisses him, but with a warning that he will keep Caleb under constant surveillance and control. The novel documents the truth of Falkland's promise, as Caleb finds himself constantly watched by Falkland's agents. Critics have interpreted the surveillance in Godwin's novel as a barely exaggerated reflection of the structures of political paranoia specific to the 1790s, reading Caleb's fear that Falkland's "colossal intelligence" has the power to "reach through all space, and his eye penetrate every concealment" as indicative of opposition fears about British government power in the wake of the French Revolution.[35]

Caleb Williams is not simply *about* paranoia, however; it is, in addition, a novel that itself instantiates the logic of paranoia.[36] Like Trotter, Godwin sought to depict a world in which all possibilities for individual action, and even self-consciousness, were ruled by a system. However, where Trotter's account implied that the author of the account had somehow located a position unaffected by the system that he described, Godwin's novel presented a first-person account of the effects of such a totalizing system. It thus faced directly the question that Trotter himself had elided, namely, from what position is it possible to speak the truth of a system of complete control? If Falkland's system depends upon secrecy, and if it is as overarching and controlling as Caleb presents it, then how has it come to pass that Caleb is able to describe the system in the way that he does in this novel?

Godwin's difficulty in answering this question is highlighted by the two quite different endings he composed for the novel. The second ending—and the one he chose to include in the published version of the text—suggests that Caleb's ability to unveil the system is a consequence of the collapse of the system itself. Thus, the "official" version of the novel concludes with Falkland's miraculous courtroom admission of guilt, apparently provoked by Caleb's (unprecedented) rhetorical mastery in a judicial context. Having finally succeeded in bringing Falkland before a magistrate, Caleb is surprised to find himself moved by the obvious desperation of his tormentor and curiously regrets having brought him before the law. Caleb's ambivalent verbal testimony results in a mutual sympathy between Falkland and Caleb:

> When I expressed the anguish of my mind . . . [Falkland] seemed at first startled and alarmed, lest this should be a new expedient to gain credit to my tale . . . But as I went on he could no longer resist. He saw my sincerity; he was penetrated with my grief and compunction. He rose from his seat, supported by the attendants, and—to my infinite astonishment—threw himself into my arms!
>
> "Williams," said he, "you have conquered! . . . And now,"—turning to the magistrate—"and now, do with me as you please." (335)

Given Falkland's power over Caleb prior to this event, and Caleb's earlier failures to combat Falkland's system by means of rhetoric, this scene reads more like a fantasy of omnipotence on Caleb's part than a conclusion that has any causal precedent in the story itself (an impression highlighted by Williams's own "infinite astonishment"). The published ending of the book thus does not in fact show us *how* an individual could successfully oppose a system but rather simply emphasizes the obvious point that if a system falls apart by itself, an individual can retrospectively describe its former operation.

Where Godwin's published conclusion accounted for Caleb's power to speak the truth about the system through the rather unsatisfying premise that the system had, for uncertain reasons, collapsed, the first conclusion to the book that Godwin composed kept Caleb within the logic of system but at the cost of undercutting the authority of his narrative. In Godwin's original conclusion, Caleb has apparently become completely unhinged, reduced to a state in which, by his own admission, "wild and incoherent visions perpetually succeeded each other" and he is "subject to wanderings in which the imagination seems to refuse to obey the curb of judgment" (343). Caleb finds himself at the mercy of his digestive tract—"As soon as I eat, and drink, I fall asleep again" (345)—and the novel ends with a disjointed ramble within which even Falkland's reality is questioned:

"I wonder who that Mr. Falkland was, for every body to think so much about him?—Do you know?" (346). The problem with *this* ending is that by calling into doubt the reliability of the narrator, it casts doubt retrospectively over the entire narrative that preceded, which calls into question the existence of the system that Caleb has purported to describe. In his attempt to reveal the overarching power of system, Godwin thus found himself on the horns of a narratological dilemma: either it is unnecessary to expose a system, since the capacity to expose it means that the system has already collapsed on its own (published ending), or the system retreats as an identifiable object of analysis, in which case the system *may* be an objective reality, but it may also be a subjective illusion, and it is impossible to locate a criterion that would allow one to opt for one or the other interpretation (unpublished ending).

While Godwin's unpublished ending is arguably even less satisfying for readers than the published one, its emphasis on the digestive tract nevertheless helps us to discern the chylopoietic logic of Godwin's approach to the relationship between systems and individuals, for Caleb's subjective reduction to the expressions of the digestive tract is in fact simply the end point of a narrative trajectory that is already marked by pivotal experiences of nausea. As Caleb describes his discoveries about the extent and power of Falkland's system of surveillance and influence, each of these illuminations about the system of which he is a part is announced by a physiological "revolution." When Caleb first begins to suspect that Falkland might be guilty of murder, he feels a "total revolution" of his "animal system," though at this point still in the mode of an exhilarating surge:

> While I thus proceeded with hasty steps along the most secret paths of the garden, and from time to time gave vent to the tumult in my thoughts in involuntary exclamations, I felt as if my animal system had undergone a total revolution. My blood boiled within me. I was conscious of a kind of rapture for which I could not account. (135)

In this passage, Godwin confuses the distinctions between Caleb's body and its surrounding medium by creating a parallel between his internal animal economy and his external perambulations. Both Caleb's internal animal economy and his external wanderings are indices of complexity, for the "secret paths" of the garden mirror the "secret paths" within his mind that enable previously isolated pieces of knowledge to become linked to one another. Yet his exultation is simply the first moment of a larger process, for just a few pages later, he finds himself nauseated. As Caleb, in a sort of "trance," attempts to open the trunk that might confirm his bodily hunch about Falkland's guilt, Falkland himself comes into the room, and realizes that Caleb suspects him. The two leave the room separately, but shortly afterward Caleb suddenly becomes aware of the terrible position in which he now finds himself. When Falkland sends for him later, this "message roused me from my trance. In recovering, I felt those sickening and loathsome sensations which a man may be supposed at first to endure who should return from the sleep of death" (140). This "epoch was the crisis of my fate, dividing what may be called the offensive part from the defensive, which has been the sole business of my remaining years" (139). Nor is this Caleb's sole experience with nausea, for whenever he discovers a new realm into which Falkland's power extends, this effect returns.[37]

Though we might be inclined to naturalize Caleb's nausea as a normal response to very bad news, such an observation would leave unexplained *why* Godwin chose to emphasize Caleb's corporeal reactions. However, in *Enquiry Concerning Political Justice*, published a year before his novel, Godwin had outlined an epistemology that explained why nausea could result from unexpected information. In the *Enquiry*, he contended that what we take as natural ideas "are regularly generated in the mind by a series of impressions, and 'digested' and arranged by association and reflection" (101). Godwin's claim was not as physiologically specific as the notorious assertion of French physiologist P. J. G. Cabanis that "the brain in some sense digests impressions . . . [and] organically secretes thought."[38] Nevertheless, for Godwin as for Cabanis, the term "digest" emphasized a materialist interpretation of mental activity, denoting both the specific work carried out by the brain on sensory impressions—the brain "digests" impressions in the etymological sense of separating them into elements—and the product of that work (i.e., by separating sensory impressions, the brain is able to create an abstract, or "digest," of sensory impressions).[39] Yet precisely because impressions are digested in the first place, it is also always possible, with the assistance of media such as speech and books, to "un-digest" earlier associations in order to create new patterns of mental associations.

In the calm and placid philosophical style of the *Enquiry*, Godwin emphasized the rational telos by means of which readers ought to orient themselves in their efforts to disgorge, gently and slowly, their previously assimilated systems of ideas. *Caleb Williams*, by contrast, explores the effects of sudden mental indigestion. In his novel, Godwin also presented the power of systems in terms of mental digestion, explaining Falkland's ability to exonerate himself of Tyrell's murder as a consequence of his ability to generate "so well-digested a lie as that all mankind should believe it true" (141). Caleb's nausea exemplifies what occurs when a lie is un-digested too quickly: the nausea initiates an uncontrollable process of feedback between Caleb and Falkland, as the revolution of his animal system provokes Caleb to act precipitously, which in turn provokes Falkland to react, which in turn provokes further nausea in Caleb, which in turn encourages him to act precipitously, and so on. Yet such a process of positive feedback cannot go on ad nauseam, and thus Caleb's submission to his digestive tract in the original ending of the book is coincident with both his and Godwin's discovery that there are simply no more available lines of flight in this narrative of systems. The "message" of *Caleb Williams* is in this sense in alignment with the emphasis on moderate and temperate reform outlined in Godwin's *Enquiry*: the novel offers a negative example of mental un-digestion, positioning nausea and paranoia as the result of the too hasty revelation of social systems. Though *Caleb Williams* may present a narrative of paranoia at the level of content, it aims, by means of negative example, to discourage paranoia in its readers by illustrating the problems with overly speedy mental un-digestion.

Yet Godwin's difficulty in determining how to conclude *Caleb Williams* highlights his only partial success in achieving the goal of communicating this message. Where the paranoid narrative of *A View of the Nervous Temperament* resulted from the conflict between Trotter's desire to treat language as a transparent medium and his need to position books and reading (i.e., media that employ the medium of language) as causes of nervous illness, the paranoia of *Caleb Williams* is the result of a tension in Godwin's more complex, but still problematic, understanding of the importance of communication media. Godwin had far more faith than Trotter in the capacity of books to oppose the dynamics of systems within which individuals find themselves. At the same time, he was clearly worried about granting books too much power. In the *Enquiry*, he struggled to endow books with the virtues of speech, suggesting that while "political justice" may require books as a sort of memory storage device for truth, speech remains the proper means of information gathering and dissemination. "Books have by their very nature but a limited operation," Godwin contended, for they do not engage the *force* of reason. Books have "a sort of constitutional coldness. We review the arguments of an 'insolent innovator' with sullenness, and are unwilling to expand our minds to take in their force. It is with difficulty that we obtain the courage to strike into untrodden paths, and question tenets that have been generally received" (288–89). Conversation, on the other hand, engages the force of reason, for "it accustoms us to hear a variety of sentiments, obliges us to exercise patience and attention, and gives freedom and elasticity to our disquisitions" (289).

Caleb Williams highlights the theoretical difficulty that can result from this desire to position speech as more virtuous than writing. Like Godwin himself, Caleb is committed to rhetorical mastery in the medium of voice, for he believes that it is through spoken discourse—what he calls "the unpremeditated eloquence of sentiment"—that he can build a consensus and escape Falkland's system (220). However, Caleb's expectations are continually disappointed because he finds himself constantly outpaced by the dynamics of a print system characterized by (in Godwin's words from his *Enquiry*) "the easy multiplication of copies, and the cheapness of books," two developments that ensure that "everyone has access to" print productions.[40] In a text-based society, relying on speech to catalyze mental un-digestion is simply not a viable position, for print productions are capable of occupying every position in advance, and "the unpremeditated eloquence of sentiment" always comes too late. The system of print consistently overwhelms Caleb's attempts to establish a community of sympathetic orality, and the final result is either complete submission to the digestive tract (unpublished ending) or fantasies of absolute sympathetic omnipotence (published ending).

Trotter's and Godwin's difficulties in containing collapsurgence were consequences of the tension between their materialism, which rendered compelling the case that the reach of systems was invidious and totalizing, and the implication that textual and linguistic media were in some unspecified way able to overcome systems. We can see in these difficulties the origins of the "performative contradiction" that Jürgen Habermas tirelessly diagnosed as being at the heart of the accounts of modernity developed by Max Horkheimer and Theodor Adorno, Michel Foucault, Jacques Derrida, and others: that is, the attempt to use the "medium" of rational discourse to free one *from* a system that is itself positioned as at least in part a consequence *of* that same medium of rational discourse.[41] What both Trotter and Godwin lacked, in short, was an account of language able to explain how collapsurgence is possible, how the nausea engendered in the wake of one's sudden awareness of a constraining system can also serve as a route for a possible escape from the system rather than simply serving as an affective echo of the fact of one's containment.

COLERIDGE'S CORPOREAL REVERSALS
AND THE MEDIUM OF LANGUAGE

We find an approach to language better able to account for collapsurgence in Samuel Taylor Coleridge's critique, in the mid-1790s, of British patterns of consumption. Coleridge articulated his critique directly and publicly in his "Lecture on the Slave-Trade," which he gave in Bristol in 1795 and then published, in slightly altered form, in the fourth number of his short-lived periodical, *The Watchman*. The audience for Coleridge's lecture probably consisted primarily of fellow travelers: Unitarians and others already opposed, or inclined to be opposed, to both the slave trade and slavery. As Lewis Patton and Peter Mann note, Coleridge in his lecture promoted abolition "by two means: seeking the enactment of [antislavery and anti-slave-trade] laws and cutting the revenue of the West Indian plantations through a boycott of sugar and rum."[42] His speech thus had an explicitly hortatory function, encouraging his audience to accept two different kinds of responsibilities: a public responsibility for altering the laws that determined collective definitions of allowable behavior and a private responsibility to monitor and adjust patterns of commodity purchases.[43] Yet where Trotter's and Godwin's efforts to rouse their readers into action were troubled by the authors' tendencies to treat both voice and print as neutral media for the transmission of information, Coleridge was far more interested in, and attentive to, the ways certain forms of language use could mark out "events." Though such events were related to the casual mechanisms by means of which systems controlled subjects, they could not be reduced to those causalities. As a consequence, Coleridge's method of mental rumination was not susceptible to the charge that within the terms of his own argument, he was simply perpetuating, rather than opposing, the system against which he wrote.

At the time that he gave his lecture on the slave trade, Coleridge was struggling to reconcile competing philosophical commitments to the materialism of the eighteenth-century philosopher David Hartley and to an approach to language that was implicitly anti-materialist. In his *Observations on Man, His Frame, His Duty, and His Expectations* (1749), Hartley had sought to describe how thought emerges from "vibrations" of "the medullary Substance of the Brain," and he had provided an extended account of how such vibrations between distinct sensory impressions could produce "associations" between ideas.[44] *Observations on Man* was a pivotal text for Coleridge; he declared in a 1794 letter that he not only subscribed to Hartleyan associationism but went "farther than Hartley and believe[d in] the corporeality of *thought*."[45] Yet Hartley had developed a peculiar version of materialism, for he argued that his doctrine of vibrations and mental association was fully consonant with, and in fact supported, Christian scripture and doctrine. It was this aspect of Hartley's thought that seems to have especially attracted Coleridge's interest. Coleridge described Hartley as "the great master of *Christian* Philosophy," and the figure of Hartley seemed to serve the young Coleridge primarily as a sort of promise that one could link, at least in principle, an interest in the material basis of bodies and thought with the freedom of will and thought demanded by Christian doctrine.[46] Thus, even as Coleridge's claim of allegiance to Hartley and associationism seemed to align him with the materialist account of the relationship between bodies and thought advocated by authors such as Godwin and Trotter, Coleridge tended to employ Hartleyian terms in ways that worked against the materialist aspects of Hartleyianism. Just a year after claiming to be a more thoroughgoing materialist than Hartley, Coleridge contended in his 1795 lecture *Conciones ad Populum* (also given in Bristol) that "*to emancipate itself from the Tyranny of Association*, is the most arduous effort of the mind"—this despite the fact that within Hartley's system, "emancipation" from mental associations was, strictly speaking, impossible.[47]

Coleridge's struggle to reconcile his interest in materialist explanations of mental phenomena with his Christian belief in free will was reflected in his attempts to formulate a philosophy of language.[48] He was attracted by both the figure and philosophy of John Horne Tooke, an outspoken opposition political figure and etymologist. For Tooke, politics and etymology were intimately connected, for he sought in his publications on language to reveal how confusions about words directly contributed to inequitable political institutions. However, as James McKusick has noted, Tooke was also fundamentally a Lockean materialist, with the consequence that his etymologies traced words back to their origins in sensory impressions of objects; thus, for Tooke, nouns were the primary elements of language (33). Coleridge sought to link Tooke's approach to political change as a matter of language use to a philosophy in which the formation of words was expressive of a mental *act*, not simply a passive index of external impressions; he thus tended to see verbs as more primary than nouns. For example, though Coleridge in his 1795 *Lectures on Revealed Religion* "repeatedly criticizes the abuse of language perpetrated by his 'aristocratic' opponents," providing "several examples of names arbitrarily imposed upon objects or events for a nefarious purpose," he also contextualized this interest in the "true" origins of words within an explicitly religious framework that was quite alien to Tooke's materialist approach to language.[49] Though Coleridge would not attempt to formulate a philosophy of language in earnest until roughly 1800, the tension between his commitment to materialist associationism and his belief that language expressed human free will was already evident in his 1795 lecture on the slave trade.

This tension came to the surface in Coleridge's use of figurative language in his lecture. In advocating for a boycott of slave-produced goods, Coleridge employed the trope of "blood sugar," suggesting to his listeners that a "part of that Food among most of you is sweetened with the Blood of the Murdered."[50] As Timothy Morton has noted, Coleridge was not alone in suggesting that British sugar was sweetened with slave blood; his use of this image was an instance of a more widespread "aversive topos" of antislavery discourse, by means of which authors sought, through means of literary representation, to render "sweetened drinks of tea, coffee and chocolate . . . suddenly nauseating by the notion that they contained the blood of slaves."[51] The trope of blood sugar was intrinsically oriented toward a vision of system, for it produced its rhetorical effect by folding together the two ends of the economic system, production and consumption. To understand the trope of blood sugar was to be nauseated by it, and this nausea depended upon the ability of auditors or readers to hold together two temporally and spatially disconnected events, juxtaposing the human suffering incurred during sugar production with the commodity that was then ingested by the consumer.

However, Coleridge linked this more widespread trope of blood sugar to another image, that of a resurrected Jesus working for the antislavery cause. Were the Savior to return to earth today, Coleridge claimed, his miracles would not consist in the transformation of an inedible substance into an edible one, such as changing nonpotable water into wine. Rather, he would reveal to our "fleshy eye" those bodily torments of slaves that make our sweet foods possible. Jesus, Coleridge wrote, would "convert the produce into the things producing, the occasioned into the things occasioning! Then with our fleshy eye should we behold what even now truth-painting Imagination should exhibit to us—instead of sweetmeats[,] Tears and Blood, and Anguish."[52] What Coleridge could accomplish only through the image of "blood sugar"—that is, the linguistic condensation of the representation of produce (the sugar consumed by the British) and the representation of production (the laboring slave)—Jesus would produce in the flesh by literally turning sweetened foods into laboring and bleeding slaves.

In his perceptive analysis of the blood sugar metaphor, Morton notes that Coleridge's supplementary image of Jesus reveals the peculiar understanding of "reality" upon which these images of blood sugar and an antislavery Jesus are based. On the one hand, by referring his listeners to the second image of Jesus making real the truth of slavery, Coleridge suggested that figurative language simply *reflects* reality: the metaphor of blood sugar mirrors the reality of how sweetened foods are produced. Reality, from this perspective, has its own intrinsic power to convert the unbelieving and the indifferent; to look upon reality is to be freed from what Coleridge described as "artificial wants." However, lacking Christ's powers of literal making-real, Coleridge was forced to work with the less efficacious medium of language, which can only mirror or echo the powers of conversion that reality itself affords.

On the other hand, Morton notes, Coleridge was speaking to an audience who was, in all likelihood, *already* abstaining from slave-produced products and who already knew all about the reality of the slave trade. To address *this* audience with

the image of blood sugar and the supplementary image of Jesus transforming sug-
ary foods into laboring, bleeding slaves renders the role of figurative language in
Coleridge's speech much more complicated:

> The [blood sugar] topos does not simply reveal a "Real" of slavery underlying
> the figure of sugar; [rather,] the materiality of the figure itself is at stake. It is
> an apocalyptic rhetoric that decodes the slave-trade, but its unveiling process
> draws attention to the materiality of the very veil which has been torn away.
> Abstinence in itself appears not to be an adequate means of resistance. It is
> not disillusioning enough; it becomes questionable whether abandoning "false
> consciousness" is a way of subverting ideology.[53]

To speak metaphorically about the reality of the slave trade to those who already
knew its reality was, in other words, to treat language as something *more* than
simply a medium for mirroring or transmitting the force of this extralinguistic
reality. It was to treat language—and, more specifically, the nonreferential capacities of language, which enabled a speaker such as Coleridge to "refer" to things
that both he and his listeners knew did not actually exist—as possessing its own
power, a power quasi-independent of those realities of the slave trade with which
Coleridge's audience was already familiar.

Morton's reading can be further refined, for Coleridge's use of the trope of
blood sugar points not just at *a* "Real" of slavery; rather, it alludes to the material reality of slavery while also expressing *another* kind of reality. On the one
hand, the trope of blood sugar acknowledges the kind of causal reality that occurs
among bodies, such as the physical bodies of slaves and slavers, the movements
through space and time of ships that carry those bodies, the injuries to slave bodies produced by disease, deprivation, extended work, and so on. Language could
be used in its referential capacity to point to the locations and movements of
these bodies in space and time and the causal relationships among those bodies.
Exhaustive documentation of these spatio-temporal and causal relations was the
goal of many antislavery advocates, including Thomas Clarkson, upon whom
Coleridge relied in writing his lecture. Coleridge cited, for example, Clarkson's
claim that slave ships remained seaworthy for only half as long as other ships
because the close confinement of so many bodies rotted "the very timbers of the
vessel."[54]

On the other hand, the trope of blood sugar emphasizes that this use of language to refer to causal relationships among material bodies did not, on its own,
allow one to grasp the *meaning* of the slave trade. One grasped the slave trade as
the *slave* trade—rather than, for example, simply a neutral movement of bodies

through space and time—only when one acknowledged another kind of reality, a reality that, though it was linked to these causal relationships among bodies, nevertheless occurred solely within and through language. This linguistic mode of reality was expressed, for example, in legislative decisions that had enabled and perpetuated the slave trade, as well as that desired future legislation by means of which antislavery advocates hoped to put an end to that same trade. Language did not express this second mode of reality by means of referentiality; it was not, in this second case, a matter of using language to refer to causal relationships among material bodies. Were Parliament to declare the slave trade "illegal," for example, such a declaration would indeed relate to relationships among various kinds of bodies—ships, money, British slavers, African slaves, American plantation owners—but it would not point to any corporeal or causal aspect of those bodies. (Nor would this declaration refer to a Lockean "secondary" quality that, though not truly in the object, seemed to be there as a consequence of the ways our human senses engaged the world; the "legality" or "illegality" of the slave trade, in other words, is not analogous to the "redness" of a sunset.) To declare something "illegal" was, in short, to use language in a quite different way than when one employed propositions referentially as a way of pointing out qualities of material bodies, such as the dimensions of slave sleeping quarters on ships or causal relationships among bodies, or such as the effects of human effluvia on ship timbers.

This second, nonreferential use of language is partially captured by the concept of "performative speech." Certain uses of language—for example, a judge's declaration that a defendant is "guilty," a minister's proclamation that two people are now "husband and wife," or a legislative vote that trading in humans shall henceforth be illegal—express a kind of reality that occurs in a different register from language referring to extralinguistic causal relationships among material bodies; performative speech acts do not use language to refer to an existing property or causal relationship but rather use language to *do* something, to create an event.[55] Such speech acts, of course, require some sort of enabling frame. If a judge is disbarred, for example, his or her proclamations of "guilt" or "innocence" no longer have the power to create events. Yet the need for an enabling frame should not obscure the fact that language is not being used referentially in performative speech acts.

The difference between the use of language to refer to causal relationships among bodies and the performative use of language is captured by the distinction Gilles Deleuze and Félix Guattari draw between "states of affairs" and "events."[56] States of affairs refer to the material realities of bodies and the causal relationships among these bodies, while events refer to realities expressed only in and through language (and which do not, at least directly, produce any changes in material bodies). Deleuze and Guattari suggest that this distinction is necessary insofar as one cannot explain our experience of the world solely in terms of causal relationships among material bodies. One may certainly posit, along with many eighteenth-century materialists—and even some idealists, such as Immanuel Kant—that causal relationships among bodies are completely determined by physical laws (e.g., the law of gravitation and laws of chemical relations) and that an observer with complete knowledge of all the material properties of all bodies could use language to describe completely all causal relations between bodies. The problem with this completely referential understanding of the relationship between language and reality, though, is that it cannot satisfactorily explain, in terms of material properties and causal relationships, those events that are of such interest to human beings. A fully referential theory of language has trouble even with the examples favored by speech act theorists, which tend to focus on transformations that occur in isolated individuals (e.g., proclamations of an individual's guilt or declarations of marital union), and the situation is even more complicated when we consider events such as wars or social movements. To which specific constellations of matter and which causal relationships, for example, is one referring when one speaks of "the English Civil War" or the "French Revolution"? To which bodies and causal relationships does a phrase such as "the slave trade" refer—especially if, as Coleridge suggests in his lecture, some of those people who understand themselves to be opposed to the trade are nevertheless unwittingly supporting it? Nor are these isolated examples, for we constantly use language to express events—"the English Civil War," "British slavery," "the French Revolution," "Waterloo"—that, though they certainly refer, in a minimal way, to material bodies, nevertheless do not derive their meaning from that reference.

Causal relations between material bodies do not explain events in part because there seems to be a different kind of causality, what Deleuze and Guattari call a "quasi-causality," operating among events. Thus, one event can "inspire" another, and each event seems to make sense only when it is understood as drawing its resources from another event. There are relationships of resonance, for example, between the British legislative decision to end the slave trade and legal proclamations made by the French revolutionary government that "freed" its citizens from aristocratic oppression, as well as between both of those events and the events of the much earlier English Civil War. Or, to take another example, Coleridge alludes to one event, Jesus turning water into wine, in order to inspire his audience to make personal resolutions to abstain from ingesting products produced by slaves and then to work toward another event, antislavery legislation. This distinction between the causal relations of states of affairs and the quasi-causality of events constitutes neither a dualism nor a parallelism: that is, states of affairs and events are not to be understood as two distinct, autonomous series that miraculously mirror one another. Rather, Deleuze and Guattari write, "the independence of the two lines [of states of affairs and of events] is distributive, such that a segment of one always forms a relay with a segment of the other, slips into, introduces itself into the other."[57] To account for the world we experience, we must assume a real distinction between states of affairs and events, between bodies that are causally determined, on the one hand, and expressive instances of language that cannot be fully understood in terms of the causal determination of bodies, on the other. At the same time, we must not lose sight of the fact that every such expressive instance of language is itself an attempt to intervene in states of affairs, an effort to constitute an event by creating new "relays" between states of affairs and events.

The power of Coleridge's speech, its simultaneous power to nauseate and encourage action, is a consequence of acknowledging the reality and interrelation of both states of affairs and events. When he contended that a "part of that Food among most of you is sweetened with the Blood of the Murdered," the words *you, food,* and *blood* referred to specific material objects, or classes of objects: a subset of the auditors in front of him as well as possible readers of the text; the food these auditors or readers had eaten or would eat; the blood of actual slaves. The word *sweetened,* by contrast, referred to a fundamentally different kind of reality than the nouns and pronouns of the sentence. Unlike these latter, *sweetened* was not a quality that was located in any actual body—one could not, even by careful observations, locate particles of sweet dried blood within food items (nor, as I noted above, could it be taken to refer to a Lockean "secondary quality" such as "red," which is produced by the effect of the material composition of an object on human senses). Nor did the term "sweetened" refer to an imaginary object: food sweetened with blood was not to be understood as an unlikely or impossible imaginary compound of real objects that were each possible on their own (e.g., a "centaur" as an imaginary combination of a horse and a human, or a "square triangle"). Rather, Coleridge used the term "sweetened" to express that an event had occurred. The force of Coleridge's sentence, its ability to induce nausea even in those who had already ceased consuming food produced by slaves, depends upon the expression of an event that has already happened and cannot be undone.

The event that is expressed through language imposes upon a listener or reader a different kind of imperative than statements that refer to states of affairs. Because an event cannot be reduced to a relationship among bodies, but rather takes place only within and through language, Coleridge's auditors or readers could not rid themselves of the event expressed by the word *sweetened* by means of literal digestion: had the problem been that food had literally been sweetened with blood, one could simply wait for a day or two, after which time the foreign substance would either be absorbed or excreted. The event expressed by the word "sweetened" could not, in fact, be addressed solely at the level of bodies: simply abstaining from slave-produced goods was not enough. Instead, this event that had already happened called upon Coleridge's auditor to produce another event in the future, such as a legislative end to the slave trade. This latter event could address, though not erase or causally affect, the earlier events to which it was related. It was true, of course, that such a future event would likely, albeit indirectly, alter relationships between bodies: the legislative end of the slave trade would mean, for example, that it was no longer legal to capture individuals in one place and transport them to another for the purpose of selling them to other people, and such a legal change would likely discourage many from engaging in the slave trade. Yet the reality of these events remained different and distinct from that of states of affairs. States of affairs and events relate to one another but not in a causal fashion; rather, these two modes of reality constantly traverse individuals, and changes in bodies or the production of new events alter the speed and duration of one's shuttling between the modes, or "relays," to use Deleuze and Guattari's term.

Though Coleridge's distinction between two modes of reality had precedent in Christian theologies with which he was familiar, his distinction eluded these earlier models. The natural philosopher and Unitarian theologian Joseph Priestley, for example, had also distinguished between the reality of material bodies, which were ruled by strict causal necessity, and the reality of "miracles," which explicitly exceeded causal necessity. As Douglas Hedley notes,

Priestley envisages miracles as providing empirical evidence that Jesus of Nazareth is the Messiah of men. Such infringements are not only compatible with but presuppose a causally determined universe: the mechanics of Newton present the laws of what is habitually the case and the biblical story describes the unique infringements of them. The strictness of the former highlights the religious and metaphysical significance of the latter: if God can infringe the laws of the universe for his adopted Son, it is equally plausible to believe that God can raise believers in his Son to eternal life.[58]

Yet Coleridge's *hypothetical* use of the image of Jesus emphasizes that the relevant distinction in his lecture on the slave trade is not that between material causal reality and divine miracles. Coleridge's image of Jesus converting "the produce into the things producing" was not intended to have referential force—he was not claiming that this miraculous act really had or would happen—but rather underlined the power of language to express events that occur only within language.

Coleridge sought to think the intersection between these two kinds of reality (states of affairs and events) through the figures of the stomach and digestion. In his lecture on the slave trade, he described the activities of the mind in terms of eating, arguing that the mind must "busy itself in the acquisition of intellectual aliment."[59] This suggested that the mind was more like a stomach than an eye, and as a consequence, "consumption" and "digestion" could not be understood simply as lower-order bodily functions but rather as fundamental processes of transformation that occur in both body and mind (or, more accurately, processes that vex the distinction between the low body and the elevated mind). Though Coleridge's metaphor of intellectual aliment recalled Godwin's account of mental digestion as a process of abstraction, Coleridge used this metaphor to stress the ruminatory, rather than abstracting, capabilities of the mind. He saw mental abstraction as one of the primary causes of social injustice, criticizing those "Illuminated" Deists and atheists who were able to maintain a neutrality concerning slavery by "soar[ing] above the vulgar superstitions of the Gospel" (248). Coleridge suggested that these atheists and Deists so privileged rapid and dizzying intellectual ascent that they could not truly digest the arguments and images that should lead them to oppose slavery actively. "It is not enough," he wrote in the "Introductory Address" of *Conciones ad Populum* (1795), "that we have once swallowed . . . Truths—we must feed on them, as insects on a leaf, till the whole heart be coloured by their qualities, and shew its food in every the minutest fibre" (49). Where Godwin had suggested that intellectual digestion was inextricably linked to processes of abstraction, Coleridge contended that mental digestion involved complex processes of incorporation, and should one seek to avoid these processes, truths would be excreted as quickly as they had been gulped down.

Coleridge's more complex account of mental digestion made it difficult to tell where literal ingestion ended and metaphorical ingestion began, for he positioned each as a condition—or better, a relay—for the other. In "A Moral and Political Lecture," a lecture also delivered in 1795 and published in the same year,

Coleridge suggested that literal consumption of food preconditioned mental digestion of truths. In this lecture, he contended that Freedom was menaced by a small group of misguided "Friends of Liberty," who, though they were "possessed of natural Sense . . . and . . . natural Feeling," were so lacking in the material necessities of life, including food, that they had become susceptible to manipulation. As a consequence, in the presence of "some mad-headed Enthusiast," these literally hungry friends of liberty "imbibe . . . Poison, not Food, Rage not Liberty" (9). A full belly, Coleridge suggested, was a precondition for thinking clearly. Yet mental digestion also determined patterns of literal consumption, as his implicit call for further boycotts of slave-produced food (and other items) made clear. Rather than seeking to make a firm distinction between real and metaphorical consumption, Coleridge positioned them as two moments within a recursive process in which each mode of digestion could encourage or discourage tendencies in the other.

While Coleridge's digestive epistemology diverged significantly from Godwin's, it nevertheless pointed toward a model of political action that was quite similar to that proposed by the author of the *Enquiry*. For both Godwin and Coleridge, to emphasize mental digestion was to encourage a politics of moderation. *Pace* the radical position that one ought to assimilate truth quickly and then immediately act, both Coleridge and Godwin contended that truths had to be slowly digested, and one ought to act only after lengthy consideration. Coleridge also agreed with Godwin and Trotter that a hypervigilant mode of self-monitoring was essential in the fight against problematic social systems. Trotter focused more on the literal diet of Britons, while Coleridge and Godwin emphasized mental digestion, but all three used the figure of digestion to encourage Britons to exert control over their patterns of literal and metaphorical ingestion.

At the same time, Coleridge's commitment to a distinction between two kinds of reality distanced his contribution to chylopoietic discourse from that of Trotter or Godwin. Trotter and Godwin had struggled to coordinate literary form with their implicit suggestion that readers could successfully monitor their patterns of consumption (literal consumption, in Trotter's case, and mental, in Godwin's). The problem with Trotter's account was that the form in which he cast it—a book designed for laypeople—relied on the very textual medium that he described as one of the contributing factors to the spread of nervous illnesses; as a consequence, his effort to hold his readers' attention by inducing nausea and panic seemed more likely in the terms of his account to increase nervous illness than to stem its tide. Trotter failed to recognize this problem because he treated language as a transparent channel that simply reflected the reality of bodies and could be used as a tool to communicate thoughts from author to reader, a belief that allowed him to exempt his own, well-intentioned publication from the systemic effects of books and reading he had illuminated in his account of the causes of nervous illness. Godwin was far more attentive to coordinating literary form and content, and in *Caleb Williams* he presented nausea and panic as a problem, rather than a solution. Yet in seeking to locate the redemptive force of language in face-to-face speech rather than books, Godwin's novel implied either that publications, including his own, were powerless in the face of speech or that printed texts created a system from which it was impossible to escape. The twin endings for *Caleb Williams* indicated the two horns of this dilemma but did not allow for its resolution.

Coleridge's distinction between two kinds of reality allowed him to understand the un-digestion of social systems in a way that avoided the performative contradictions Trotter and Godwin encountered. Coleridge acknowledged that there was a materialist causality that operated at the level of bodies, and which could be more or less specified at various levels. At the level of individual human bodies, for example, one could determine quite specific causal relations (ship captain X had transported Y number of slaves from point A to point B on date D), and though it was more difficult in practice to specify precise causal relations at the level of what Hartley had called "the medullary Substance of the Brain," one could nevertheless assume that medullary particles interacted in causally determinate ways. And like Trotter and Godwin, Coleridge's rhetoric was intended to show that a "system" determined his listeners and readers in their most intimate habits: words, dietary desires, and even patterns of thinking were all means by which systems took hold of an individual. Yet Coleridge's lecture was premised on the principle that there was, in addition, a register of "events" that occurred only in language and, though they expressed relationships among bodies, were neither caused by nor could have any direct effect on those states of affairs. Coleridge's image of blood sugar expressed this freedom of events from material causality, for it initiated relationships of resonance among events rather than referring to any real material quality of a body. This distinction between the reality of bodies and the reality of events thus provided Coleridge with that which Trotter and Godwin lacked: namely, the possibility of a perspective on "the system" that was not causally determined by that system. Coleridge's image thus also rendered thinkable the possibility that one could "un-digest" a system by acknowledging and producing events. Producing events meant using the mouth not for the ingestion of food but rather for the creation of a linguistic "surface" that hovered over states of affairs like a mist or haze (e.g., Coleridge's invocation of blood-sweetened sugar that existed nowhere yet encompassed completely the slave trade and its products).[60] It is from this perspective of the production of events that we ought to interpret Coleridge's hortatory rejection of "artificial" wants; it was grounded not simply in a Puritan distaste for the body but also in Coleridge's recognition that the production of events was linked to the question of food and digestion.

The opacity of the material body and its digestive apparatus thus played a paradoxical role in Coleridge's early thought and in his lecture on the slave trade. Both Trotter and Godwin enlisted the increasingly evident complexity of digestion in support of their materialism. Because the body was made out of matter and digestion involved complex actions of transformation and assimilation (and thus could not be understood as simply a matter of changing the scale of food), it was at least plausible, perhaps even likely, that what one ate could determine what one thought. For Trotter and Godwin, the opacity of digestion—the fact that it was not exactly clear how digestion worked—rendered the universal reign of material causality plausible. Coleridge, by contrast, drew on the opacity of digestion to emphasize a dimension of reality that was distinct from that of material causality. Thus, even as Coleridge declared himself in his correspondence to be more materialist than Hartley, his image of digestion did not refer solely to the material causality of bodies but also to a noncausal relationship between bodies and language. Hence the strange effect of Coleridge's image of blood sugar, which does not aim simply to *disgust*, since that would lead to prophylactic rejection of particular products rather than sustained engagement in the antislavery cause, but instead aims at something like *ruminatory nausea*, which would enable a slow working toward a new system.

Unfortunately, Coleridge's interest in ruminatory nausea proved short lived. Even in the version of his slave-trade lecture printed in the *Watchman* the next year, Coleridge subtly altered his alimentary logic. While he retained the metaphor of blood sugar, the image of Jesus transforming produce into producer, and the critique of food items such as sugar, rum, and coffee, he deleted his opening reference to "intellectual aliment." In place of that image, which emphasized the connection between digestion and language, Coleridge cited a passage from his poem "Religious Musings" (1796), in which he described an "ascent" of the faculties by means of which "sensual wants / Unsensualize the mind."[61] In place of stressing rumination that was part and parcel of collapsurgence, Coleridge reverted to a more traditional emphasis on abstraction.

Coleridge's movement away from his early alimentary logic was further encouraged a few years later when he encountered the work of Immanuel Kant. As Coleridge later wrote in *Biographia Literaria*, Kant "took possession of me as with a giant's hand," and reading Kant encouraged him to shift his understanding of the relationship between states of affairs and events.[62] Drawing on Kant's distinction between the phenomenal world—the world of material bodies and strict causality that could be investigated and described by science—and the noumenal realm, which was the source of moral judgments but could not be explained in the terms of the forms of intuition (space and time) or the categories of phenomenal world (matter, causality, and so forth), the events expressible by language became for Coleridge simply expressions of the moral will, which was itself guaranteed by God. Rather than understanding states of affairs and events as establishing relays with one another, Coleridge instead increasingly came to view the two categories as in orthogonal relationship. The theoretical interest of language, for him, was as a consequence to be found less in its capacity to express and connect events than in its capacity to punctuate the phenomenal world of closed causal relationships. As we saw in my account of his later interest in theories of life (chapter 3), this shift in his understanding of the powers of language did not put an end to his interest in the relationship between language and vitality, but it did discourage further interest in nausea and collapsurgence.

NAUSEA AND DISGUST

As suggested by the literary forms of the case studies I have discussed here—a medical account intended for lay audiences, a novel intended to draw out the implications of a political philosophy, and a lecture on the slave trade—Romantic-era efforts to induce collapsurgence did not aim at a properly "aesthetic" goal. These texts did not seek to produce a space of reflection on the basis of which a judgment of taste (or distaste) could be produced; rather, they aimed at a more complex, intimate, and paradoxical sensation. I have used the term "collapsurgence" to describe this "strong" sensation, which was a compound of both nausea (encouraged by the recognition that one was completely animated by a contemporary system) and, at the same time, an invigorating sense that one *could* break the hold of this system. In *Thus Spoke Zarathustra*, Friedrich Nietzsche provided a particularly striking image for, and a late Romantic echo of, the dynamics of collapsurgence, describing a shepherd into whose mouth a "heavy black snake" has crawled. Because the snake has "bit itself fast" into the man's throat, simply gagging is not enough to expel the offending object; rather, one must channel nausea (*Ekel*) into action and bite through the snake, for it is only thus that one can "embar[k] with cunning sails on unexplored seas."[63] As Nietzsche's image suggests, collapsurgence does not produce judgments of taste but rather existential transformations, and this in turn requires an ability to ruminate, thus keeping the nauseating object within oneself.[64]

It is this extra-aesthetic telos that links, but at the same time fundamentally distinguishes, Romantic-era collapsurgence and the concept of disgust. As Winfried Menninghaus has demonstrated, the concept of disgust is bound up with the emergence, in the eighteenth century, of a quasi-independent field of "aesthetics."[65] Within both eighteenth-century and Romantic aesthetic theory, disgust was never the opposite of the beautiful (that pole was occupied by the ugly) but rather a vexing link between experiences of art and experiences that had to be excluded in order to maintain the frame of "art." Thus, for eighteenth-century theorists of the aesthetic such as Johann Adolf Schlegel and Moses Mendelssohn, as well as for proto-Romantic theorists such as Kant, disgust named that which could not be transformed into art—"Disgust alone is excluded from those unpleasant sensations whose nature can be altered through imitation."[66] At the same time, however, disgust could also result from aesthetic experience that was *too* purely aesthetic; Mendelssohn argued that "pure sweetness soon leads to *Eckel* [disgust]."[67] Thus, as Menninghaus stresses, disgust was both the other of aesthetic experience in general and the danger that always threatened "pure" aesthetic experience from within.[68]

As Romantic theorists such as G. W. F. Hegel and Friedrich Schlegel introduced a historical dimension into aesthetics, the disgusting was understood less in terms of timeless objects of abjection and more as a lure that introduced a dynamic tendency into art practices. Menninghaus notes that

> Schlegel sees the turn toward the disgusting as the natural tendency of an art whose unceasing and fully self-supporting striving for otherness—otherness being the mark of "genius"—exhausts the "old stimuli" and almost invariably seizes on "ever more violent and penetrating ones" . . . the disgusting thus figures as an extreme subspecies of a modern aesthetics of the "shocking," and by 1798, it is recognized as the almost inescapable vanishing point, as the negative eschaton of the modern art system's accelerated *drain on stimuli*. (8)

Disgust, for both eighteenth-century and Romantic aesthetic theorists, though in differing ways, was therefore a figure that remained outside aesthetic experience and, paradoxically, also a tendency that emerged from within aesthetic experience proper.

Though this confusion of the inside and outside of aesthetic experience makes disgust easy to confuse with what I am calling "collapsurgence," the two are fundamentally distinguished by the aversive principle that motivates the former.[69] Disgust names that which has to be rejected, or at least held at a distance, for aesthetic experience to be possible. For this reason, it has often been linked to the maintenance of a social system—disgust names "matter out of place," to use Mary Douglas's phrase, in the sense that it is a reaction to an object (or person) that seems to threaten one's way of life and from which one must thus take one's distance.[70] Collapsurgence, by contrast, names a revulsion directed toward that from which one *cannot* distance oneself: the nervous illness that Trotter sees everywhere; Falkland's system that perpetually outpaces Caleb's attempts to flee; the "blood sugar" that has already spread over everything and thus cannot be expelled. Nietzsche's image of the thick black snake that one cannot vomit forth is again useful here, for collapsurgence is not a matter of rejecting "matter out of place" so that one can ensure the possibility of pure aesthetic experience (or, for that matter, any other kind of pure experience, such as scientific or moral experience); it is, rather, an experience of the lack of any pure discourse, sphere, or system into which one might escape.[71] Thus, collapsurgence is not simply a valorization of whatever those in power hold to be disgusting (e.g., the embrace of the grotesque body by popular culture, as described by Mikhail Bakhtin), nor is it a satiric utilization of the disgusting in order to shock readers back onto the "right" path (e.g., Jonathan Swift's use of fecal or cannibalistic imagery in his satirical texts as a means for shocking readers so that they would collectively respond to this disturbance by returning to stable, homeostatic social structures).[72] As Trotter's, Godwin's, and Coleridge's texts make clear, the nausea of collapsurgence is based on feeling that "the system" has extended its grip into the body and its digestive processes, thus ultimately into life itself. As a consequence, one cannot expel the offending object, but one must instead ruminate both upon and within it.

ROMANTICISM, THE AVANT-GARDE, AND THE LITERATURE OF EMBODIED SYSTEMS

Holding on to this important distinction between the aesthetic concept of disgust and the extra-aesthetic concept of nausea allows us to recognize that the key link between Romantic literature and the twentieth century is captured less by the rubric of the "avant-garde," though that link indeed exists, than by what we might call the literature (and art) of "embodied systems." Theorists of the avant-garde, from Clement Greenberg to Renato Poggioli to Peter Bürger, locate the origins of the early-twentieth-century avant-garde in Romanticism, though accounts of this relationship vary. For Greenberg, Romanticism initiated a tendency toward "kitsch" (namely, imitation of the effect, rather than the cause, of art) that was the

mirror image of the avant-garde; for Poggioli, Romanticism first initiated that oppositional, experimental approach to art the avant-gardes later embraced; while
for Bürger, we find in Romanticism the origins of that separation of the field of
"the aesthetic" from other spheres of life that was finalized in late nineteenth-
century aestheticism, which served as the motivation for the emergence of the
avant-garde.[73] However useful such narratives may be individually or collectively,
they all tend to focus attention on the world of art and to privilege experiences
of shock and disgust. As a consequence, these accounts are less able to address
modes of discourse and practice that, as in the case of Trotter's, Godwin's, and
Coleridge's texts, are not explicitly artistic and which aim at nausea rather than
disgust.[74]

Focusing on collapsurgence, by contrast, reveals a different lineage between
the Romantic era and the twentieth century, one that encompasses not only "artistic" productions, such as the literature of Burroughs and Pynchon, but also a
whole field of documentary exposés of the health effects of global systems of commodity production. Though few contemporary documentary accounts would
adopt the relatively direct understanding of the effects of food or environmental influences on mental capabilities that underwrote Trotter's account of the
rise of nervous illness, the performative contradiction that plagued his account
nevertheless hovers perpetually at the borders of contemporary representations
of the bodily insults to which our corporately controlled environments subject
us.[75] Rachel Carson's landmark *Silent Spring* (1962) and Robert Kenner's recent
documentary *Food, Inc.* (2008) are marked by that same tension between an attempt to induce bodily panic and paranoia in the reader or viewer by revealing the
complete and universal reach of a system (which suggests that one cannot change
this system) and the activist inspiration of the work (which implies that one *can*
change the system). The scientific models of natural causality upon which these
documentary accounts draw are far more complex than those Trotter and Godwin employed in the late eighteenth century, for these contemporary accounts
induce paranoia through the atmospheric vector of "risk" rather than the more
coarse materialism of the eighteenth century.[76] Yet the apparent epistemological
modesty of the notion of risk—the admission that causal relations are so complex
that we cannot even in principle map them all out—does not overcome the basic
paradox at work in Trotter's and Godwin's accounts but simply diffuses it. Insofar
as the contemporary concept of risk maintains the premise of strict (even if often
unverifiable) material necessity, it recognizes only *one* mode of reality (namely,
materialist causality), which in turn leads us back to that same double-bind of
panic-inducement and paralysis that marked Trotter's and Godwin's accounts.

If much documentary work finds itself caught in the same shallows as Trotter's and Godwin's texts, we can nevertheless also find contemporary work that adopts a much more Coleridgean approach, emphasizing both the material reality of "the system" that entraps subjects and the linguistic reality of events that cannot be reduced to this material reality. Pynchon's *Gravity's Rainbow*, for example, though admittedly not often understood as a neo-Romantic novel, is nevertheless motivated through and through by the strategy of collapsurgence. The centrality of systems and paranoia in *Gravity's Rainbow* is difficult to miss; as Leo Bersani notes, paranoia is "the narrator's most cherished word and concept" and the book obsesses over the term to such an extent that it "even gives birth to a new English verb": "to paranoid."[77] Moreover, *Gravity's Rainbow*, like *Caleb Williams*, confuses any easy distinctions between "things as they are" and literary representations of reality, for, as Friedrich Kittler stresses, Pynchon's "text . . . is essentially assembled from documentary sources, many of which—circuit diagrams, differential equations, corporate contracts, and organizational plans—are textualized for the first time."[78] This confusion between real life and literary representation is intended to encourage paranoia in readers by suggesting that the global and even supernatural plots and conspiracies partially uncovered within the novel spill over into—or, more accurately, begin in—the real world. This confusion of textual representation and things as they (really) are means that Pynchon's novel does not in the end seek to exempt itself from the real world: "far from holding out the promise of a postexegetical superiority to the world that it represents, Pynchon's work permanently infects us with the paranoid anxieties of its characters," suggesting that "literature is never merely an agent of resistance against networks of power-serving knowledge" but is rather "one of that network's most seductive manifestations."[79]

Yet the affective power of Pynchon's novel depends, like Coleridge's image of blood sugar, on a tension between states of affairs, which bear upon bodies, and events, which are only expressible in language. The plots depicted in the novel depend on various forms of bodily causality—most prominently, a Pavlovian stimulus-response logic, in which the bodies and consciousness of characters respond automatically to stimuli. Yet the novel itself is an attempt to name events: the Zone, the Rocket, the White Visitation. Like Coleridge's blood sugar, these nouns are not exhausted in their reference to specific configurations of bodies but rather express events that hover like a haze over states of affairs. If Pynchon's novel seeks to confuse the distinction between real and represented life, it does so less to assist identification of the real conspiracies that control readers' bodies and consciousness than to serve as a means by which readers can create new events by linking themselves to earlier events. Just as "sweetening" indexed for Coleridge the reader's power to introduce points of inflection (events) into both personal and political states of affairs, "paranoiding" names for Pynchon not a capacity that facilitates identification of the true causal relationships of states of affairs but rather a power to express events. As a consequence, reading Pynchon does not result in a feeling of restraint but rather, as in the case of Coleridge's essay on the slave trade, a sense of ecstatic freedom, a sense that the world *can* be otherwise, despite its plots and systems.

These Romantic-era and postmodern examples emphasize that collapsurgence is a textual strategy that emerges between aesthetics and science, and depends upon exploiting tensions immanent within both. Thus, collapsurgence draws—in fact, it depends—upon scientific discourses that validate claims about the causal relationships of bodies, such as the scientific claims about digestion that underwrote Trotter's, Godwin's, and Coleridge's texts (or Pavlov's claims about stimulus and response, in the case of *Gravity's Rainbow*). Collapsurgence also finds its greatest support in scientific discourses that directly describe relationships between the human body and mind or explore how living bodies incorporate the external world. At the same time, collapsurgence exploits the fact of *competing* scientific accounts of a given phenomenon. In the Romantic-era, this meant exploiting the fact that though digestion had increasingly become a scientific object of study, it was still not clear what mechanisms were responsible for its operations. By locating and exploiting these tensions, these points of indecision or potential inflection, within scientific discourses, collapsurgence is able to draw on both the sense of reality that attends scientific discourse and the reality of language. In so doing, one produces a text that is neither solely documentary nor solely literary but is instead itself a point of inflection, a production of an event.

CHAPTER 5

The Media of Life

> He comes to himself out of his sensuous slumber, recognizes himself
> as Man, looks around and finds himself—in the State. An unavoid-
> able exigency had thrown him there [*warf ihn hinein*] before he
> could freely choose his station.
>
> *Schiller*, On the Aesthetic Education of Man, 28

Romantic-era poetry and prose is, in many ways, literature devoted to what Martin Heidegger was later to call *thrownness*; that is, literature that assumes that human existence only begins when one finds oneself already cast, both mysteriously and irremediably, in the middle of things. Whether it is the unreferenced and unknown "this" with which William Wordsworth began, mid-line, the 1798 version of what would become *The Prelude* ("Was it for this / That one, the fairest of all rivers, loved / To blend his murmurs with my nurse's song") or the now-forgotten curse thrown forth in the distant past by the protagonist of P. B. Shelley's lyrical drama *Prometheus Unbound*, Romantic-era verse is replete with beginnings that can only gesture toward, rather than name, a past cause or surrounding and determining context. Even G. W. F. Hegel's *Phenomenology of Spirit*, the Romantic-era philosophical narrative that so notoriously purports to account for the entirety of existence, begins—and is premised on the principle that one cannot but begin—right in the middle of things, with a consciousness that finds itself amid concrete "heres" and "nows." Though the technique of beginning a story in medias res long predated the late eighteenth century, literature of the Romantic era transformed a literary device into an existential axiom; Romantic-era literature is, in both its content and form, an interrogation of whether it is possible ever to get beyond one's thrownness into the middle of things.

This shift in what it meant to write in medias res depended, I suggest, upon a transformation of the concept of medium, a newly vitalized sense of what it meant to be amid-things. The importance of media is especially evident in Hegel's oeuvre, for his historical accounts of nature, spirit, law, and aesthetics are essentially nothing but narratives of the transformation of mediation (*Vermittlung*) into immediacy (*Unmittelbarkeit*) and immediacy into mediation. There is no doubt that Romantic-era authors still depended heavily on an earlier sense of the term "medium," one generated by seventeenth- and eighteenth-century natural philosophers, who had used it to denote the material spaces that connected otherwise disconnected points. Francis Bacon and Isaac Newton, for example, had described the transmission of light and sound as dependent upon the qualities of "media" such as air and water. Yet Romantic-era authors vivified this earlier sense of the term, presenting media not just as vehicles of transmission but also as conditions of possibility for the life and growth of living beings. This newer sense of the term was especially evident in biological discussions of the solids and fluids that surrounded living beings (for example, the soil that sustained plants or the water upon which marine animals depended for their existence). Romantic-era authors also often linked older and newer understandings of media, describing the ways media of transmission enabled a growth, transformation, and progress of human existence, and many hoped to locate in "cultural" media both the origins and possible sublation of the thrownness that was entailed by human biological existence. For many Romantic-era authors, in other words, even if the medial condition of humanity meant that it was impossible to begin otherwise than in medias res, the dynamics of media promised to deliver humanity to a telos that lay beyond the limits of particular media.

Twentieth- and twenty-first-century readers have tended to lose sight of this vitalist dimension of Romantic-era understandings of media. This was a consequence, I suspect, of the fact that around the middle of the nineteenth century, the term "medium" underwent a disciplinary mitosis that parsed out "life" and "culture" to different fields of knowledge production. Mid- to late-nineteenth-century biologists and physiologists appropriated the vital understanding of "media," using the term to describe the nutritive fluids or solids they employed in experiments to isolate and keep alive cells and organisms. At the same time, those in what would come to be called the humanities increasingly came to think of media as solely a cultural phenomenon, and thus used the term to describe the material substrata (e.g., printed paper or photographic film) by means of which ideas, images, and sound were stored and transmitted from one place or time to another. Significantly, the last several decades have witnessed a partial recrossing of the two cultures, for as both biological and humanities research have become increasingly computer-oriented, understandings of life have for their part come to be parsed through notions of communication and transmission (e.g., the genetic "code"), while at the same time the field of synthetic biology employs communicational media as a means for literally creating new forms of life.[1] Taking advantage of the opening offered by this contemporary re-hybridization—this capacity to think anew the chiasmic crossing of life and communication—I return in this chapter to the earlier Romantic vitalization of the concept of medium, for we can find in this period ways of thinking about the vitality of media, and the media of vitality, that still remain lost (or, more charitably, implicit) in contemporary discussions of "biomedia."

I especially wish to stress the *generative* capacities of time in Romantic vitalizations of the concept of media. That Romantic authors were interested in temporality is, of course, not news, for the Romantic-era has long been known as the period that gave birth to the concepts of historicism, geological "deep time," and time as a subjective form of intuition rather than an external reality.[2] However, I emphasize here—and it is already implicit in the serial form in which I have cast these examples of interest in rethinking time—that in the Romantic era, connections established between time and media tended to function as concept generators. To think about the relationship between life and media in the Romantic era was also to rethink the ways in which time insinuated itself in life, soliciting change and transformation, and it was thus also to rethink time more generally.

In this chapter I document a series of particular accounts of the relationship between time, life, and media, as well as the ideational and institutional dynamics that provided a milieu for these variant understandings. Romantic-era authors invented two different—and ultimately incompatible—narrative schemata in their efforts to understand the vexing relationship between media and living beings. On the one hand, authors found themselves solicited by narratives of "mediality," that is, narratives in which every apparent end could be transformed into a means—a medium—for another end. In its purest form, this process of mediality was one that foreclosed the possibility of any final end. On the other hand, many of the same authors fascinated by mediality also sought to contain the disseminative logic of this schema within hierarchical narratives of "perfectibility." Within this schema, mediality had a directionality and telos; biological media, for example, were the means by which God or Nature produced successively more perfect living beings, with humans and their social institutions—including discursive accounts of life—invariably positioned as the final cause of this process.

The tension between these two narratives produced two quite different series of thought about media and life. We find what I call (with a nod to media theorist

Friedrich Kittler) an "official" Romantic-era series, which implicitly understood the tension between these two narratives as a problem to be solved. This first series, which I follow through three discursive fields—the biology of Jean-Baptiste Lamarck, the lay-oriented medical writings of British surgeon Richard Saumarez, and the philosophy of G. W. F. Hegel—presented media in terms of fluidity, positioned fluidity as the precondition of perfectibility, and described perfectibility in terms of organic unity. This approach to media encouraged a hermeneutic conception of media, one that valorized a discourse about media and life to the extent that it could reflect upon—and, ideally, account for—its own origins. However, separating itself from this official Romantic-era narrative of media was another, "unofficial," account, one that treated the tension between mediality and perfectibility not as a problem to be solved but rather as a generative source of variations (a problematic rather than a problem, we might say). This second approach also began with the premise of fluidity, but fluidity was in this case understood as the precondition for a generative spacing that produced constant variation and self-difference. Where the official Romantic-era series encouraged a hermeneutic understanding of media, the latter approach suggested that human institutions do not function as the final nodal points of organic unity but are instead themselves vectors for producing populations of variations within both nature and culture. I track this approach across a variety of discourses as well: the dark idealist philosophy of F. W. J. Schelling, the teratological biology of Étienne Geoffroy Saint-Hilaire, and Mary Shelley's variation on the Godwinian novel, *Frankenstein*. I emphasize that rather than seeking to sublate the thrownness of life through the concept of organic unity and the form of a philosophical system, these latter authors sought to develop poetic forms of discourse aimed at encouraging further vitalist forms of experimentation.

THE RISE OF ROMANTIC MEDIA (FROM TRANSMISSION TO TRANSFORMATION)

Romantic-era conceptions of media owed much to seventeenth- and early-eighteenth-century natural philosophy, within which "medium" emerged as a key term of mechanics, the science of the movement of matter in general.[3] For natural philosophers, a medium denoted the material space that enabled the transmission of forces or particles between distant points. Francis Bacon described the effects of different media, such as air and water, on the propagation of sound, the transmission of odors, and the flow of magnetic force, while Isaac Newton discussed the effects of "rarer" and "denser" media on the refraction of light.[4]

This concept of medium as a means for transmission or communication was then taken up in the mid- to late eighteenth century by authors interested in thinking psychological events in terms of—or, in its stronger formulation, as dependent upon—the phenomena described by physics. In his *Enquiry Concerning Political Justice*, William Godwin applied this conception of media to human psychology and physiology, arguing that "thought" is "the medium through which the motions of the animal system are . . . carried on," while the human body is "our medium of communication with the external universe."[5] This understanding of medium as a means of transmission also provided a way of understanding psychological error and delusion: in his *Lectures on Natural and Experimental Philosophy*, George Adams cautioned that a false philosophical system, "once fixed in the mind, becomes as it were the medium through which we see objects; they receive a tincture from it, and appear of a different colour from what they do when viewed by the pure light."[6] More generally, wherever eighteenth-century authors saw communication or transmission, they tended also to find media: canals that connected distant geographic locales, for example, were described as "medi[a] of communication," while another writer described Parliament itself as a "medium" by means of which the whole of Britain could be addressed.[7] And in what was to become a famous appropriation of the concept, Adam Smith described the institution of paper money as a "medium" that allowed debt and credit to circulate in the British colonies while at the same time saving "as much as possible the expence of so costly an instrument of commerce as gold and silver."[8]

Yet it turned out that one could not simply explain physiological, psychological, and social phenomena by means of a natural philosophical conception of medium without altering and shaping the concept of medium itself. Thus, the more that "medium" was applied to living bodies and concerns, the more something other than transmission and communication came to the fore, for the term also came to connote vital transformation and development. We can detect a sort of mutual infection of the concepts of media and vitality as qualities of living beings migrated into descriptions of media and, conversely, qualities associated with external media slipped into descriptions of living beings. When Godwin, for example, described the effects of the "medium of unrestricted communication," he turned to metaphors of growth, rhetorically asking "With what delight do we contemplate the progress of intellect, its efforts for the discovery of truth, the harvest of virtue that springs up under the genial influence of instruction, the wisdom that is generated through the medium of unrestricted communication?"[9] In similar fashion, if canals could serve as a medium of communication, it was because they drew on what Adams described as the tendency of bodies of water to encourage the growth of human commerce and invention:

> Water serves to the art and navigation of man, as air serves to the wings of the feathered species. It is the easy and speedy medium, the ready conduct and conveyance, whereby all redundancies are carried off, and wants supplied. It makes man as it were a denizen of every country on the globe. It shortens every distance, and ties the remotest regions together. It carries and communicates the knowledge, the virtues, the manufactures and arts of every clime to all. It gives springs to industry, energy to invention.[10]

As both of these examples show, a medium, when applied to the context of human affairs, tended to point to two different functions. A given medium enabled a particular kind of activity, such as communication between individuals or transportation of goods from one place to another. This particular activity, however, was then itself understood as a medium, or means, for enabling the emergence of a more final "end" or value. Communication between individuals, for example, was the means for producing wisdom, while transportation of goods from place to place was a means for encouraging industry and invention. In the context of human affairs, in short, the capacity of a medium to enable communication of something from one point to another was often understood as simply the means by which the medium generated something new and valuable.

If the sense of medium as transmission had been most fully expressed in natural philosophy, this new, vitalized view received its strongest expression in biological approaches to living beings.[11] The British surgeon Richard Saumarez noted in *A New System of Physiology* (1799) that plant seeds can remain in a "torpid state" for long periods of time, but as "soon as these seeds are placed in media fitted for their action," such as rich soil, then the "living principle" will again begin to act, producing growth and transformation, while the German biologist Gottfried Reinhold Treviranus noted the variety of media within which plants could exercise their powers of vegetation ("jedem Medium, worin Pflanzen vegetiren können").[12] Nor were discussions of vital media restricted to plants. French biologist Jean-Baptiste Lamarck argued that the specific shape of animals was a consequence of the medium in which they lived: fish possessed fins and lacked necks, for example, as a consequence of the "dense medium" (*milieu dense*) that they inhabited.[13] Plants and animals, in other words, might encounter a variety of media, such as different soils, bodies of water, and atmospheric gases, but only certain of these constituted a true medium for a given plant or animal, for only certain of these determined the conditions of survival and growth. From this perspective, a medium was less a channel between two points than what *encompassed* a living being, allowing it to survive and thrive.[14] Edward Jenner's famous 1798 account of smallpox vaccination stressed that human beings could improve their chances for surviving and thriving by treating domesticated animals as media capable of transforming some kinds of infectious matter, for effective vaccination against smallpox could only be ensured when "a disease has been generated by the morbid matter from the horse on the nipple of the cow, and passed through that medium to the human subject."[15] In Jenner's account, media were not simply what surrounded living beings; they could also penetrate into the interior of the living being itself. The Romantic-era political writer John Thelwall drew on the work of surgeon John Hunter to argue that blood was a "medium, by which alone the stimuli necessary for the production and sustainment of Life can be absorbed and properly diffused through the organized frame."[16] For Thelwall, blood functioned as a communication medium, in the sense that it transported and diffused stimuli throughout the body—yet *what* it communicated was precisely the stimuli of life itself. From the perspective of the living body, the fact that a medium allowed something stable to be "communicated" between two points was of secondary importance, for what media truly enabled were the transformative processes of life itself.

Where the natural philosophical understanding of medium in terms of transmission might have seemed relatively straightforward, this newer understanding of medium as a condition of possibility of concrete living beings proved more difficult to think. On the one hand, as Saumarez noted, life depended upon a relationship of "fit" or harmony between the living being and its medium—in the absence of soil, for example, a seed would eventually "crumble to decay."[17] On the other hand, the relationship between the living being and its medium was one neither of identity nor of mirroring. For Lamarck, both media and living beings were essentially just combinations of fluids and solids, but living beings were distinguished by the fact that they were more intensively "canalized" than the media in which they existed (that is, they contained more channels and connections between parts). The importance of this difference between living beings and their media was especially evident in the case of temperature, for though the living body depended upon the heat it received from its surrounding medium, it nevertheless could not simply mirror or match that temperature; consistent identity of the temperature of a plant or animal body with its medium was just another name for death. As British surgeon William Lawrence noted, we "well know what happens to the body after death: its heat is lost, and it soon reaches the temperature of the surrounding medium."[18] Thinking the relationship between life and medium thus required concepts capable of handling what we would now describe as processes of recursion and feedback, for while the medium of a living being provided the latter with its condition of possibility, the relationship between medium and living being nevertheless also required constitutive and productive forms of difference between the two.

THE OFFICIAL ROMANTIC-ERA ACCOUNT OF MEDIA: LAMARCK, SAUMAREZ, HEGEL

Two approaches to this difficult thought of mediated life emerged in the Romantic era. In both, there was an effort to hold on to two ultimately incompatible narrative schemata for understanding the relationship of life to media. On the one hand, we find a narrative of "mediality," within which every seeming end could always be transformed into a means for another end (a process that foreclosed the possibility of any final end). On the other hand, we find a hierarchical narrative of "perfectibility," within which mediality indeed had a unified directionality, telos, and final end. Romantic-era authors interested in media invariably felt themselves solicited by both of these narratives, and the accounts they produced were thus inevitably hybrids. Yet my goal here is teratological rather than simply descriptive, for I want to outline two different logics—two different paths—that Romantic-era authors pursued by means of their hybrid textual inventions. This section describes what I call the "official" Romantic-era account of mediated life, an account aligned with institutions, such as botanical gardens and state-sponsored educational institutions, that in principle understood hybridity and interminable mediality as problems and either transmission of pre-existing form or ascent to perfection as their solution. In an attempt to take this logic seriously, the order in which I consider these three projects—Lamarck's zoological philosophy, Saumarez's "new system of physiology," and G. W. F. Hegel's absolute idealism—has itself the form of a dialectic, with each author progressively clarifying the stakes of a solution that understands interminable mediality as a problem to be resolved.

Lamarck's Sociobiology

I begin with Lamarck because of these three thinkers, his zoological philosophy was most fully under the sway of the narrative of mediality, making only a perfunctory nod toward the narrative of perfectibility (though, as we shall see, his attempt to accommodate his account to the demands of propriety and perfectibility came at an intellectual cost). Jean-Baptiste Lamarck was born into the

French aristocracy and became a government naturalist, a member of the French Academy of Sciences (which he joined in 1779), a key figure in the French royal botanical gardens (Jardin et Cabinet du Roi) and, following the French Revolution, professor of zoology at the National Museum of Natural History (Muséum national d'histoire naturelle), a position he maintained until his death in 1829.[19] Today, Lamarck is best known as a pre-Darwinian evolutionist, one who asserted that living beings could, through their own efforts, modify their organic structure and then pass on those changes to their progeny (a view corrected by Darwin's emphasis on the culling processes of natural selection).

Central to Lamarck's zoological philosophy is his claim that living beings are nature's means (*moyens*) for creating complexity.[20] For Lamarck, "nature" is a complicated term, and in *Zoological Philosophy* (1809), he defined nature as

> the totality of objects comprising . . . (1) all existing physical bodies; (2) the general and special laws, which regulate the changes of state and position to which these bodies are liable; (3) lastly, the movement distributed at large among them, which is continually preserved or being renewed, has infinitely varied effects, and gives rise to that wonderful order to things which this totality embodies. (183; PZ, 359–60)

Nature thus refers to the collection of fluids, solids, and gases that make up the universe and the invariant laws of physics and chemistry that govern those fluids, solids and gases. At the same time, it also signifies the "wonderful order" produced in these solids, fluids, and gases by natural movements and laws. One of Lamarck's primary goals was to convince his readers that this order has a temporal dimension, for he claimed the passage of time produces increasingly perfect beings. The naturalist himself was an instance of the most perfect being (humans), for, Lamarck asserted, "it is clear [*evident*] that since the organisation of man is the most perfect, it should be regarded as the standard for judging of the perfection and degradation of the other animal organizations" (73; PZ, 142). The task of the naturalist is thus "to grasp the order which [nature] everywhere introduces, as well as her progress, her laws, and the infinitely varied means which she uses to give effect to that order" (9; PZ, 1). *Zoological Philosophy* was intended to illuminate nature's progressive movement toward perfection as she built upon simple organisms such as infusorians and polyps to arrive at more complex—and thus, Lamarck claimed, more perfect—organisms such as birds, mammals, and eventually, human beings.

What distinguished Lamarck's account of nature as an order of increasing perfection from earlier eighteenth-century descriptions of a divinely instituted great

chain of being was his insistence that this order had to be produced *immanently*—
within nature, within time, and through natural means (*moyens*). Lamarck insisted that nature, understood as the physical world and the laws that governed
it, was not capable of creating, all at once, a diversity of different species of plants
and animals.[21] Rather, he contended, a diversity of species could only be created
slowly, over time, as the result of differences between the internal needs of animals and the external "circumstances" in which they found themselves: "nature
has little by little fashioned the various animals . . . by the aid of much time and
an infinite variation of environment [*une variation infinie dans les circonstances*]"
(40; PZ, 66).[22] As circumstances change, the needs of living beings change, and in
striving to meet these changed needs, the organs of living beings are strengthened
or atrophied, processes that in turn lead to the emergence of new kinds of organs.
These organic changes are passed on to progeny, and the result over time is the
emergence of new species. However, because nature has as its goal the creation
of ever-greater complexity, this slow process of change produces an "order to progression," leading from simple plants and animals to much more complex, and
perfect, mammals.

Insofar as nature denotes both the means by which organic perfection is
achieved, and this goal of organic perfection itself, living beings end up bearing
a double relationship to their media. On the one hand, as Georges Canguilhem
notes, living beings for Lamarck exist in "distressful and distressed" relationship
to their media, for the living being is continually forced to try to "stick" to an ever-
changing and indifferent medium:

> The life and the medium [*milieu*] that is unaware of it are two asynchronous
> series of events. The change of circumstances comes first, but it is the living
> itself that, in the end, initiates the effort to not be let go by its medium [*milieu*].
> Adaptation is a repeated effort on the part of life to continue to "stick" to an
> indifferent medium [*milieu*].[23]

Canguilhem's gloss highlights the deep pathos of Lamarck's account, for it is
a story of needful, desiring beings forever bound, through series of picaresque
variations, to try to adapt themselves to media that can never be anything but
indifferent to them. Though a medium supplies a living being with its conditions
for existence, in the form of food, temperature, and other necessities, the medium
nevertheless "does nothing *for* life" (13, my emphasis). On the other hand, the
unceasing efforts of living beings to stick to their media are themselves "means"
(*moyens*) through which nature is able to complexify itself. Thus, Lamarck writes,
"Living bodies . . . constitute, by their possession of life, nature's principal means

[*le principal moyen*] for bringing into existence a number of different compounds which would never otherwise have arisen."[24] It is this *dynamic* relationship between milieux and multitudes of living beings that distinguishes the Lamarckian milieu from the much more classical conception, proposed at least as early as Hippocrates, that external environments in some way "influence" living beings.

Yet precisely because of this double reference that life bears to media, it is not at all clear why mammals in general, and humans more specifically, should necessarily constitute the *final* stage of nature's ascent to perfection. Lamarck clearly hoped to head off this question by means of the rather peculiar way he narrated the order of nature. For rather than tracing the ascent of nature from the most simple to the most complex forms—replicating narratively the temporal progression he claimed nature itself had employed—Lamarck's account followed nature's "order in the inverse direction," beginning with the most complex animals, then tracing what he described as the "progress of degradation" downward to the most simple animals (72; *PZ*, 139). This narrative strategy made it easier for Lamarck to position humans as the pinnacle and standard of organic complexity, thus effectively foreclosing at the outset the question of whether nature might create new, even more complex organic systems than those possessed by mammals and humans.

Yet the *biological* criterion of complexity to which Lamarck's account commits itself gives no positive grounds for taking very seriously the purportedly self-evident axiom of human perfection. Lamarck contended that the "progress" of living organization was essentially a process of increasing canalization within living bodies; that is, greater complexity was created to the extent that "the movements of the fluids in the supple parts of the living body which contains them . . . cut out paths and establish depots and exits," creating "canals and afterwards various organs" (189; *PZ*, 374). Lamarck's emphasis on canalization as the criterion of complexity suggests that he understood complexity solely in morphological rather than in functional terms: that is, a more complex organism is one with a greater number of discrete and interconnected parts rather than, for example, an organism that (perhaps because of its more complex inner organization) has a greater capacity to "adapt" to changing milieux.[25] However, if nature aims at biological complexity, understood as an increase in biological canalization, then it is by no means clear why we should accept that the human system of organs should be the end or goal of this process of canalization. While it may indeed be the case that humans cannot imagine anything more complex than themselves, Lamarck had taken pains to point out at the start of his account that human cognitive limits are not necessarily adequate to a full understanding of the natural order.[26]

Thus, in locating nature's goal of complexity in a biological register, Lamarck begged the question of how his own discursive account was able to grasp fully, and thereby determine the limits of, biological complexity.

Saumarez on Final Causes and Perfectibility

The goal of justifying the claim that human discourse is able to grasp and explain biological complexity motivated the account of the relationship between media and life in the work of Richard Saumarez. Though Saumarez, a British surgeon, has now been largely forgotten, he developed a vitalist political philosophy that served as the basic template for the later vitalist conservativism of John Abernethy, and his virtues were extolled by S. T. Coleridge in *Biographia Literaria* (1817).[27] Saumarez developed his account of the relationship between biological and discursive media in *A New System of Physiology* (1799), a lengthy two-volume work that purported, via its subtitle, to "comprehen[d] the laws by which animated beings in general, and the human species in particular, are governed, in their several states of health and disease." Directing his work toward a general readership, Saumarez not only discussed physiological phenomena and comparative anatomy but also considered the "media," such as universities, through which the knowledge produced by these sciences was disseminated to society as a whole.

Like Lamarck, Saumarez traced a progress in organization through the natural world from simple to complex living beings, and Saumarez, too, positioned humans as the goal and apex of this process. However, where Lamarck presented each stage in the progress of organization as a means by which nature fulfilled its only real goal, increasing complexity, Saumarez developed a complicated system of multiple "final causes." He argued that the variety of organizations of living beings "arose from the difference in the end each had separately to fulfill."[28] Thus, vegetables "are destitute of organs of sense or of sensation" because "the *propagation of the species alone* [is] the final cause of their existence" (1:ix), while "brutes" possess sense and sensation because their final cause is both "*the gratification of the appetite and the propagation of the species*" (1:x). Humans possess both the means of propagation and sense but, in addition, are able to exercise "higher faculties of cognition and of reflection," and Saumarez thus concluded that the final cause of a human being is "the perfection of his mind" (1:xii). This final cause was itself characterized by two linked "ends," namely, "adoration of the Deity and the exercise of the moral virtues" (1:xiii).

Saumarez's emphasis on final causes allowed him to avoid the deep historicism so central to Lamarck's understanding of means and ends. For Lamarck, nature had no preexisting *plan* it instantiated over the course of time; rather, it simply had a *goal*, to increase complexity. The diversity of living beings was the means by which nature sought to achieve this goal, but because nature had no atemporal perspective upon its own operations—nature, in a sense, stumbled incessantly forward, under severe limitations—it made no sense to speak of a "final cause" of any particular kind of living being. Saumarez's account, by contrast, implied a fundamentally static conception of nature, one in which there was indeed a preexisting plan for each level of living being (vegetables, brutes, humans). God designed each of these modes of life for a specific purpose; thus, it did indeed make sense, within his system, to speak of the final cause of a living being.

Saumarez's emphasis on final causes and a transcendent designer also allowed him to situate the *study* of life within life itself. He described "education and instruction" as the "necessary *media*" that "connect the dawn of reason to the full perfection of it," thereby leading humans from the "dormant," "vegetable existence" of savage societies to the perfection of civil societies (1:xiii). Saumarez implicitly presented education and instruction as biological activities, in the sense that they were actions dictated by the organic structure of humans, and it was thus by means of education and instruction that humans fulfilled the final cause of their organic structure, perfection of mind. Moreover, the premise of Saumarez's "new system of physiology" was that the study of living beings ought itself to be part of education and instruction—that is, his text was to function as one of the media by means of which one learned to "live properly" by learning about the nature of living beings. By understanding the hierarchical order of nature and the different final causes of its various stages, humans would be able to perfect their minds. This was a powerful premise, for it suggested that at stake in popular discussions of the nature of life and living beings was not simply the truth about other species but the very essence of humanity; to misunderstand the true nature of the living world was to risk returning humanity to the "vegetable existence" of savage life. This premise also justified the *regulation* of public discussions of the nature of life and different kinds of living beings, for dissemination of inaccurate claims about these topics would, from this perspective, encourage a more general decline of civil life. The authority to disseminate information about the nature of life, and to conduct further research into living beings, Saumarez suggested, must be delegated to particular social institutions, such as universities, which he described as "the means by which the final cause of human existence is attained" (1:169).

Yet even as Saumarez's emphasis on final causes justified his interest in regulating education about the nature of life, it introduced a tension into his concept of medium. Saumarez usually presented media as communication channels that facilitated the movement of something already formed; media were the material spaces through which immaterial forms and principles moved from one location or time to another. In the case of biological conception, for example, Saumarez proposed a variant of the ancient Aristotelian explanation of generation, arguing that

> the sensible qualities of [semen] constitute the vehicles only through which the energy and specific power are imparted, and are the media only through which it is exerted. Semen is the recipient by which the paternal character is received, and the penis constitutes the medium by which it is conveyed to the ovum which the female evolves. (1:423–24)

In their role as media, both semen and penis pass on a form (the paternal character) that is generated elsewhere. The event of conception in humans then initiates yet another series of mediations as the immaterial "principle of life" begins to "evolve" the brain, such that an equally immaterial individual soul can eventually employ the brain as a medium for communicating already-formed acts of will to the external world.[29] Just as semen and the penis (and, apparently, the mother) serve only as material vehicles for the transmission of form from father to child, so the brain functions only as the material medium by means of which the mind transmits preformed acts of will to the world.

Saumarez clearly wanted to limit the function of media to the communication of form through space and time, yet his emphasis on perfectibility also demanded that media be able to help *structure* those forms they purportedly only transmitted. Saumarez presented the perfection of mind as the final cause of organic existence, but achievement of this perfection required the assistance of the "media" of education and instruction. Education and instruction were not simply a matter of teachers inscribing truths on the blank-slate minds of students; they were, in addition, a matter of allowing students' minds to restructure themselves such that truths would stick. The possibility of perfecting oneself through the media of education and instruction, in other words, meant that at least some media must do more than simply transmit "content"; they must also be able to transform the receiving mind.[30]

Given the pedagogical aspirations of Saumarez's New System of Physiology, one might have expected him to provide a criterion of perfection, a mark by means of which an individual would know whether the recursive processes of a particular educational regimen were contributing to or leading away from the perfection of mind. That Saumarez provided no such criterion no doubt reflected his belief, implicit in his authorship of a "new system of physiology," that he was himself much nearer to a state of mental perfection than were his readers. Yet his failure to describe the marks by means of which perfectibility might be known makes it difficult to determine whether Saumarez believed that *all* individuals were capable of a medial process of self-transformative education, or whether he understood self-transformative education as a process reserved for a small but glorious band of individuals like himself, who would then use media in their more limited "broadcast" function to transmit preformed mental contents to the rest of us.

Rather than resolving this problem at the level of the individual, Saumarez reproduced it in a historical register. Having argued that the possibilities for mental perfection depended upon the organic development of the brain, he suggested that the historical progress of humanity from savage to civil society had depended on the relative "magnitude of the brain" of different groups of humans. Although qualifying that "the subject is still open for particular enquiry," Saumarez contended that we

> possess facts sufficient to draw a conclusion that the most rational systems are endued with a greater proportion of brain than the less rational, and that the brain of rational beings in general is proportionably larger than the most instinctive. The skull of a white is considerably larger than that of black, of an African than of a monkey, of a monkey than of a dog. (1:160–61)

Where it had remained unclear in Saumarez's account of educational institutions whether he believed that every individual had the capacity for progressive enlightenment, his claim that science had established a quantitative hierarchy of brain media across types of humans clarified his position: since some entire groups of humans were less capable of enlightenment than others on account of their limited brain matter, enlightenment was not an option open to all.

Though Saumarez intended his claims about the dependency of mind upon the development of the brain to justify his assertion that different groups of humans possessed different mental capacities—an assertion that in turn was to support his account of the progress of human society from savagery to the civilized states of Europe—this is not the conclusion toward which his text actually points. The problem, in brief, is that Saumarez's criterion for *mental* development—brain size—is too meaty to serve his purpose. Even were one to accept his claim about the relative sizes of brains of Africans and Europeans, this by no means foreclosed the possibility that some other (non-African and non-European) group might have larger brains than Europeans or that future groups of humans might grow larger brains. In other words, insofar as one accepted his claim that the biological medium of the brain imposed limits on the exercise of the mind, there was no apparent reason to exempt his own expression of mind from these limits, no reason to conclude that he (and his European peers) represented the "most rational system" and full enlightenment. The dilemma of Saumarez's position was that he presented the interaction between brain and mind in terms of a *changing limit* the brain imposed on the mind; as a consequence, even as his concept of final causes allowed him to avoid Lamarck's open-ended "progress of complexity," his claim that mental perfection depended upon the nondiscursive register of biological media meant that he was, in principle, unable to justify the claim that his discursive account of life was of necessity true or unsurpassable.[31]

The racism of Saumarez's claims, so evident now, may encourage modern readers to dismiss his new system of physiology, recognizing in it not the progressive philosophy of spiritual ascent that it claimed to be but rather a regressive ideology of racial and class superiority. Though I share this assessment of Saumarez's system, it is nevertheless important to understand *how* to understand this assessment, for his book also dramatizes—albeit unintentionally—the problem of "justification" that continues to plague accounts of the relationships between human bodies and mental limits. Saumarez's claims about relative brain size are simply false, but as Troy Duster has noted, nineteenth-, twentieth-, and now twenty-first-century physicians and scientists have been perfectly happy to adjust the *specific* physical criteria by means of which they account for purportedly racial differences (with genetic risk factors being the most recent candidate) and have even seen in these changes evidence of the progress of science, its capacity for self-correction.[32] In response, we in the humanities have tended to assert that the differences between groups of humans that Saumarez and more recent scientists have ascribed to biological "race" are in fact entirely social in origin; that is, they are differences that depend upon perceptions of others, assumptions about social values, access to educational and employment opportunities, and so on. From this perspective, Saumerez's error—and the error of his modern epigones, such as Richard Herrnstein and Charles Murray, of *Bell Curve* fame—was not simply that of picking the *wrong* physical criteria to account for what they see as mental differences but rather in picking physical criteria in the first place.[33] That is, Saumerez's account of mental limits was itself limited by his too-embodied theory of media: had he simply stuck with the claim that media, including human bodies, are neutral channels that transmit form, then he could have avoided his racist views. Or, to put this another way, one must diagnose and overcome Saumerez's limits by adopting a *disembodied* theory of media.

While one admittedly avoids the racism of Saumerez's claims about brain-based mental limits by assuming that human bodies are neutral and equivalent containers for cultural content, such a position finds itself hard-pressed to explain those physical inequities that result *from* racism (e.g., higher levels of disease and mortality due to both stress and lack of access to medical resources that are part and parcel of racism).[34] The deeper problem, though, is that of justification: how does one *justify* a disembodied theory of media? One *can* employ the sciences to justify the more limited point that Saumerez's particular claims about embodiment and mental limits are incorrect, but—as Duster notes—the sciences do not provide much help in justifying a disembodied theory of media. What, in other words, *justifies* the view that disembodiment, rather than embodiment (of whatever variety) is the proper criterion for assessing the limits of enlightenment?

Media and Auto-Justification in Hegel

G. W. F. Hegel's philosophy of absolute idealism presents arguably the only consistent answer to these questions. Hegel developed his account of life and media in a discursive context quite different from that of Lamarck or Saumarez: though well read in German and French physiology and natural philosophy, Hegel was trained in law, theology, and philosophy, rather than the study of living beings.[35] Yet he also developed his philosophy in a context dominated by the critical thought of Immanuel Kant, which had both highlighted the difficulties inherent in trying to account for a progressive "enlightenment" of humanity within the terms of a mind-matter dualism and pointed to the study of living beings as a possible solution to bridging the relationship between the "phenomenal" and "noumenal" sides of human existence. While several German biologists adopted Kant's dualism as the foundation for their work, Hegel took another route.[36] Seeking to overcome the problems of dualism entirely, Hegel developed an idealist account of the progress of "spirit" (*Geist*), arguing that human beings ought to understand both themselves and nonhuman nature as expressions of an Absolute Subject able to transform every apparent limit imposed by recalcitrant materiality into media for the further expansion of spirit. By stressing the activity of mediation (*Vermittlung*) rather than focusing on media as substances or spaces, Hegel was able to establish a *discursive* criterion for the perfection of spirit, for he argued that the ability to articulate in language the history of the progress of spirit was itself the proof that the final stage of spirit's perfection had been reached.

Hegel's most extended discussion of the medial nature of spirit occurs in his first major work, *Phenomenology of Spirit* (1807).[37] The work begins with a discussion of instruments and media, which functions as an implicit critique of Kant's "critical" approach to epistemology. In his *Critique of Pure Reason* (1781), Kant had argued that further progress of knowledge was possible only *after* we have determined the nature and limits of the mental faculties through which we come to cognize the world. Hegel implied in the preface to *Phenomenology of Spirit* that such an analytical endeavor had more in common with anatomy than with the study of the "fluid nature . . . of an organic unity" that philosophy required, and in his introduction he contended that understanding cognition as either a sort of "instrument" (*Werkzeug*) or "a more or less passive medium [*Medium*] through which the light of truth reaches us" is fundamentally problematic if our goal is knowledge of the Absolute.[38] If we understand cognition as "the instrument for getting hold of absolute being, it is obvious that the use of an instrument on a thing certainly does not let it be what it is for itself, but rather sets out to reshape and alter it," while if we understand cognition as a medium, "then again we do not receive the truth as it is in itself, but only as it exists through and in this medium." Yet, he contended, *pace* Kant, that it does not help to determine the limits of cognition, whether cognition is understood as instrument or medium, for this simply takes us back to our starting point. To "remove from a reshaped thing what the instrument has done to it" or to subtract the distorting "refraction" from a medium is in either case to take away all that cognition itself provided, leaving us with nothing. Thus, whether we understand cognition as an instrument or a medium, "we employ a means [*Mittel*] which immediately [*unmittelbar*] brings about the opposite of its own end." If we seek to know the Absolute, "what is really absurd is that we should make use of a means [*Mittel*] at all" (46; W, 3:68–69).

For Hegel, these apparently insurmountable problems do not suggest that knowledge of the Absolute is impossible; rather, they reveal a deep and fundamental misunderstanding of the concept of medium and its relationship to the process of mediation. The problem with the Kantian conception of cognition as an instrument or a medium is that, in either case, we "assum[e] that there is a *difference between ourselves and this cognition*" such that "the Absolute stands on one side and cognition on the other, independent and separated from it" (47; W, 3:70). Hegel implied that though this conception of cognition as a medium seems to be based on an analogy with physical media, in fact it diverges significantly from them. In the case of physical media, the "medium" denotes a fundamentally homogenous material space (that is, the points or elements of this space differ only in degree or position, not in kind). So we can speak of a sound passing from point A to point B within a fluid medium only if A and B are already composed of the same "kind" of thing.[39] However, as Hegel points out, the situation is fundamentally different when one speaks of *cognition* functioning as a medium. In this case, cognition is by definition supposed to connect two fundamentally different kinds of things—for example, a "subject" and an "object" or a particular consciousness and the Absolute. Thus, the very premise upon which the physical concept of medium depends—namely, that a medium connects points or elements that are the same—has been implicitly abandoned when one purports to understand cognition in terms of media. In the case of consciousness, the term "medium" seems to name a magical act of bridging, which somehow partakes of both kinds of being and translates each into the other.

Hegel's solution to this problem has two aspects. On the one hand, he suggested that mediation was possible only if there is nothing but *one* universal medium. It is true, of course, that we encounter the world as apparently made up of diverse kinds of beings: rocks, plants, animals, stars, elements, other people, and so on. However, Hegel's philosophy is intended to reveal this as a limited view; from the perspective of his absolute idealism, everything that seems "objective," as well as everything that seems unique to a particular subject, is in fact a misrecognized part of the absolute subject, which encompasses everything that is. The true idea of a medium is thus not that of a "third" that connects two different kinds of beings but rather that which allows a *self*-relation, a "pure relating of self to self" (68; W, 3:94). A medium, in other words, denotes that which enables elements to come into contact with one another *as* a unity (rather than as a mere paratactic collection of qualities or things that bear no essential relation to one another).[40] The task of each of Hegel's books is thus to document the different "intermediate media" through which the absolute subject slowly has come to recognize that it is the universal medium, and that what it mediates is nothing but itself. *Phenomenology of Spirit*, for example, documents the different forms of consciousness by means of which the absolute subject has come to recognize itself as itself, beginning with a consciousness that is in medias res, surrounded by concrete "heres" and "nows," and ending with the form of self-consciousness that is able, through the medium of Hegel's philosophy, to recognize itself as the concrete universal medium that holds together as a unity all apparently separate subjects and objects.

Yet even as Hegel argued that there was one all-encompassing medium, he also stressed, on the other hand, the *activity* of mediation over an understanding of medium as a substance or material space. For Hegel, mediation (*Vermittlung*) can never be thought apart from immediacy (*Unmittelbarkeit*), and both must be understood in relationship to movement. Every movement begins with immediacy, in the sense that immediacy is that point of apparent solidity that enables one to make an initial movement forward. Mediation, by contrast, allows one to turn back toward the starting point. In *Phenomenology of Spirit*, for example, Hegel begins with what seems like the most immediate form of consciousness—a single consciousness aware of a preconceptual manifold of concrete things that are simply "here" and "now"—because this starting point allows a movement outward (i.e., a movement by means of which consciousness will come to recognize that this apparently immediate starting point is itself actually mediated). Moreover, the final stage of consciousness in *Phenomenology*, "the idea [*Begriff*] as pure knowledge," itself serves as the immediate starting point, or presupposition, for Hegel's subsequent *Science of Logic*.[41] There is thus no essential contrast between what is immediate and what is mediated, only a contrast that is relative from the point of view of movement. Or, as Hegel, puts it, "*there is* nothing in heaven or nature or spirit or anywhere else that does not contain just as much immediacy as mediation, so that both these determinations prove to be *unseparated* and *inseparable* and the opposition between them nothing real."[42] Rather than understanding a medium as something prior to, and in principle separable from, the activity of mediation that the medium purportedly enables, Hegel instead implied that the very concept of a stable, preexisting medium is a naïve "substantialization" of the more primary activity of mediation.

The medium of idealism—Spirit—is thus not to be understood as a neutral, objective space in which the movement of mediation can take place or not, but rather as a univocal field that enables elements to come into contact with one another *as* a unity. Individual media are "fluid" to the extent that they enable a kind of turning, and turning back, which itself results in a *self*-relation, or "pure relating of self to self." History is a process of Spirit-ification, Spirit's increasing ability to turn what had seemed like opaque fixed points into self-reflections. The history of art, for example, narrates the movement of art through various media, or what Hegel terms "elements" (*Elemente*): art begins with architecture, which employs fixed and resistant media such as stone or metal, and then locates increasingly fluid media, such as the articulated dancing body of the human being or the even more articulated language of poetry.[43] An artistic medium is "fluid" insofar as it allows each element of a work to be oriented toward its origin in the freedom of Spirit, rather than stressing Spirit in its self-limiting forms, such as the invariable tensile strength of a certain kind of stone or the effects of a pigment on the human eye. A state of pure fluidity is reached when philosophy enables thought itself to reflect all of the more bound materials of art within itself.[44]

Hegel contended that it was only this self-mediation of the absolute subject that deserved to be called "life." "Life" is a complicated term in Hegel's philosophy, for in some contexts it denotes that which distinguishes living from nonliving beings; in others, the process of self-mediation of the absolute subject. Hegel considered living beings at length in his *Philosophy of Nature*, tracing the evolution of nature's external forms from their origin in abstract time and space to "vitality in its true form," animal life.[45] Within this account, a living being is that particular kind of natural being that is able to externalize itself into an object—a living body—"in order to return into itself" in the form of feeling and thought (275; W, 9:340). Yet from a broader perspective, this capacity of living beings to externalize themselves in order to return to themselves is simply an expression of the self-movement of the absolute subject itself. Feeling and thinking living beings function, in other words, as the final medium of a process through which the absolute subject has evolved (itself) from a merely implicit unity of all that exists into a unity able to recognize itself *as* this unity. Life, from this broader perspective, is not fundamentally a property that distinguishes one kind of being from another; rather, it is that *movement* of the absolute subject through different media to the point where it is able to recognize itself in the discursive capacities of a particular kind of living being, namely, humans.

While Hegel's conception of the distinction between living and nonliving beings owes something to Kant, his conception of life is fundamentally Aristotelian. For Aristotle, living beings have the capacity for "movement" (*kinesis*), though this term refers not simply to physical change of location but to any kind of change or alteration. Hegel follows Aristotle in linking life to movement, and though he argues that the highest forms of animal life are indeed characterized by their capacity for physical movement—the plant represents a lower form of animal life precisely because it cannot "freely determine its place, i.e. *move from the spot*" (305; W, 9:373)—the life of the absolute subject is defined not by physical displacement but by its historical evolution of objective and subjective forms. The absolute subject thus "moves" in the sense that it transforms the relationships among its parts, making explicit relationships that were initially only implicit.

This Aristotelian understanding of movement allowed Hegel to move beyond the "transmission" conception of a medium that dominated natural philosophical discussions of physical phenomena. Because Hegel, like Aristotle, understood movement as any kind of change, his conception of medium had to be able to account for changes beyond simple shifts in physical location. Physical movement, for Hegel, was an epiphenomenon of the primary mode of movement, growth. However, Hegel insisted that growth was possible only on the basis of *unity*, and thus a medium could enable growth only because it established unities. Media might facilitate "transmission" and "expression," but only because they first—and more fundamentally—established organic unities within which elements could then be put into communication.

On the one hand, Hegel's idealistic account of media resolves the problem of justification inherent in Lamarck's and Saumarez's accounts of the perfectibility of life. Where all three authors shared the belief that one could discern in different kinds of living beings a goal-driven growth in complexity, Lamarck and Saumarez tied this process of perfectibility to the nondiscursive register of biological media, which implicitly called into question whether their own discursive accounts of this process could in fact be the final word on perfectibility. Hegel, by contrast, insisted that there was only one true medium, that of absolute spirit, which *could* articulate itself (via Hegel's philosophy), and which had reached perfection when it was able to announce itself discursively as this universal medium.

On the other hand, even as Hegel resolved the problem of justification that plagued Lamarck's and Saumarez's accounts, he did so at the cost of plunging the absolute subject into a state of post-vitality, for his ability to explain the life of the absolute subject meant that the absolute subject had moved beyond life. The life of the absolute subject is its movement, its evolution within itself of objective and subjective forms. Yet the absolute subject moves, and thus lives, only when it remains incomplete—that is, prior to the point at which it is able to recognize itself as itself.[46] Once it has recognized itself as itself through the medium of Hegel's philosophy, it by definition no longer grows or develops; it has reached the end of history. This means that the absolute subject no longer moves and, thus, no longer lives. Of course, the fact that absolute subject no longer lives does not mean that it dies; rather, it has entered a state of post-vitality, beyond both life and death.[47]

MEDIA AND VARIATION: GEOFFROY, SCHELLING, AND SHELLEY

Media as fluidity, fluidity as precondition of perfectibility, and perfectibility as organic unity. I have described this as the "official" Romantic-era narrative of media in part because Lamarck, Saumarez, and Hegel—republican scientist, self-appointed advocate for the British constitutional monarchy, and government educator, respectively—each aspired to represent the voice of the state in one form or another, in part because their emphases on perfectibility, hierarchy, and organic unity could so easily be appropriated by the state (which, like the institutions of which it is composed, seeks to transmit its form unchanged over time). This shared rubric of official narrative harbored a hierarchy of differences, of course. Lamarck, whose account of life and media was the least able of the three to account for itself, had the most tenuous relationship to centralized authority.

Though Lamarck did indeed continue to "stick" to the French state through-out its various changes in the late eighteenth and early nineteenth centuries, and though his botanical position was intended to consolidate and regulate the plant world for the French state, his zoological philosophy became increasingly liminal, and by his death, in 1829, it had been more or less completely eclipsed by Georges Cuvier's "functional anatomy," which was premised on the stability of biological forms over time.[48] By contrast, Hegel—able to account for his phi-losophy by subordinating mediality to the perfection of organic unity—began his career in liminal teaching positions but ended up as one of the official voices of the German educational mission.[49]

Notwithstanding these differences, however, it should not come as a surprise, nor is it a coincidence, that the accounts of media developed by all three authors seem quite at home within the account of the "discourse network" (*Aufschrieb-system*) of 1800 developed by media theorist Friedrich Kittler. Drawing on Mi-chel Foucault's concept of a general episteme that was articulated throughout what might otherwise seem like unrelated late-eighteenth- and early-nineteenth-century institutions and discourses, such as proposals for "reformed" prisons and educational systems, Kittler shows the techniques by means of which this epis-teme was "personalized" (that is, inscribed in structures of subjectivity). Kittler's account emphasizes the newly privileged nuclear family, the emergence of the figure of the poet as a mediator between nature and the citizen, and the de-velopment of all-encompassing philosophical "systems," such as those of Fichte and Hegel, which were intended to allow citizens to locate their place within the linked systems of nature and state. The wager of Kittler's project—and what makes it particularly relevant here—is his claim that the links between the per-sonalizing techniques upon which he focuses and the impersonal institutions of Foucault's account become visible only if we attend to the *media*—namely, handwritten and printed texts—through which what he calls "data, addresses, and commands" were circulated in the Romantic era.[50] Kittler argues that the dominance of handwritten and printed texts in that era encouraged an interpreta-tion of media as *hermeneutic conduits*, channels that enabled a continuous flow of meaning from a source (Nature/Spirit) through the channel of a handwritten text (which linked the mind and the word through the intermediary of the writing hand) and from there to the printed book, which passed on to readers the flow of meaning. Perhaps surprisingly, given the centrality of communicational media to his account, Kittler does not address the explicit interest in the term "media" in the Romantic era that I have outlined above. However, it would not be difficult to position what he describes as the "hermeneutic" premise of the discourse

Hegel (media as condition of organic self-relation)

↑
╱

Saumarez (media as transmission of form)

↑
╱

Lamarck (media as instigator of organic change)

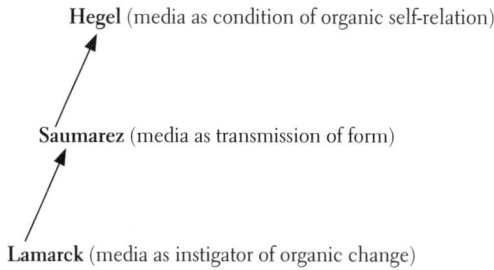

Figure 1 The Media Dialectic of Perfectibility, from Lamarck to Hegel

system of 1800 as the milieu within which Lamarck's, Saumarez's, and Hegel's accounts of media thrived, and to see Hegel's version as simply the theoretical clarification of this premise.

Yet what ought to make us hesitate before arriving at this conclusion is the *diversity* of theories of media in the Romantic era. Kittler's account is intended to distance us from a hermeneutic understanding of media by contrasting the discourse system of 1800 with the discourse system of 1900, in which new media-inscription technologies, such as phonograph and film, as well as psychophysical research instruments, suggested that the processes of "meaning" that a subject can grasp are in fact epiphenomena of unconscious, asemantic, and corporeal processes of communication. But Kittler's disinterest in the actual, and diverse, uses of the term "media" in the Romantic era, combined with his own reliance on a cybernetic understanding of media (that is, media as systems made up of sender, code, channel, and receiver), suggests that we may want to reconsider whether all Romantic-era theories of media in fact tend toward hermeneutics.

What if, for example, we were to see my dialectical series Lamarck-Saumarez-Hegel as itself part of a population or space of variation? I premised this series on the notion that each author represented a moment of a progressive movement by means of which one-sidedness and partiality were overcome (see figure 1). Thus, Lamarck's understanding of media as instigator of organic change explains the *possibility* of an organic ascent to perfection, but it cannot justify that humans have reached the final stage of this ascent. This problem is partially solved by Saumarez's understanding of media as transmission of form, which would explain the final stage of the ascent yet still cannot *justify* the claim that any humans have yet arrived there. Hegel's understanding of media as condition of organic self-relation both explains and justifies why humans occupy the final stage of perfection.

Yet if this dialectic were in fact simply a part of a wider environment of theories of media (see figure 2), then what had appeared initially as an inexorable dialecti-

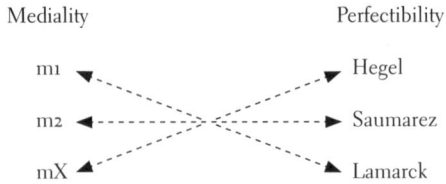

Figure 2 Two Poles for Understanding Media—Mediality and Perfectibility

cal movement running from Lamarck to Hegel would now appear to be simply one way for us to link three media theories that aim at perfectibility, but at the cost of being able to explain other media theories (m1, m2, etc.) that take as their telos not perfectibility but rather mediality.

From this latter perspective, Lamarck would not be understood as the starting point of a series that moves inevitably to Hegel but rather as a node that is linked both with authors interested in perfectibility *and* with authors interested in pure mediality. Embedding our initial dialectic in this way would in turn demand that we attend to *two* quite different textual criteria. If the criterion that generates the line that culminates in Hegel is the ability of a text to account for itself, the criterion that generates the movement toward a spectrum of expressions of pure mediality would be the *generative* ability of a text; that, its ability to facilitate another node in the series, or another genre of expression. (And this would mean, as a consequence, that the two "ends" of the series I have plotted above should not be understood as equivalent, or homologous, for where the series on the right truly ends with Hegel, the left end is disseminative and interminable.)

Darkness and the Primal Fluid: Schelling

We can begin to test this alternate schema for understanding Romantic-era narratives of media—that is, to see these theories in terms of an effusive series rather than an upwardly inclined dialectic—by beginning with the theory of media that emerged in the work of Hegel's dark twin, Friedrich Wilhelm Joseph Schelling.[51] Hegel and Schelling were fellow travelers during the 1790s and the first half-decade of the nineteenth century, and Schelling, like Hegel, initially committed himself to the premise of a universal vital medium that came to self-consciousness by unbinding fixed forms. Schelling was initially more interested than Hegel in the history of the objective aspect of Spirit, developing a "nature philosophy" that began from the "point from which nature can be posited into *becoming*," a starting point that, Schelling argued, had to be "the *most primal fluid*."[52] In this primal fluid, all the particular natural forms that would later come into being within time and nature interpenetrated one another indifferently. The *history* of nature was for Schelling the result of a tension between two drives: a drive toward specificity, which produced the determinate forms in which we now encounter nature—chemical atoms, planets, plants, and animals, for example—and a continual infinite striving, which kept the whole of nature in a constant state of production. Schelling used the figure of the whirlpool as a synecdoche of this state of productive tension:

> *Example*: a stream flows in a straight line forward so long as it encounters no resistance. Here there is resistance—a whirlpool forms. Every original product of nature is such a whirlpool, every organism. The whirlpool is not something immobilized, it is rather something constantly transforming—but reproduced anew at each moment . . . Certain points of inhibition in Nature are originally set up . . . [b]ut at each moment comes a new impulse, as it were, a new wave, which fills this sphere afresh. (18n)

The goal of nature philosophy, then, was to produce within human consciousness an awareness of this infinite striving within the products of nature, allowing the philosopher to intuit the otherwise invisible activity that made possible "fixed" forms. In complementary fashion, Schelling's philosophy of *subjective* spirit was intended to reveal those same tensions within intellectual products.[53] From this perspective, Hegel and Schelling seemed to share the premise that both the worlds of nature and culture had to be understood in terms of media and mediation, with media itself understood in terms of a dynamic fluidity that led upward to perfectibility and organic unity.

Yet even as Schelling committed himself to the premise of perfectibility, he nevertheless remained suspicious of the implication that contemporary philosophy was unfolding in a post-medium condition. For Hegel, philosophy, understood as systematic thought, enabled a condition beyond mediality, a condition in which *Geist* could grasp its organic unity precisely because thought had become freed from the material limits of earlier media. For Schelling, by contrast, philosophy could only be truly systematic—that is, could only be led back to, and thus justify, its starting point—if it remained bound to materiality and limitation. Schelling's emphasis on materiality is highlighted by his understanding of the philosophy of nature as complementary to, rather than a precursor of, the philosophy of spirit, and in his contention that *art* remained a necessary "organ" to philosophy. Where Hegel had argued that the materiality of artistic media allowed *Geist* only a limited grasp of its unity, Schelling maintained that the materiality of art enabled an intuition, an intuition that philosophy itself could not provide, of the unity of nature and spirit. In his *System of Transcendental Idealism* (1800), Schelling argued that "it is art alone which can succeed in objectifying with universal validity what the philosopher is able to present in a merely subjective fashion."[54] In his account, the work of art is able to unify objectivity and subjectivity less because art allows spiritual form to be imposed on recalcitrant matter than because the artwork appears to us as the result of "genius," that is, a combination of subjectivity and something that is beyond subjectivity, the "indwelling element of divinity in human beings."[55] The genius, in other words, "links finitude with infinitude," presenting us with something spiritual that nevertheless cannot be fully grasped (84). Philosophy could not, and should not, attempt to overcome this ungraspability, which meant that it did *not* develop within a post-medium condition. Rather, philosophy should aim at thinking this necessary relation between finitude and infinity, limitation and the unlimited.[56]

Schelling's suspicion of the post-medium condition was a consequence of the significant differences between his and Hegel's conceptions of media. For Hegel, medium was simply a way of naming as a substance Spirit's centripetal self-turning—the reflection of Spirit back upon itself—which had as its necessary end, or telos, a condition of static organic unity beyond both movement and life. For Schelling, by contrast, the telos of complete organic unity could not be reached, and the universal medium should not, as a consequence, be understood in terms of reflection or turning. Rather, what media enabled was a form of ecstatic self-differentiation. In the case of living organisms, for example, Schelling argued in *The First Outline of a System of Natural Philosophy* (1798) that the development (*Bildung*) of increasingly more perfect organisms had to be understood as "an infinite process of formation [*Bildung*]," rather than something that could ever be completed.[57] One had to understand existence as a medium for difference, a medium within which concrete organisms were "variations" (*Abweichungen*) on an ideal archetype that could never be fully instantiated (*Werke*, 2:64). Where Hegel understood the fluidity of media in terms of a schema of spatial unity—media, for him, enabled a centripetal self-turning—Schelling, by contrast, understood the fluidity of the primal medium as that which enabled an effusive process of seeking to exhaust possibilities.[58]

From this perspective, the goal of philosophy was not, as it was for Hegel, a recounting of the coming-to-unity of Spirit but rather the invention of concepts and narrative schemata able to understand and to participate in this infinite process of variation. Thus, where Hegel's philosophy became increasingly monogeneric, in the sense that he sought to situate all his writings within the category of the "encyclopedia," Schelling pursued his project through sketches, outlines, systems, dialogues, a history of the origin of time, and more. In the generic variety that constitutes his oeuvre, he consistently presented human beings as that node within existence that sought to ensure its own capacity for variation—its own freedom—by grasping how existence as a space of variation became possible.

Media as Space of Variation: Geoffroy

My account of the official Romantic-era narrative of media took the form of a dialectic that began with zoology (Lamarck), moved upward through physiology (Saumarez), and concluded with philosophy (Hegel)—a disembodying ascent that traces a movement away from a literary genre obsessed with the obscure and dark effects of matter on bodies and toward a genre committed to the clarity of spirit. The movement that would allow us to link Lamarck and Schelling, by contrast, does not take the form of a dialectic or trajectory but must rather be understood as the emergence of a space of variation, a space that emerges as part of a common effort to invent literary forms that remained open to the vitalizing powers of the fluidity of media and matter. In Lamarck's case, this was the genre of zoological philosophy, which was arguably far more interested in grasping the transformative effects of media on organisms than in tracing the ascent to "perfection," while for Schelling, it was the search for a genre able to reflect upon and participate in the "infinite process of formation [*Bildung*]" that the medium of will enabled.

Étienne Geoffroy Saint-Hilaire's "teratology," or study of monstrosities, opened up a space between Lamarck and Schelling, effectively hybridizing elements of the two. A disciple and then colleague of Lamarck, Geoffroy owed his biological career to the institutional reorganizations that occurred during the French Revolution. In 1793, both he and Lamarck were appointed as professors of zoology at the newly founded National Museum of Natural History. However, Geoffroy, almost thirty years younger than Lamarck and just at the start of his career in the 1790s, accepted an invitation in 1798 to serve as a scientist on a trip with Napoleon Bonaparte to Egypt. This trip kept Geoffroy away from Paris until 1802 and arguably delayed his research and writing. At the same time, it introduced Geoffroy to new species and allowed him time to work out the theoretical implications of these discoveries. He pursued many of these implications in *Philosophical Anatomy* (*Philosophie anatomique*), published in 1818, a work that alluded, in both its title and content, to Lamarck's earlier *Zoological Philosophy*. However, by that date, Lamarck's zoology, which even initially had received a mixed reception, had been eclipsed by Georges Cuvier's far more empirical "functional anatomy," premised on the stability of biological forms over time.[59] As a consequence, while

Geoffroy wrote from an official institutional position within the state, his explicit support of Lamarck's account of biological change increasingly relegated him to a minority position within this institution.[60]

Though Geoffroy adopted Lamarck's account of media as instigators of corporeal transformations, he reconceptualized this account in such a way that media were positioned as a space of variation, rather than a "stage" for perfectibility, as had been the case for Lamarck. Lamarck had seen organic change as an entirely immanent process of transformation, in the sense that new organic canalizations depended upon the specific organic form from which they were a departure. Geoffroy, by contrast, argued that differences between species could be explained only if one assumed an "abstract animal" that was never expressed as such but took on particular forms as a consequence of the media within which it emerged. This "transcendental" approach encouraged Geoffroy to look for "homologues" between different kinds of animals: that is, similar kinds of connections between parts in what were otherwise widely different species. Thus, as Gilles Deleuze and Félix Guattari note in A Thousand Plateaus, for Geoffroy,

> Anatomical elements may be arrested or inhibited in certain places by molecular clashes, the influence of the milieu, or pressure from neighbors to such an extent that they compose different organs. The same formal relations or connections are then effectuated in entirely different forms and arrangements. It is still the same abstract Animal that is realized throughout the stratum, only in varying degrees, in varying modes.[61]

Both Geoffroy and Lamarck stressed the dependency of organic form on media, but where Lamarck understood this dependency in terms of a continual filiation of change (the organic structure of an earlier animal or species serving as the basis for subsequent animals), Geoffroy proposed a model in which organic innovations relied upon a perpetual "return" to an abstract animal.

A vestige of Lamarck's notion of perfectibility remained in Geoffroy's account, but it was transformed to such an extent that it instead functioned as a principle of variation, rather than perfection per se. Instead of understanding the variety of existing and extinct species through Lamarck's schema of gradually increasing perfection (the latter understood as increasing canalization), Geoffroy argued instead that the "development" of one set of organic connections in a species was always offset by a reduction in development in another set of organic connections. Bones, for example, reached a "maximum of development" in fishes, and the scientist could, as a result, give a name to particular arrangements of bones based on the function they played in this kind of animal. Yet this was the maximum of development of one *component* of animals, not of animals themselves, and other components might be more developed in other species. Geoffroy's concept of an abstract animal allowed him to think of organic change as a process in which components expanded, contracted, and drifted from one location in an organism to another.[62]

Because Geoffroy understood all real, particular animals as grounded in the abstract, "transcendental" animal, he invented a new experimental approach and genre—teratology, which for Geoffroy meant both the production and study of monsters—with which to investigate and document possible permutations of animal form. Focusing primarily on the model organism of the Romantic era, the poultry egg, Geoffroy sought to produce changes in the exchanges between eggs and their surrounding media by covering the eggs with more or less permeable substances, such as varnish or wax, or by soaking eggs in different fluids, such as salt water.[63] The purpose of these experiments was to determine *which* changes in the milieu of an egg altered fetal development. Geoffroy's premise was that the fetal development of mammals and poultry depended in part upon the milieux that surrounded each:

> It is known that the fetus in the womb feels the same alternatives of health and disease as its mother, for [the fetus] is not just in an incubation-pocket, but in a milieu in which the surfaces that are in contact with [the fetus] provide it with the elements of its nutrition; it is thus quite natural that its regular or irregular development depends on the good or bad qualities of these same elements that it draws with his mother.
>
> We can apply the same reasoning to a fetus that has emerged from the shackles of uterine life in the manner of chicken fetuses, since the fluids that will be transformed into organs are in the case of the ovipares gathered together at a time when the existence of the fetus is not yet perceptible to our senses. (271; my translation)

Since a poultry fetus developed outside the body of its mother, the experimentalist could more easily influence the key elements of the milieu, such as heat:

> The transformation of the molecules contained in an egg, and the transformation of molecules of a corpse have the same beginning—or, more generally, all changes of animal substances occur under the action of heat as the first cause. Prevent the tendency to motion excited by this cause and the organization will stagnate; and indeed, cool—or, what comes to same thing, control the caloric—and there is more incubation or more putrefaction. (272; my translation)

While Geoffroy seems to have been largely unsuccessful in his attempt to create monsters experimentally, his interest in the logic of monstrosity represented an important modification of Lamarck's evolutionary claims. Rather than attributing organic changes to the grown organism's active effort to "stick" to its changing medium, Geoffroy instead argued that media produced organic changes directly in developing fetuses, without any need for effort on the part of the adult organisms. As a consequence, for Geoffroy "monstrosity" did not mean, as it had in the eighteenth century, a falling away from perfect form; rather, monstrosity was a means for revealing the disseminative powers of vitality that were at play in every instance of life.[64] Moreover, where Lamarck's evolutionary claims for continuous organic variation relied solely on comparisons between extant species and the fossil record of past species, Geoffroy's approach was premised on the principle that the experimentalist could create a space of variation in the present, soliciting new organic transformations by altering the media within which organisms developed.

Media Spaces: Shelley's Frankenstein

I noted above that the textual criterion that marks a movement toward expressions of pure mediality is not the ability to account for one's account but rather generativity, the ability to facilitate a next experiment. Schelling exemplified generativity in his perpetual search for a philosophical genre capable of accounting for the relationship between time, life, and media; for Geoffroy, the science of teratology was founded on the analogous principle that one must continually solicit, through new experiments, the creative powers of nature. In this subsection, I consider a final unofficial Romantic-era account of media, Mary Shelley's *Frankenstein* (1818), which hybridizes Geoffroy's emphasis on biological monstrosity with Schelling's interest in the ways cultural media, such as art, enable a space of freedom. If Hegel's idealism systemized the official Romantic-era account of communicational media by subordinating nature and materiality to a self-reflexive centripetal turning of spirit, Shelley's novel deepened the alternate, variational approach, exploring the ways in which cultural and biological media cross one another in the figure of chiasmus, each serving as medium for the other.

Insofar as *Frankenstein* centers around Victor Frankenstein's de novo creation of a living being, the link between life and media is clearly central to the novel. However, as many readers have noted, Shelley was remarkably reticent about describing the particular medium Victor employed to create a living being. While it is certain that Victor visited charnel-houses during a period of initial research that led to his discovery of "the cause of generation and life," it is less clear whether his creature was constructed from human body parts that he appropriated during

these visits or from a non-organic material such as clay.[65] Shelley's relative disinterest in the medium Victor employed in constructing his creature suggests that
the significance for the novel of the creature's technical origin is not to be found
in the particularities of its bodily medium. Rather, the significance is that it allows
Shelley to imagine a *variation* on the human form: the creature is very much like,
but not precisely the same as, a human being (he is larger and faster than human
beings, for example).

The creature's status as a variation on the human form has been partially
acknowledged in psychoanalytic readings that emphasize the uncanny doubles
and triples of the novel: for instance, Robert Walton, the arctic explorer with
whom we begin the novel, is a "double" of Victor, who produces yet another
double by creating a creature.[66] However, insofar as such accounts seek to reduce
a multiplicity of characters to a single origin—reading Walton and the creature,
for example, as psychological projections or multiplications of Victor—they also
leave unexplained the ontological difference that the creature introduces into this
multiplicity. This lacuna is problematic, for if Mary Shelley had been interested
solely in psychological projections or doublings, presumably there would have
been no need to introduce an ontological variation on the human form. Understanding these characters as variations, rather than as doubles of one another or
as distorted projections of an original, allows us, by contrast, to take into account
the creature's peculiar ontology, his nonhuman humanness.

Reading Victor, Walton, and the creature as variations of one another also
helps us to recognize that the key media in the novel are not the bodies of characters per se, but rather the fluid external media within which these characters find
themselves, and which solicit each individual in slightly different ways. Shelley's
interest in fluid media is signaled at the start of the novel as Walton describes
his dependency on the flows and currents of the ocean, his "inspiration" by the
"cold northern breeze" that blows down from the North Pole, and his quest to
discover the source of the magnetic flows that cover the earth.[67] Recalling Lamarck's conception of *milieux*, Walton's quest emphasizes the way media solicit
active responses from individuals: neither Walton nor the culture of which he is
a part simply accept their given relationship to the seas and magnetic field of the
earth but instead seek to alter their relationships to these media through technical
means. Significantly, Walton understands this attempt to change human relationships to environmental media in terms of *cultural* values, positioning his quest
to regulate oceanic and magnetic orientation as a means for resituating himself
within the social field, either by finding himself a friend or by earning him the
"glory" that he would then deserve.[68]

The parallelisms established in the novel among Walton, Victor, and the creature map out a space of variation within which we can locate different ways of trying to cross cultural and natural media with one another. If Victor's creation of a creature is "like" Walton's quest to discover the Northwest passage and to locate magnetic north, this is because both seek to achieve cultural values—for example, glory—by transforming human relationships to natural media. And where Victor and Walton attempt to translate discoveries about natural media—for example, the location of magnetic north or the nature of life—into cultural achievements, the creature, by contrast, is caught at the intersection of natural and cultural media, solicited equally by each yet without any means of controlling these calls. The creature presents his early biography as that of a mobile heliotrope, driven along the heat and light differentials established by the rising and setting sun; after escaping Victor's dwellings, the creature orients himself toward the sun, responding to light and dark by closing his eyes, and he later goes in search of food by following the "setting sun."[69] His heliotropism draws him into a social milieu, as he is solicited by the emotional attachments of the De Lacey family, finding pleasure and meaning in his capacity to "sympathize in their joys."[70] When his efforts to integrate himself into this human community fail, he again returns to his heliotropic orientation—"But how was I to direct myself? . . . the sun was my only guide" (164)—and the creature eventually leads Victor to the North Pole, where, as Walton notes at the start of the novel, "the sun is for ever visible; its broad disk just skirting the horizon, and diffusing a perpetual splendour" (49). It is in this region in which the dominant media of the novel—light, water, and magnetic fluid—meet and become fixed that the paths of the three protagonists converge.

Victor finds his way to the North Pole via a Paracelsian approach to media. Where Walton tended northward as a consequence of his efforts to change culture through discoveries about natural milieux, and the creature is drawn there by the sun, Victor's voyage to the pole has its origins in his attempts to confuse the distinction between cultural and natural media, using each as a generative matrix for the other. Like Walton and the creature, Victor is solicited by both natural and cultural media: as a child, his experience of lightning—"a stream of fire" that destroys a tree near his house—had a decisive impact on his educational interests even at the same time that he thrived within a cultural milieu of "mutual affection" with Clerval and Elizabeth (71). Like his alchemist hero Paracelsus, though, Victor sees no fundamental difference between natural and cultural media, and searches for those points of vital generation within media that enable each to cross into the other.[71] For Victor, the creation of a living being is to be a means for eventually achieving two ends that each confuse the nature-culture distinction: he desires the gratitude of a newly created species, and he hopes to renew past social relationships by reanimating dead bodies. Perhaps not surprisingly, then, his successful animation of his creature transforms what was to have been a means into its own end, for when this object that was to have facilitated further research begins to move from within itself ("a convulsive motion agitated its limbs" [85]), it is clear that it has now become a subject, something that lives in and for itself.[72] Perhaps even more disturbing, what was to have been a technical means now becomes a determining medium, an element of the environment that is always there in advance of Victor and which determines his possibilities for action. By the novel's end, Victor recognizes that his actions are fully determined by a tropism established by his creation, for he finds himself bound to his creature even when his desire for vengeance wanes: "I pursued my path towards the destruction of the daemon, more as . . . the mechanical impulse of some power of which I was unconscious, than as the ardent desire of my soul."[73]

Though it is tempting to read Victor's itinerary through a didactic lens, interpreting his pain and eventual death as evidence of his arrested or imperfect moral development, Shelley's novel, like Geoffroy's experiments with monsters, has more of an analytic than an evaluatory goal. Insofar as Frankenstein treats the fluidity of media as a space of variation, questions of greater or lesser complexity or perfection are largely irrelevant. Though Victor selected a human form for his creature because of what he describes as the greater "complexity" of humans relative to other animals (81), the relationships among Walton, Victor, and the creature all occur within the same rung of creation, so to speak. Walton, Victor, and the creature establish a field of variations rather than a hierarchy of being, and thus what distinguishes these three characters from one another is not their complexity relative to one another but rather the differing degrees to which each can grasp—or, on the contrary, is held by—the natural, cultural, and communicational media that surround them. The creature is able to function as part of Victor's milieu, populating each of the novel's geographies in advance of his creator, because the harsh environments in which the two find themselves have less of a grasp on the creature than they do on Victor. When Victor and the creature meet on Mont Blanc's Sea of Ice, for example, those same frozen waves that presented Victor with a serious locomotive impediment—"The field of ice is almost a league in width, but I spent nearly two hours in crossing it" (124)—do not hold back the creature, who moves through this sea like a fish in water: "I suddenly beheld the figure of a man, at some distance, advancing towards me with superhuman speed. He bounded over the crevices in the ice, among which I had walked with caution" (125). At the same time, though, the creature finds himself constantly outpaced by a communicational medium such as writing: as I discuss at more length below, the books the creature reads, which include both published belles lettres and Victor's laboratory journal, exert a tremendous hold upon him, determining completely his sense of his place in the world. The creature is grasped by, and cannot himself grasp, the medium of writing: though he is, like Caleb Williams, a powerful speaker, he remains at the mercy of print. In the space of variation that the novel opens up between Walton, Victor, and his creature, in other words, the novel focuses our attention on the *differing* extents to which each character presents a point of action for various media—and, conversely, the extent to which each character can grasp and exert control over different media.

Frankenstein thus suggests that the differences among kinds of media (natural, cultural, communicational) are a function of the extent to which each kind of medium presents a collection of individuals with "handholds," points at which the medium can be grasped and made to speed up or slow down actions. A medium counts as *natural* when most individuals are grasped by, but can never grasp in return, its fundamental dynamics: the natural medium, in other words, is one that acts at too small or too large a scale, or acts too swiftly or slowly, for most individuals to act upon it. Walton and his crew, for example, are able to use the seas to move from one point to another on the globe, yet they ultimately cannot sail swiftly enough to reach the North Pole before the water freezes, and once the polar seas have frozen, winter passes too slowly for them to wait for the ice to melt. Though they can cover themselves with furs and make fires to protect themselves from the cold, and in this way change the medial conditions immediately around their bodies, these measures have no power to grasp and affect the fundamental dynamics of the medium itself (the vast stretches of air and frozen water around and beneath them). *Cultural* media, by contrast—the media Edmund Burke had described as a kind of "second nature"—are those in which the fundamental dynamics *can* be partially grasped by most individuals (they are on the border of too-fast- or too-slowness).[74] Cultural media include those narrow paths that humans create in the midst of natural media (e.g., canals that facilitate transportation within a country or the shipping lanes across the seas that connect different countries), institutions such as modes of government (e.g., monarchy, democracy, etc.), and affective fields (e.g., the field of sympathy within which the creature seeks to find his place). In *Frankenstein*, what Clerval describes as "the progress of European colonization and trade" may move slowly, but it neverthe-less occurs at a rate that Clerval can imagine "material[ly] assist[ing]"; by the same token, cultural prejudices act just slightly too quickly to prevent Justine's ex-ecution.[75] Finally, *communicational* media, such as speech and writing, are those that can be directly grasped by most individuals. Victor can grasp and respond immediately to the creature's speech on the Sea of Ice, and Robert can respond to the creature's speech at the end of the novel.

Yet by emphasizing ontological differences among the novel's central charac-ters, Shelley also points to the fact that what counts as a communicational media for one individual can function as a cultural or even natural medium for another. The creature, for example, grasps speech but is always outpaced by writing, while Victor grasps writing but cannot keep up with his creature in (what is for Vic-tor) an inhospitable Sea of Ice.[76] Media differences are not absolute; rather, they mark relative relationships of speed and grasp, and a given medium solicits each individual in a slightly different way. These differences need not prevent us from making generalities: both Victor and Walton, for example, respond to, and grasp, writing in more or less the same way. However, Shelley's novel suggests that such generalities represent not a "proper" or ideal response but only a dense grouping of individual responses that also shade off into "outliers" that fall far beyond the norm (e.g., the creature's response to writing).

Understanding media differences in terms of speed and grasp also makes it pos-sible to contextualize the central claims of the other, official Romantic-era media narrative. From this perspective, Hegel's understanding of language as the most fluid medium, the medium in which absolute organic unity became possible, was the result of misattributing to one medium (writing) a dynamic that in fact depended upon *differences between* media. Writing may indeed have facilitated fluidity in the Romantic era, but this was a consequence of the fact that—as both Foucault and Kittler have demonstrated—it enabled new relationships of speed between different social institutions. The feeling of fluidity that underwrote Hegel's commitment to philosophical language as the most elevated medium was really a feeling of intensive *difference*, a feeling of acceleration, that writing facilitated among institutions and between institutions and nature. However, as Schelling emphasized, this feeling of fluidity does not come to an end but rather persists, precisely because non-organic difference itself persists. From this per-spective, earlier media forms are not sublated by later forms but rather persist as points of potential differentiation and tension. Though Victor's embodied cre-ation comes to self-awareness through belles letters, it is precisely his biological embodiment that allows him to outpace Victor, and though Walton comes to wisdom through the medium of the epistle, these communications nevertheless must lag behind the "old" media technology of the ship.

NOVELS, DEVIATION, AND SPECIES

Yet why, we might ask, should this second, unofficial, approach to the connections between fluidity, media, and life emerge with especial force in a *novel*, especially since the unofficial account of media had such a troubled fate in philosophy (many of Schelling's texts remained in draft form) and in biology (Geoffroy "lost" to Cuvier, at least in the short term)? This is especially puzzling if we consider that the early-nineteenth-century novel, unlike philosophical literary forms such as Lamarck's zoological philosophy or Hegel's philosophical encyclopedism, was certainly under no obligation to give an account of itself. As an explicitly fictional form, the early-nineteenth-century novel was not required to explain either its real conditions of possibility—that is, an individual novel did not need to account for the natural processes that had led humans to the form of the novel—or its telos (that for the sake of which it existed). A novelist was, of course, always free to make claims about the conditions of possibility and telos of novels, and Percy Shelley's original preface to *Frankenstein* arguably attempted to address both of these issues by suggesting that the novel's purpose was to inculcate domestic virtues. Yet contra the biological, medical, and philosophical genres I have discussed, such claims neither were necessary elements of the early-nineteenth-century novel nor, even when they appeared in a preface, could have the weight of claims presented in a "realistic" genre such as biology, medicine, or philosophy. Nor have early literary critics detected much correspondence between the plots and characters of late-eighteenth- or early-nineteenth-century novels and the vital aspects of media and milieux that intrigued Schelling and Geoffroy. If anything, literary histories have tended to emphasize the parallels between Hegel's official conception of media and Romantic-era creative literature by pointing to the correspondences between the Hegelian model of organic development and the *Bildungsroman*, Wordsworth's *Prelude*, and Goethe's *Faust*.[77] If, as these accounts suggest, Romantic-era creative literature was more generally focused on organic models of character development, how then can we understand the fact that at least one novel also served as a medium for expressing the unofficial conception of media—that is, media as a fluid space of vital variation?

As a first step toward an answer, it is worth noting that in the eighteenth century, the form of the novel was itself sometimes presented as a milieu for generating madness, understood as deviation from a proper state. Foucault notes that for eighteenth-century critics of the novel, this literary form was understood as establishing an "artificial milieu" that was especially "dangerous to a disordered sensibility" (and thus, especially dangerous for women). The novel "constitut[ed] a milieu of perversion," for "it detaches the soul from all that is immediate and natural in feeling and leads it into an imaginary world of sentiments violent in proportion to their unreality, and less controlled by the gentle laws of nature."[78] Novels solicited the interests of their readers, thereby—in Lamarckian fashion— altering the physiology of these readers, leading to illness and madness. From this perspective, Shelley's emphasis in *Frankenstein* on the disordering effects of fluid media would be an act of willful reappropriation, of folding into the content of her novel earlier concerns about this literary form.

The publishing context of the early-nineteenth-century novel also encouraged reflections on the ways that novels could be understood in terms of species and variations. This period saw massive growth in the number of novels published, increasing anonymity of the consumers of novels, and a consolidation of the review system for assessing novels. These three factors encouraged reviewers to describe novels in terms of genres or "species"; as Walter Scott noted in his anonymous review of *Frankenstein*, "This is a novel, or more properly a romantic fiction, of a nature so peculiar, that we ought to describe the species before attempting any account of the individual production."[79] Thinking novels in terms of species and genres enabled reviewers (and presumably, by extension, readers) to identify and thus assess the criteria that ought to guide a reading and assessment of a particular novel. Many contemporary reviewers, for example, understood *Frankenstein* as an example of the "Godwinian" novel established by Mary Shelley's father, William Godwin.[80]

Yet as *Frankenstein* demonstrated for these same reviewers, one could take neither a Cuvierian nor even quite a Lamarckian approach to "species" of novels, for they exhibited neither stability nor an ascent to perfection. Instead, one had to think species of novels as a field of variations, some of which might result in new literary species. Thus, even as many reviewers of *Frankenstein* felt certain that they had read a Godwinian novel, they were also aware that this particular text represented a *new* kind of reading experience. Some sought to understand this novelty in terms of monstrosity, attributing the peculiarity of *Frankenstein* to its hyperbolic stretching of existing literary forms. The reviewer for *The Edinburgh Magazine and Literary Miscellany* began by claiming that *Frankenstein* represented "one of the productions of the modern school in its highest style of caricature and exaggeration," going beyond even the Godwinian novel, to which it was formally related, and described the central premise of the novel as one of "those monstrous conceptions" produced by "the wild and irregular theories of the age."[81] Others suggested that *Frankenstein* was a new species of literary form, one that (according to Walter Scott) "excites new reflections and untried sources of emotion" and thus "enlarge[s] the sphere" of the "fascinating enjoyment" of reading novels.[82]

Whether they understood *Frankenstein* as a monstrous variation of an existing species or as an example of speciation itself, these reviewers implicitly positioned the novel as a medium, a kind of "strange nature" that solicited unprecedented experiences of intensity from readers.[83] Moreover, however much some of Shelley's reviewers wanted to describe the novel as simply an *effect* of its milieu—as a monstrosity, for example, produced by the wildness of the contemporary age— they also implied that this novel in particular (and by extension, novels more generally) had the power to alter those social institutions or dynamics of which it was purportedly only an expression (if only by further facilitating the wildness of those institutions or dynamics). Anticipating twentieth-century accounts that understand novels less as *reflecting* social dynamics than as accelerating or decelerating social change, *Frankenstein*'s first reviewers presented the novel as a vector for the transformation of social institutions.[84]

THE ONTOLOGY OF POPULATION

How might we understand this conception of media, common to Geoffroy, Schelling, and Shelley, and even to some of Shelley's reviewers, as a force that not only solicits variations but also produces new kinds of groups, such as species, by means of these variations? The concept of "population" provides a way of synthesizing these elements, and also of understanding them at a deeper level. Michel Foucault has noted that the term "population" underwent a semantic shift in the Romantic era. Where earlier authors had used the term as the opposite of "depopulation"—that is, population referred to the processes by which a "deserted territory was repopulated after a great disaster, be it an epidemic, war, or food shortage"—by the late eighteenth century, the term had come to provide a conceptual framework for determining specific facts about large collections of people (e.g., all the inhabitants of a country). For Romantic-era authors, a population was "a set of elements in which we can note constants and regularities even in accidents, in which we can identify the universal of desire regularly producing the benefit of all, and with regard to which we can identify a number of modifiable variables on which it depends."[85] To return to the example of Jenner's smallpox vaccination, for example, the concept of a population allowed eighteenth-century physicians to determine a "normal" level of deaths that occurred in a population due to the inoculation procedure itself, to compare that with the death rate due to inoculation for a given subsection of the population (e.g., children), and then to determine how to reduce the larger of those two death rates (61–62).

Formalizing Foucault's description, we can say that in order for any "set of elements" to constitute a population—whether a population of living beings or a population of novels—there must be (1) a *source* of constant variation; (2) a *plastic plane* within which those variations can be embedded and held for some period of time, and which will allow these variations to move in relationship to one another; and (3) a series of *selective forces* that traverse the plane and destroy some, but not all, of those variations. The "vital" concept of media that interested Geoffroy, Schelling, and Shelley names the conjunction of these three principles. That is, the term "medium" indicates a plastic plane able to hold variations long enough for some of these to be destroyed, and some amplified, by selective forces. This means that, rather than thinking the topology of media in terms of channels—as material means for moving thoughts, goods, debt, and so on, from one point to another—we must think it in terms of a sticky surface with two sides, one oriented toward a source of variation and the other toward selective forces. Both the source of variation and the selective forces remain out of reach of human intervention, and thus always seem to come from beyond; they are what produces that sense of "thrownness" with which I began this chapter. In the biological realm, the source of variation is often understood as "chance," while in the cultural realm it might be "inspiration," and the selective forces can be variously understood as originating from "nature," "the market," or some other source. Though the source of variation and the selective forces cannot themselves be influenced, one *can* alter the plastic plane in ways that will take these forces into account.

Conceptualizing media in terms of a population ontology allows us to see that common to Lamarck's and Geoffroy's milieux, Schelling's primal fluid, and Shelley's natural, cultural, and communicational media is the question of *where* and *how* populations of variations come into contact with selective forces. For Lamarck, the term "milieu" denotes not simply a natural element such as water or air, or even those natural elements (or assemblages of natural elements) that allow an *individual* animal to exist. Rather, milieux are those assemblages of natural elements that allow *multitudes* of animals to exist long enough to reproduce and, equally significant, respond differently to changes in the milieux (i.e., to try to "stick" to their changing milieux, to use Canguilhem's term). Geoffroy followed Lamarck's basic approach: milieux were always spaces for multitudes of variations. However, Geoffroy distinguished between the milieux in which the *development of organs* took place and the milieux in which the *matured living being* existed. He sought, moreover, to determine experimentally the precise limits of particular media: around which threshold values, for example, would changes in the temperature of the atmosphere around an egg produce differences in organic development? Schelling's primal fluid was a space for variation as well, one that enabled that "infinite process of formation [*Bildung*]" by which possible variations were explored.

Yet *Frankenstein* went further in its articulation of a population ontology, for Shelley sought to specify not only those plastic planes within which populations of variations encountered selective forces but emphasized as well the ways natural and cultural media each served as sources of variation for the other. As I noted, Walton, Victor, and the creature are not doubles of one another but rather variations, a perspective that allows us to see how their differences are amplified by the cultural media to which each is attracted. If we consider the books that each read in his formative years (or formative months, in the case of the creature), we find that though each was drawn to a "realistic" genre—a genre understood as presenting an account of things as they really are—there were nevertheless significant differences in their responses. Walton was drawn to travel narratives, Victor read books of magic, while the creature read what he took to be "histories," and the slight differences between these realistic genres encouraged very different personal trajectories.[86] For example, the totalizing ambitions of the epic—that is, its claim to represent all possible positions in the social milieu—played for the creature a pivotal role, for, not having found anyone in Milton's epic like himself, the creature felt banished from *all* social milieux and thus forced to become a part of the natural milieu.[87] In order to have understood his differences from others in a way that would have allowed him to find a place *within* social milieux, the creature would have had to have read a literary form that emphasized populations rather than totalizing hierarchies of existence; he would have had to have read novels rather than histories.[88]

From the perspective of the population ontology that underwrites the vitalist concept of media, that media often enable *communication* between different geographic points turns out to be a secondary issue. Or, more precisely, communication between different geographic points must be understood as a way, though not the only way, populations of variations can emerge. This point was arguably more evident to eighteenth-century authors than it is in contemporary discussions of mass media, for as I noted at the start of this chapter, authors such as George Adams, Adam Smith, and William Godwin explicitly connected the communicational aspect of media such as canals, money, or printed texts to fields of difference: global fields of "invention," in Adams's case; transcolonial fields of fluctuating debt and credit, for Smith; and international "harvests of virtue" in Godwin's terminology. For these authors, a communicational medium was a means of establishing a plastic plane sufficiently large that populations of variations (e.g., a multitude of attempted "inventions," in Adams's case) could be captured and exposed to selective forces.

POPULATION, COMMUNICATION, MEDIATION

Kittler's late work on the ontology of communicational media arguably gestures toward this more fundamental understanding of media in terms of population. His starting point is that communication cannot be understood in terms of a traditional philosophical ontology of "form" that is imposed on "matter," since such an ontology always misses the essential *distance* upon which communicational media depend; there is only a need for communication when distances are involved.[89] In place of the ontology of form and matter, Kittler proposes that we conceptualize communicational media in terms of an ontology of *encoding*, *transmission*, and *storage*: that is, the creation and holding in place of a field of addresses, between the points of which commands and data circulate in feedback loops. Kittler's concept of "storage" partially captures what I have described as the plasticity of the medial plane, while the concept of "encoding" partially addresses the fact of selective forces (i.e., communicational media store only what they can encode, and encoding procedures always frame what is stored in certain ways). Yet however much Kittler's ontology of encoding, transmission, and storage represents an improvement over the form-and-matter schema, his ontology is ultimately unable to think either the importance of variation or the fact of populations. Kittler simply takes for granted that there are sources of variation that produce different instances in a given communicational media—his Romantic-era discourse network, for example, consists not of many copies of one single book, but many copies of many different books—and as a consequence he fails to recognize or account for the *form* of those differences (namely, populations).

Hegel's concept of "mediation" is arguably even less able to take into account the population ontology that enables media. As I noted above, Hegel suggested that the concept of mediation had two advantages over the concept of media. First, the concept of mediation does not mistakenly substantialize, or make a "thing," out of media but instead emphasizes the *processes*—the turnings and movement, which depend upon what Kittler calls "distance, absence, and nihilation" (25)—that media enable. Second, the concept of mediation keeps in view the *discursive* aspect of all media and mediation: that every claim about an

instance of mediation, whether such an instance concerns nonhuman living beings or human social processes, is always made *by* a subject *for* the sake of some
telos or end. For Hegel, all observations about specific mediations were made by
the absolute Subject (Spirit) for the sake of Spirit's recognition of itself as all that
is, including those "things" that formerly seemed foreign to it. (For dialectical
materialists, by contrast, claims about mediation are made by human observers
for the sake of overcoming class conflict.) Yet insofar as the concept of mediation
seems inseparable from the concept of a single, unified subject—whether that of
absolute idealism's Spirit or historical materialism's "species being"—it is not at
all clear how a concept of mediation can deal with the fact of *population*, which
cannot be sublated into a unity or a subject. It is true, of course, that a claim about
population is still a claim, and one made for a discursive end. However, a "self-
conscious" claim about population emphasizes that there can be no final word
on populations, and the telos of such a claim is to enable further claims about
populations.

From this perspective, the way to avoid substantializing media lies not through
the concept of mediation but rather through the vitalist concept of media I have
outlined in this chapter. The vitalist concept of media also emphasizes that media
are not things but processes, but it further specifies the nature of these processes
through the notion of plastic planes that allow sources of variation to come into
contact with selective forces. The vitalist concept of media does not lose sight
of the discursive nature of concepts, including the concepts of "media," "life,"
and "populations." However, because it begins from the premise of a population,
rather than an unified meta-subject, the vitalist concept of media understands
concepts as means by which individuals change (often unintentionally) the speed
of relations among the elements of multiple kinds of media (natural, cultural, and
communicational).

MEDIA, TIME, AND HISTORY

The fact that we can find both a dominant, official Romantic-era media narrative *and* a series of unofficial variant theories means that explaining media in the
Romantic era requires an approach that recognizes the space of variation within
which these series of theories of media proliferated. This means that we cannot
rely on the official Romantic-era media narrative for our explanation, for such
an account only understands variations as means by which a narrative moves
toward a complete and final conclusion. It is only by adopting a more vitalistic
approach to media—that is, approaching media in terms of a non-organic generativity, rather than either transmission of form or organic unity—that we can

understand the multiplicity of Romantic-era theories of media. Both the content and reception of Shelley's *Frankenstein* indicate that we must turn to a population ontology to understand vital media: that is, we must see them as plastic planes that enable populations of variations and expose them to selective forces. Though some of those who predicated their work on a population ontology—for example, Lamarck—nevertheless continued to think of selective forces as producing "progress" and "ascent," Shelley's novel stresses that a population ontology demands a much more complicated conception of the temporality of media. Media are not that which allow an atavistic past to be sublated into the present but rather that which confuse temporal directionality by suspending temporal determinations. Media allow what seemed "past" to be perpetually folded back into the present—a re-layering of past, present, and future that is emblemized in Victor's ability to arrive at the future of science by drawing on its magical "past."

This vitalist approach to media also allows us to reconceptualize our understanding of the concepts of vitality and generativity in the history of nineteenth-century science and to question especially our tendency to chart a path from eighteenth-century mechanism to a Romantic vitalist reaction and then back to mechanism. In place of such a narrative, we can observe that the figure of a primal generative fluidity was used by *both* "mechanists" and "vitalists" in the eighteenth and nineteenth centuries. The premise of primal fluidity, in other words, was not tied to a specific "-ism"; rather; it was productive whenever the thought of origin was linked to time scales beyond those of human experience. In the eighteenth century, for example, Immanuel Kant, William Herschel, and Pierre Simon Laplace developed early versions of the "nebular hypothesis," intended to provide a mechanical account of the origin of galaxies from swirling clouds of matter. This theory gained widespread currency in Britain beginning in the 1830s, and often has been understood by historians of biology as a conceptual resource that supported belief in biological evolution by drawing on the authority of astronomy.[90] However, as Simon Schaffer has noted, there was not one nebular hypothesis but a series of such hypotheses that served multiple, and sometimes competing, purposes in Britain, including (1) relatively narrow and discipline-specific "astronomical accounts of the construction of the nebulae"; (2) "cosmogonic stories about the origin of the Solar System . . . [which could be used] to illustrate the compatibility of divine creation and supposed natural laws"; and (3) "general models of the universal progressive development shown in the heavens, on Earth and in human society."[91] Schaffer is interested primarily in describing how competing social groups exploited the nebular hypothesis for their own ends, but his account also thereby ends up highlighting the social conditions of possibility—namely, competition between different social groups—that helped to produce a multitude of theories of a primal generative fluid.

A vitalist approach to media also illuminates for us the uncanny similarity between Hegel's Romantic-era and Kittler's information-age understandings of media. Kittler is as committed to the premise of the univocity of media as was Hegel, for neither can imagine media as spaces of variation. For Hegel, the limitations of all past media have been gathered up and overcome by the fluidity of philosophical discourse, and the fluidity of the latter is a consequence of its ability to cleave to only that which was spiritual in all previous media forms. Kittler is no Hegelian, for he emphasizes the *dependency* of discourse and thought on the materiality of the dominant communicational media. Yet his account is nevertheless as univocal as Hegel's, for precisely because he understands media solely in terms of communication, his account ends up seeking to demonstrate the regimented unity, the lockstep, that the materiality of media entails. Thus, though Kittler intends to demonstrate the extent to which *media* produce the experience of thrownness—the Kittlerian subject can never catch up to itself, for its ends and means are always determined by the communicational circuits that enable subjectivity itself—he understands that thrownness in rigidly linear terms. Where Hegel sought to overcome thrownness by employing the fluidity of media as a means of getting behind the subject, revealing the subject both as what is thrown and what throws itself, Kittler insists that though the subject is indeed part of a circuit, it can never move swiftly enough to catch up to what throws it.

The alternate media narrative I have plotted here approaches the thrownness that media entails quite differently. Rather than understanding media through the figures of communication and transmission, Schelling, Geoffroy, and Shelley saw media through the figure of life, as generative matrices. As both Geoffroy's and Shelley's works make clear, the figure of life was not equivalent to the figure of the organism; rather, life is what produces variations. Living one's thrownness, as a consequence, was not a matter of expressing an essence, of seeking to overcome the condition of living in medias res by reaching the standpoint of static eternity; instead, it was a matter of functioning oneself as a medium by producing tensions and differentials between other people, other institutions, and other forms of life.[92]

Finally, while we can agree with the now commonplace claim that the ability to theorize media depended historically upon the fact of at least *two* dominant media—that is, it was only the difference between two media that could make the fact of mediation phenomenologically available—the vital conception of media I have outlined forces us to understand this claim quite differently from how it is usually intended. Most accounts of the emergence of the media concept point to the existence in the late nineteenth century of multiple *communicational* media, such as photography, film, and gramophone, which collectively—because of their technical differences from print—emphasized the fact of multiple, fundamentally incompatible media.[93] Yet my analysis suggests that the concept of media emerged not because of the technical differences between multiple forms of communicational media but rather because of the more fundamental difference in ontological assumptions between vital and communicational media. In other words, one needs the difference between at least two media in order to think the concept of media, but these two media must be of different orders or registers. The difference that makes a difference is not that between two kinds of communicational media (e.g., the difference between print and film) but that between communicational *and* vital media, for it is only from the perspective of the latter kind of medium that the differences between media are possible and can be thought.

CHAPTER 6

Cryptogamia

The whole plant realm will . . . appear to us as like a vast sea, as
necessary for the contingent existence of the insect as the actual seas
and rivers to the contingent existence of fish.

Goethe, "On Morphology," 84

For better or for worse, Romanticism has come to be associated with the love of
plants. Nor need one look hard for examples to justify such an association: William
Wordsworth's claim that his heart "dances with the daffodils"; the German
romantic fascination with the unattainable "blue flower"; J. W. Goethe's researches into the metamorphoses of plants; the articulate Lilly of William Blake's
The Book of Thel; the trembling and panting mimosa of P. B. Shelley's "The
Sensitive-Plant"—these are just a few instances of the dense vegetative network
that sprouts throughout Romantic literature. The Romantics' exuberance for
plants has often struck later poets and critics as a naïve, and even embarrassing,
enthusiasm, an affection bordering on affectation that (depending on the critic)
either must be excused or ought to be mocked as overwrought pathos. Even a generation or two ago, a scholar of Romanticism such as Robert Maniquis was willing
to concede that "the Romantics sometimes loaded too many overwhelming feelings on defenseless flowers," and this charge could no doubt also be extended to
all Romantic-era representations of vegetation (154).

Ecocriticism, to its credit, was the first school to recognize this fascination with
plants as an *event*, as the emergence of something new within literature and, as
a consequence, the emergence of a new function for literature. The Romantics
were not, of course, the first to represent plants in poems and stories, nor were
they even the first to link plants in creative literature to botanical research. But
in place of what we might call the agricultural emphasis of earlier georgic and
pastoral poetry—an understanding of plants, that is, as existing fundamentally

for the sake of human ends—the Romantics were the first to link representations of plants to what Karl Kroeber has described as a "new biological, materialistic understanding of humanity's place in the natural cosmos."[1] Rather than functioning as symbols for human concerns and ends, Romantic plants are for ecocritics pathways leading readers to an awareness of their embeddedness within larger ecological and cosmic processes; by stressing the cyclical links between plants and clouds, for example, Romantic poems promote a "proto-ecological vision of the hydrological cycle."[2] From this perspective, if some readers have found the Romantic love of plants excessive and embarrassing, it is only because they have not fully assimilated that new ecological sense of propriety to which the Romantics committed themselves.

Without contesting the fact that the Romantic turn to plants was an event within (and for) literature, this chapter will nevertheless take a slightly different approach to Romantic flora. Like all obsessions, the Romantic love of plants was ambivalent, grounded as much in a sense of vertiginous loss as of devotion and respect. Plants did not serve Romantic authors solely as means of ascent into an enlightened understanding of the beauty and harmony of the cosmos; they drew them as well into darker and more cryptic forms of becoming. Thus, alongside the dancing daffodils and panting mimosas were Romantic images of a vegetative abyss: the speargrass and bindweed that silently and indifferently erase all traces of Margaret in Wordsworth's *The Ruined Cottage*; the basil plant of John Keats's *Isabella; or, the Pot of Basil*, which derives its sustenance from the decapitated head of Lorenzo buried in the pot; and the thistles, fungi, and hemlock of Shelley's "The Sensitive-Plant," which, as "forms of living death," displace and kill both the mimosa and small mammals.[3]

As these latter examples suggest, at stake in the Romantic love of plants was something more than simply an eco-friendly reverence for trees, flowers, and shrubs. The Romantic relationship to plants was not simply one of "respect," with its implications of propriety and distance, but rather a vertiginous falling-for the strange and dark life of vegetation. This seduction of the human by vegetable vitality is one aspect of what I call in this chapter "cryptogamia." In the Linnaean taxonomical system, "cryptogams" were those plants with hidden reproductive organs, but I exploit here the Greek roots of this term (*crypto* = hidden + *gamia* = generation) to index and amplify the Romantic fascination with the strange life of plants.[4] Eighteenth- and early-nineteenth-century botanists and chemists were aware, in ways earlier researchers had not been, that plants live and grow quite differently from animals, and it was this difference that solicited Romantic desire and attention. Romantic cryptogamia was thus not simply a one-way twining of the plant around the human but also an embrace of the plant by the human, a joint seduction that produced what Shelley called, in "The Sensitive-Plant," a "mutual atmosphere" between species (something, as we shall see, quite different from an "environment"). Understanding cryptogamia requires that we go beyond seeing Romantic plant poems simply as ecological fables (though they may also be that) and recognize the function of aesthetic judgments in facilitating the strange hybridizations of plants and humans that mutual atmospheres enable. Recognizing the ecstatic nature of aesthetic judgments—their tendency to produce transformation and change—will allow us to see that although Romantic cryptogamia is consilient with many aspirations and practices of ecology that claim a Romantic heritage, it is at the same time based on premises of mutability and becoming fundamentally at odds with the ecological aim of (re)establishing harmonious and reverent relationships with nature.

THE STRANGE LIFE OF PLANTS

Scholars of Romanticism have tended to focus on late-eighteenth- and early-nineteenth-century interest in *similarities* between plants and animals, emphasizing, for example, claims that plants, like animals, had sexual lives and that both animals and plants had feelings and perceived their environments.[5] As historians of botany have noted, an interest in the "analogies" between plants and animals was indeed a central part of eighteenth-century approaches to plants, but this was in part a consequence of a mechanistic understanding of the universe. As historian of science François Delaporte notes, "the concepts and methods of animal studies" could be applied to the study of plants only after "plant life had [been] given the same theoretical status as animal life"; hence, in "order to analyze plants" in the eighteenth century, "mechanistic assumptions were necessary."[6] Where the Aristotelian tradition had understood plants as *lacking* many of the capacities of animals—because plants could not move, and hence pursue either food or mates, Aristotle argued that plants lacked both "sensitivity" and "sex"—the premise that plants and animals were both mechanisms encouraged eighteenth-century natural historians and philosophers to seek plant analogues for all animal organs and functions, including circulatory systems, digestive apparatuses, and sexual organs.[7] The Linnaean "sexual system" of plant identification is one of the most well-known consequences of this analogical approach to plants—and Erasmus Darwin's *The Loves of Plants* (1789) one of the best-known literary applications of the Linnaean approach—and it is fair to say that a sense of fundamental similarity between plants and animals guided much eighteenth-century work on plants.

Yet however much an analogical approach may have guided eighteenth-century approaches to plants and animals, this sense of a fundamental similarity of plants and animals began to fracture in the Romantic era, for at each point of seeming analogy with animals, the uncanniness of plant life become more and more evident.[8] Consider a function such as "respiration," which eighteenth-century researchers such as Joseph Priestley had established occurred in both plants and animals.[9] Initially, plant and animal respiration seemed not only analogous but even complementary: animals inhaled in order to gain "dephlogisticated air" (oxygen) and exhaled "fixed air" (carbon dioxide), while plants inhaled fixed air and exhaled dephlogisticated air. This suggested a virtuous and harmonious dependence of each mode of life upon the other: at the level of respiration at least, animals and plants seemed mirrors of one another. Yet this apparent symmetry was broken by the discovery of the relationship of vegetable respiration to the sun, for as Jan Ingenhousz established in his *Experiments upon Vegetables* (1779), plant respiration depended upon solar light in some peculiar and uncertain way.[10] Plants were *not* simply inverted animals, for they "fed" from, and had access to, the sun in a way that remained fundamentally baffling until nearly the end of the nineteenth century.[11]

The differences between plants and animals that Romantic-era researchers discovered were so fundamental, in fact, that they vexed attempts to develop definitions of life that could include both of these kinds of vitality. Ancient theories had simply embedded the vegetable within the animal: for Aristotle, the nutritive or "vegetable" soul was the least developed of the three souls of living beings, concerned only with growth and nutrition, and from this perspective, vegetables were just simplified animals.[12] That animals, in general, moved themselves around, while plants tended to stay put, was not a problem for the Aristotelian theory of life. Though Aristotle defined life in terms of "movement," he understood that latter term not simply as a synonym for mobility but rather as a concept that also included phenomena such as circulation of fluids and the growth from seed to fully developed organism. However, by the late eighteenth century, theorists had come less and less to discuss life in terms of a tripartite soul and definitions of "movement" had become less capacious. As a consequence, it became increasingly difficult to fit plants into definitions of vitality that almost invariably took animal life as paradigmatic. Some eighteenth-century authors, taking it as axiomatic that since all living beings moved, there must be some kind of "spontaneous motion" within plants, found themselves forced to explain the movement of sap within a plant as a kind of "circulation" and to describe the tropism of a plant toward light as an intentional action.[13] Other authors, though, pointed out that fluids within plants did not truly "circulate" but simply moved up or down in a quasi-hydraulic fashion. They argued as well that plant tropisms could be explained as the effect of mechanical causes, which in turn implied that plants were in fact "destitute of sense and spontaneous motion."[14] While no one doubted *that* plants lived, it had become increasingly evident to late-eighteenth-century researchers that vegetable life differed fundamentally from that of animals.

Romantics authors thus had many opportunities to be aware, like no generation before them, of the essential *strangeness* of plants—the uncanny fact that though both plants and animals lived, they did so in fundamentally different ways—and it was this sense of plants as instantiating an elusive form of vitality that, I suggest, most fundamentally characterizes the interest of Romantic-era authors in vegetation.[15] It was clear, by the Romantic era, that the botanical, natural-historical, and chemical sciences could indeed make significant progress in their analyses of plant physiology and reproduction. It had become equally clear, however, that the results of these investigations were not always—or even often—likely to fit comfortably into categories and schemas developed to explain animal physiology. Though it might have seemed at the start of the eighteenth century that plant life could be understood as a less complex version of the mechanics of animal life, by the start of the nineteenth century the study of plants emphasized the need for reflection on the concept of "life" itself, since animal life could no longer be taken, at least unproblematically, as the implicit model for vitality. This was, in one sense, a situation of epistemological uncertainty, an uncertainty about which concepts and approaches were most likely to produce knowledge and understanding about the life of plants. Yet it was also, and arguably more fundamentally, a situation that seemed to demand scientific innovation, in both theoretical and practical registers, as well as in the development of a "sense for the unknown" that could guide such innovation.[16] Because the study of living beings could no longer rely solely on concepts and methods developed in the study of animal life, the development of new concepts and methods would require attunement to the *strangeness* of life. This need to attune oneself to the strangeness and hiddenness of life—in part by becoming receptive to what I am calling cryptogamia, or the differences between plant and animal life—was not precisely a part of science but rather something like an extra- or pre-scientific affective comportment—a generalized mood (*Stimmung*) rather than a specialized "thought-style" (*Denkstil*), to draw on Ludwik Fleck's terminology.[17] In this sense, what I call Romantic cryptogamia should not be understood as something that flowed *from* the sciences *to* literature and philosophy; rather, cryptogamia names a more widespread effort in the Romantic era to use the results of the sciences to amplify an initial, prescientific sense of the strangeness of life.

This perspective allows us to make sense of the fact that even in their most positive, laudatory moments, Romantic-era literary representations of plants almost invariably emphasized the peculiarity of plant life. Poets, for example, stressed the extent to which vegetable growth and reproduction occurred at time scales almost unimaginable for humans. The yew tree of Lorton Vale in William Wordsworth's "Yew Trees" grows so slowly that it escapes death; it is "a living thing/Produced too slowly ever to decay."[18] The life-span of a flower, by contrast, is so fleeting that it disturbs our customary parameters of permanence and stability: the fact that the "flower that smiles today/Tomorrow dies" produces vertigo in the narrator of Shelley's poem.[19] And for John Clare, the fearful "darkness" of creeping ivy sprouts in part from its strange, looping temporality, exemplified by its ability to grow upon its own dead stalks:

> DARK creeping Ivy, with thy berries brown,
>   That fondly twists on ruins all thine own,
> Old spire-points studding with a leafy crown
>   Which every minute threatens to dethrone;
> With fearful eye I view thy height sublime,
>   And oft with quicker step retreat from thence
> Where thou, in weak defiance, striv'st with Time,
>   And holdst his weapons in a dread suspense.[20]

Like animals, plants live and die, but they live and die in ways that are alien from a human perspective. This haunting and haunted life of vegetation is often exemplified, as in Clare's lines, as a slow torque or twining: the fond twisting of living ivy around its own ruins; the bindweed that surrounds and drags the rose down to earth in *The Ruined Cottage*; or the "serpentine/Upcoiling, and inveterately convolved" trunks of Wordsworth's yew trees, which upcoil precisely by drawing an animal image ("serpentine") down into the deep time of trees.[21]

For Romantic-era authors, this alien life of plants also harbored a dark, explosive generativity. Early-nineteenth-century scientific discussions of plants stressed the disseminative powers of vegetation: "So great are the prolific powers of the vegetable kingdom," one author wrote, "that a single plant of almost any kind, if left to itself, would, in a short time overrun the whole world."[22] This colonial tendency of plants sprouts up in Romantic literary representations, for example in the potential infinity of Wordsworth's famous "host of golden daffodils," which "stretched in a never-ending line." If Wordsworth's popular poem on the daffodils nevertheless reads as a happy and safe verse, it is both because the imperial ambitions of this golden host are distanced from the narrator by means of memory (the poem could be written and later read only because this virtual vegetable imperialism witnessed in the past did not become actual) and because, at the same time, the narrator entwines his own pleasure into this self-generating vegetative line (the narrator can imagine a never-ending series of returns to his memory of the flowers, and the reader can imagine likewise a never-ending series of returns to this poem).[23] Shelley, too, sought to entwine his verse into the peculiarities of vegetatable generativity, positioning poetry itself as a mode of language that can, or ought to, aspire to plantlike modes of dissemination. In Shelley's "Mont Blanc," the narrator's fascination with the suspended animation of plant seeds, that "trance" through which "every future leaf and flower" escapes the ravages of cold, becomes a figure for the ability of the poem to awaken dormant life, while his "Ode to the West Wind" is premised on the hope that this poem—and poetry more generally—might scatter like winged seeds on the wind, remaining potent for the future even if ignored in the present.[24]

Because Romantic authors attended so closely to the strangeness of vegetable vitality, their representations of plants generally functioned less through an associational than through a deictic logic. The Romantics had inherited a long tradition of associations between plants and human qualities from both classical and biblical traditions. The Bible established strong associations between the Virgin Mary and plants such as ambrosia, lily, rose of Sharon, and myrrh, and by at least the sixteenth century, many other plants had come to have specific Christian "meanings": red carnations stood for Christ's crucifixion (as well, coincidently, for betrothal); the iris for loyalty; the orange blossom for fruitfulness.[25] Yet even as these traditional associations continued to function in the eighteenth and nineteenth centuries, Romantic-era representations of plants sought to escape allegoric values in favor of a deictic pointing outward toward literal plants. John Clare's bean plants, crab-trees, sycamores, and red clover; Wordsworth's gorse, gooseberry trees, bindweed, and daffodils; Shelley's thistles, nettles, and darnels: these are not, at least initially, allegories to be decoded but rather deictic references to literal plant life. In some cases, this pointing is quite specific, such as in the case of Wordsworth's yew tree of Lorton Vale. However, it is more frequently a gesture toward an instance of a genus: the red clover blossom John Clare addressed in his sonnet, dated 1821, and about which we now read in the twenty-first century, is treated grammatically as the same red clover that Clare saw much earlier: "I bend musing o'er thy ruddy pride;/Recalling days when, dropt upon a hill,/I cut my oaten trumpets by thy side."[26]

This possibility of treating multiple instances of red clover as all the same is grounded neither in a form of Platonism nor in the use of plants as allegories but rather in the strange reproductive logic of plants, a reproductive logic that—as both Goethe and Hegel noted in their own ways—makes it difficult to speak of plants as distinct individuals. Hegel described plants as like geometric points (*Pünkte*) and thus as lacking any interiority, a description that emphasized the difficulty in understanding plants as individuals, at least in the sense that we habitually apply that term to humans or animals.[27] Though the specific red clover plant that Clare observed as a child is indeed different than the instance to which he much later dedicated his sonnet, the peculiar nature of plant reproduction nevertheless makes it feel as though these two instances of the plant are in some sense the same.[28] To point at the red clover is thus not to point at a specific instance of a plant but rather at the strange reproductive logic of plants.

The deictic function of Romantic representations of plants did not, of course, prevent the latter from assuming either new or traditional symbolic associations. The vertiginous temporality of Wordsworth's yew tree of Lorton Vale could indeed come to stand for something else. Tim Fulford notes that Wordsworth's poem turns "the local tree into a patriotic symbol, into a guarantor of the deep-rooted and thus enduring strength of Englishness."[29] Yet an actual tree can come to stand for Englishness only if the poem can convince readers that there is something common to both. No doubt part of the work of the poem is to transfer the status of reality from a living tree to the much more vague and illusive referent, "Englishness." But this transfer will work only if there is a vital middle term, a sense that the slow life of the tree is somehow like the liminal life of "peoples" (that is, human groupings that extend far beyond one's immediate relatives). As both Fulford's and Geoffrey H. Hartman's interpretations of the poem emphasize, this middle term is hard to think, and there is a tendency to drift toward anthropomorphized and "magical" figures of half life, such as ghosts and ghostliness.[30] Yet the poem roots us in a figure of full life, albeit one that embodies a mode of vitality quite different from that with which we are familiar. It is by pointing toward the life of real trees, rather than referring intertextually to preestablished meanings or sending us into the realm of the miraculous, that the poem solicits a sense of a disseminative cryptogamia that traverses both the worlds of plant and humans.

PLANT PEDAGOGY

Because Romantic-era literary representations of plants emphasize a deictic over an associational logic—because they point at the characteristics of real plants rather than at other texts—they also provide us an occasion for thinking anew about how literary "pointing" works in the first place. For the Romantics, such pointing was understood as fundamentally a mode of teaching: to attend closely to Romantic literary representations of plants (which representations were them-selves attentive to the strange vitality of plants) was to put oneself in a position to learn something. The pedlar's tale of the speargrass and bindweed that erase all traces of Margaret teaches the narrator of *The Ruined Cottage* (and, presumably, the reader) to "trace" that "secret spirit of humanity" (line 503) that purportedly remains in the weeds, while the specific instances of long-lived vegetation in "Yew Trees" are pointed out in lecture-like fashion, the attention-directing locu-tion "worthier still of note" (line 13) drawing the reader from a description of the yew tree of Lorton Vale to that of the four conjoined yew trees of Borrowdale. The representation of the death of the gardener and her garden in the three parts of Shelley's "The Sensitive-Plant" is followed by an explicitly labeled "Conclusion" that draws a lesson from this story, while the articulate Lilly, Cloud, and Clod of Earth in Blake's *The Book of Thel* each have something to teach Thel (and again, presumably, the reader). For the Romantic poets, in short, representations of the excessive, eccentric vitality of vegetation clearly had the power to teach. Yet *how*, precisely, does one learn from such pointing at the plants, and to what does such teaching lead?

Although these examples employ different poetic forms, in each case the peda-gogy of plants operates by means of what the narrator of Wordsworth's *The Ruined Cottage* describes as a *tracing*. In describing Romantic literary representations of plants as tracings, I am again emphasizing the desire to break merely associa-tional meanings of plant imagery by incorporating quasi-scientific details about actual plants into poems. However, I also want to stress that the true object of this tracing operation was cryptogamia, the strange and excessive life of plants. Tracing cryptogamia could mean, in part, establishing linkages between poetic representations and specific, identifiable plants with some peculiar characteristic, such as the long-lived yew trees of Lorton Vale and Borrowdale or the rent and shattered stump of Burthorp Oak (which, as Clare notes in his poem on the tree, nevertheless sprouts new leaves every spring). It could also mean the inclusion in verse of precise and detailed descriptions of patterns of real plant life, such as the quasi-documentary catalog of growth patterns of local plants in the absence of human control that Clare developed in poems such as "Cowper Green" or those details about gorse that have allowed historicist critics to identify fairly precisely the setting of Wordsworth's *The Ruined Cottage*.[31] Yet tracing cryptogamia does not demand such specific geographic referents, for it could also mean engender-ing a sense of the excessive growth of a more abstract collection of plants, such as the expanding line of Wordsworth's daisies.

If the *object* of Romantic tracing is cryptogamia, tracing itself depends, for both the author and the reader of Romantic plant verse, upon cultivating a state suspended between activity and passivity. Tracing is an activity insofar as it is never a matter of simply copying what seems to be given to the senses immediately (for example, by merely describing the physical appearance of a given plant). It is, rather, a matter of searching for that fugitive vitality of plants that escapes understandings of life based on the animal template. Thus, to learn from plants is to trace *traces* rather than simply following clear outlines. (This is, in a sense, the primary lesson of Goethe's *The Metamorphosis of Plants*, which sought to locate the primal generative form—what Goethe called the *Urpflanze* or *Urblatt*—that was nevertheless virtually "present" in all parts of the plant.) Yet insofar as tracing involves attending to a mode of life that is fundamentally other than that of animals, it also necessitates a form of passivity, a submission of one's attention and thought to those "serpentine/Upcoiling, and inveterately convolved" paths that lead away from the associational logic that had previously dominated representations of plants, and away from our sense that life is fully contained in the form of the animal. In Wordsworth's description of the entwined trunks of the Borrowdale yew trees, for example, we move through and beyond an animal form of coiling (the serpentine coil) to more expansive and generalized forms of plant coiling: upcoils that ascend heavenward, rather than, as in the case of the serpent, moving horizontally, and communal, collective coilings (convolutions). The rhythmic aspects of poetry can also become "plant-ified," as assonance and rhyme help to create slow spirals and convolutions of sound that seem equally alien to the intentionality and plot-based movement of animal life.

Precisely because plant tracing leads away from the more familiar world of animal vitality, it cannot be a matter of *mapping* human relationships onto the world of plants. Mapping is grounded in a project of mastery intended to allow a return to the familiar; tracing, by contrast, aspires to something more like the irreversible transformations of apprenticeship (a point to which I return below). Michel Serres's account of the mapping function implicit in the use of animals in literary fables helps to sharpen this point. Serres develops his account of the literary uses of animals in *The Parasite*, a book that—as its title suggests—explores the logic of inequitable relationships, relationships in which one party takes but does not give back, or gives back in a fundamentally different, less costly, register (I accept your food, but give back only stories in return).[32] Serres contends that fables emphasize parasitic relationships, describing, for example, city rats who "host" country rat cousins without cost because both are able to dine on scraps left by the tax collector or vain ravens who drop their food in order to exhibit their shiny black coats to foxes, who in turn steal the neglected food. Yet he also suggests that the *genre* of the fable itself emerged as a way to legitimate parasitic relationships between humans by means of a mapping of human onto nonhuman life:

> Man milks the cow, makes the steer work, makes a roof from the tree; they have all decided who the parasite is. It is man. Everything is born for him, animals and beings . . . Plants and animals are always [man's] hosts; man is always necessarily their guest. Always taking, never giving. He bends the logic of exchange and of giving in his favor when he is dealing with nature as a whole. When he is dealing with his kind, he continues to do so; he wants to be the parasite of man as well. And his kind want to be so too. Hence rivalry. Hence the sudden, explosive perception of animal humanity, hence the world of animals of the fables. (24)

Fables, in Serres's account, are thus solutions to a problem. The problem: though I wish to treat other humans as I treat plants and animals (by taking and not giving back), humans seem to me to be fundamentally different in kind from natural entities, for humans demand recognition and equitable exchange from me. Hence, if I am nevertheless going to establish inequitable, parasitic relationships with other humans, I must have some means for mapping my asymmetrical relationship to other forms of life — I take from plants and animals but do not give them something equal in return — onto human relationships. The solution: the invention of the fable and its talking animals. The fable first maps inequitable relationships onto nonhuman life — foxes exploit the vanity of the crow; lions break their promise to share with less powerful animals; rats feast on leftovers — and then shuttles these inequitable relationships back out into the human world in the form of "universal" morals: vanity will always be exploited; the powerful are laws unto themselves; taking a free meal is dangerous. The fable asserts that these universal morals, exemplified in animal relationships, hold as well in the human world. By clothing its lessons about humans in the skins of animals (and, much less frequently, plants), the fable is able to introduce asymmetries into what would otherwise be the reciprocal field of human relationships, while at the same time disguising its justification of the abuse of humans by humans.

While it is impossible to know whether Serres's narrative of the origin of the fable is "accurate" (whatever that might mean), and while animals appear in many literary genres other than fables, his account nevertheless articulates a basic assumption that underwrites many pedagogic uses of animal imagery in literature: if human relationships can be mapped onto animal relationships, and lessons drawn from this mapping, it is only because animal life is "close enough" to human life that it can bear this cartographic operation. It is only because animals and humans belong to the same mode of vitality, in other words, that humans can take lessons from the animal world (and, as Serres notes, adopt a "parasitic" stance toward the source of the lesson in the sense that one can learn from the animal world without having to give anything back to it).

Yet if, as the Romantics were among the first to recognize, plant life differs fundamentally from animal life, this undoes from the outset the possibility of establishing a cartographic relationship between plant and human modes of vitality.[33] For the Romantics, plants are too close to the elements to engage in those shifts from giver to receiver that Serres suggests characterize human (and animal) life. If, as Hegel contended, plants do not truly eat or drink but simply transform the elements that surround them into nutrient, it is difficult to imagine plants engaging in those quests for sustenance (literal or symbolic) upon which narratives of loss and gain depend.[34] Though plants were for the Romantics far more than affectless movements of matter, they nevertheless did not quite have "drives" (for sex or food, for example); rather, they were impelled by something more like proto-drives.[35] Moreover, plants lack voice and hence remain outside the realms of promises, informational exchange, and deception. Romantic plants may appeal, and even seem to make appeals, but this is not a matter of speaking and listening. The flowers upon Wordsworth's living carpet remain in "mute repose" (line 17), while the narrator of John Clare's "Ballad—A Weedling Wild" acknowledges that any speech attributed to this plant must in fact be intuited from "its silence" (line 12). Romantic plants, in short, made poor vehicles for the mapping and metaphor of human relations. As a consequence, tracing cryptogamia was not a matter of mapping the lives of humans onto those of plants and then drawing off lessons but rather a matter of tracing an eccentric trajectory that leads away from the sureties of animal life.

MUTUAL ATMOSPHERES

Tracing was not the final goal of Romantic plant pedagogy but simply a means for generating a movement toward vegetable vitality. This impetus toward plant life did not mean dispensing with desire, voice, and attention, but it did require that these latter mimic the life of plants by moving closer to the elements. The form of Romantic poetry itself initiated this movement of decomposition by emphasizing nonmeaningful capacities of language, such as rhyme and assonance, upon which the meaning of Romantic poetic sentences depended. However, the goal of this verse was not absolute decomposition of language into elements but rather its structuring, such that it could help establish a sur-elemental "mutual atmosphere" that linked humans and plants.

Shelley's "The Sensitive-Plant" focuses on what it means to generate (and destroy) mutual atmospheres. The poem describes an ornamental garden in which the tended flowers are each "interpenetrated/With the light and the odour its neighbour shed,/Like young lovers whom youth and love make dear/Wrapped and filled by their mutual atmosphere" (1:66–69). Shelley's use of the concept of atmosphere drew obliquely on the late-eighteenth-century discovery that plants "exhale" oxygen and "inhale" carbon dioxide, a fact that encouraged researchers to explore the effects of plants on the global atmosphere, as well as on more local atmospheres. Jan Ingenhousz presented a vivid image of "dephlogisticated air [i.e., oxygen], gushing continually" from the undersides of leaves; since this gas was "inclined to fall . . . downwards," it provided a "beneficial shower for the use of the animals who all breathe in a region of air interior to the leaves of trees."[36] Ingenhousz emphasized, though, that not all vegetative atmospheres were beneficial for animals: in the absence of sunlight, plants generated carbon dioxide rather than oxygen, and thus he had "no doubt but a great quantity of plants, kept in a close and small place during a night, or by day in the dark, may do some material mischief, and even occasion death, to any person who should be imprudent enough to remain in such a place."[37] Plants contribute to the earth's atmosphere, considered as a whole, but they also generate their own local atmospheres, only some of which are beneficial for animals.[38]

However, by thinking the plant's atmosphere in terms of *love*, Shelley underscored that the function of the atmosphere was to entice and solicit the vitality of some agents and discourage that of others. This selective function of an atmosphere is dramatized in the latter part of "The Sensitive-Plant," for after the gardener dies, the atmosphere that her love helped to maintain begins to change. Amid her beautiful plants, thistles, nettles, darnels, dock, henbane, and hemlock begin to spring up. These less visually attractive plants generate their own atmosphere, one that "stifle[s] the air" and sends forth "vapours . . . which have strength to kill" (3:57, 3:75). This new atmosphere encourages a change in the flora of the now untended garden, and "When Winter had gone and Spring came back/The Sensitive-plant was a leafless wreck;/But the mandrakes, and toadstools, and docks, and darnels,/Rose like the dead from their ruined charnels" (3:114–117). The new atmosphere that surrounds the weeds does not kill indiscriminately but rather kills only those plants formerly tended by the gardener.

Like an animal cry, an atmosphere establishes a territory.[39] However, where a cry, as a more or less articulate imperative to stay away or come near, establishes a territory through an appeal to consciousness, an atmosphere establishes a territory at the level of ontology; it affects conditions of the entire possibility of living, not simply sensing or thinking. An atmosphere cannot be kept at a distance, as can a cry or color; to sense an atmosphere is to be within in it. Its closeness—its intimacy—made the atmosphere a good figure for Coleridge when, in *The States-man's Manual*, he sought to direct his readers to the light of God by refracting its luminosity through the effects of vegetative atmospheres. Coleridge first de-scribed how, in the case of a plant, a "touch of light . . . returns an air akin to light, and yet with the same pulse effectuates its own secret growth, still contracting to fix [chemically] what expanding it had refined," and then urged his readers, armed with this botanical knowledge, to become plantlike by allowing the forma-tive light of God into their own atmospheres (72).

Because of its intrinsic intimacy, an atmosphere is difficult to consider as solely, or even primarily, an *aesthetic* phenomenon. This difficulty of thinking at-mosphere in terms of a pure aesthetics is especially evident in the work of Imman-uel Kant, for whom some form of distance was essential to all properly aesthetic phenomena. Kant was suspicious of smells, for example, because, as he noted in *Anthropology*, smelling forces us to "inhale air that is mixed with foreign vapors" (49). In *Critique of Judgment*, Kant criticized the custom of "pull[ing out] a per-fumed handkerchief" to create an atmosphere of sweet-smelling air, noting that such a gesture "gives all those next to and around him a treat whether they want it or not, and compels them, if they want to breathe, to 'enjoy' [*genießen*] [this odor] at the same time" (200). While we might find a particular atmosphere pleasant, our ontological dependence on atmospheres renders them too close for pure per-ceptions of beauty or sublimity.[40] Moreover, as Kant's example of the perfumed handkerchief suggests, atmospheres are invariably collective phenomena, linking multiple people or living beings, and for this reason too, they disable that stepping back that is central to Kant's understanding of aesthetics.

Yet if atmospheres cannot be purely aesthetic phenomena, Shelley's poem shows that aesthetic experiences can themselves become elements of mutual at-mospheres. A mutual atmosphere is not solely composed of gases, for it is the visual and olfactory beauty of the garden flowers that encourages the gardener to care for them. By being beautiful—by looking and smelling in a certain way—these plants "lure" the gardener to care for them. Mutual atmospheres thus en-able double domestications: even as the gardener domesticates the plants, the plants are also domesticating the gardener. Shelley's narrator emphasizes the doubleness of this domestication by transferring qualities suspended between ontology and aesthetics, such as "tremulousness," back and forth between plants and gardener.[41] This quality of being beautiful is not an intrinsic characteristic of the plants but rather one relative to other living beings: the garden plants are beautiful to the gardener, who (as a consequence) cares for them. As soon as the gardener dies, though, this being-beautiful ends, and the mutual atmosphere of the garden plants dissipates as well, to be replaced by another atmosphere.

EXCHANGE AND INVOLUTION

Like atmospheres themselves, the logic of the mutual atmosphere is difficult to grasp and easy to confuse with other, related phenomena. A mutual atmosphere is not equivalent to either a "milieu" or an "environment"; it is distinct as well from "ambience." As I noted in chapter 5, a milieu—at least as Romantic authors such as Jean-Baptiste Lamarck used the term—is a fundamentally *elemental* phenomenon, denoting those flows of gases and fluids within which living beings always and necessarily found themselves. As such, a milieu excludes any kind of mutuality; as Georges Canguilhem put it, living beings find themselves forever trying "to 'stick' to an indifferent *milieu*."[42] Mutual atmospheres, by contrast, are *sur*-elemental; like the plants in Hegel's description, atmospheres build and depend upon elements but are not themselves elemental. Nor is a mutual atmosphere an "environment." This term, which appeared infrequently in the Romantic era, is now generally used to denote the sum of all surrounding conditions that affect a living being.[43] Yet an environment is a set of conditions *external* to living beings themselves, while a mutual atmosphere is something produced from within and between plants and animals. Finally, "ambience"—another word used rarely, if at all, in the Romantic era—is now used to refer to something that is both much more focused on perception, and much more diffuse, than an atmosphere; an ambience is the "feel" of a specific place for an observer, rather than an ontological condition of possibility of life.[44]

The logic of the mutual atmosphere is also difficult to think because the mutual entwining upon which it depends makes it difficult to apply concepts—such as "exploitation" or "mutual exchange"—that assume independent agents. Serres's notion of parasitism initially seems to help explain the generation and destruction of atmospheres in "The Sensitive-Plant," for it suggests that the survival of the garden depends upon a chain of living entities, each parasiting the next. The mimosa parasites the atmosphere of the other plants, giving back "small fruit/Of the love which it felt from the leaf to the root"; the mimosa "received more than all, it loved more than ever, /... could belong to the giver" (1:70–73). The mimosa would from this perspective serve as the titular flower of the poem not because it is the sole parasite—the narrator also refers to other "parasite bowers" within the garden (3:48)—but rather because it serves as a synecdoche for the chain of parasitic relationships that make possible the garden as a whole. Thus, even as the ornamental garden atmosphere is parasited by the mimosa, the garden as a whole parasites the gardener, for she provides the plants with "matter" (water and soil) and receives only aesthetic experience (beauty) in return. However, this apparent exploitation of one element by another only gets us so far. It fails, for example, to address the relative *stability* of the ornamental garden, as it is not this chain of parasitic relationships that undoes the garden but rather an external, unexplained cause: "And ere the first leaf looked brown—she died!" (2:60).

One might be tempted to see the relative stability of the ornamental garden as evidence that exploitation can give way to mutual exchange between quasi-autonomous agents. Serres argues that parasitism can become stabilized by moving toward a "sur-equilibrium":

> Birth of an exchange. The parasite adopts a functional role; the host survives the parasite's abuse of him—he even survives in the literal sense of the word; his life finds a reinforced equilibrium, like a sur-equilibrium. A kind of reversibility is seen on a ground of irreversibility. Use succeeds abuse, and exchange follows use. A contract can be imagined. (168)

From this perspective, the ornamental garden of Shelley's poem, like all instances of domestication, might seem to establish an exchange between human and non-human life: just as livestock "exchange" milk and flesh for protection from predators or crops receive water and are protected from pests in exchange for their fruits and seeds, so the ornamental garden plants receive water and protection from "killing insects and gnawing worms" in exchange for their beauty. The beauty of Shelley's sensitive garden would thus seal a vital harmony, a kind of living in peace with the natural world that is both described in scripture and advocated by contemporary ecocriticism today.

What troubles this interpretation, however, is the *atmospheric* role beauty plays in Shelley's depiction. In feeling herself drawn by the beauty of her plants, the gardener may have had the sense that she was the recipient of something, but this "something" is evanescent and fleeting, not anything that can be grasped, held, or appropriated; the gardener does not "receive" an experience of beauty from the ornamental plants of the garden in the way that one can receive flesh or fruits in exchange for water and protection from insects. This discrepancy highlights in part the fact that these two cases—livestock and agriculture, on the one hand, and ornamental beauty, on the other—are not truly analogous; the care for ornamental plants is simply not a part of agriculture, at least insofar as the latter is understood as a means of obtaining sustenance.[45] More fundamentally, though, this discrepancy emphasizes that "mutual exchange" is ultimately an imprecise way to account for the role of beauty in the emergence of the mutual atmosphere between the gardener and garden.

Concepts of exploitation and mutual exchange end up falling short because both assume the formulation of the subjective space that Kant demanded in the case of aesthetics, a sort of "stepping back," which is the opposite of the movement required for the emergence of a mutual atmosphere. The narrator of "The Sensitive-Plant" describes the mutual atmosphere between plants not as an appropriation of something from a position of distance but rather as an "interpenetrat[ion]" of "light and . . . odour." This interpenetration of light and odor that creates a mutual atmosphere is a drawing together, Shelley's narrator suggests, analogous in form to the convolution of new lovers, who, each sensing a fundamental change in him- or herself, commit even further to a mutual involution:

> For each one [of the flowers] was interpenetrated
> With the light and the odour its neighbor shed,
> Like young lovers whom youth and love make dear
> Wrapped and filled by their mutual atmosphere. (1:66–69)

Exploitation and exchange require a space within which transfers can occur; mutual atmospheres, by contrast, emerge and persist only if each party is initially generated, and thereafter allows itself to be further solicited, by a tightening plenum, a desire for ever less distance and space, which tends toward (even if it can never reach) the limit of producing one from two. It is this tightening plenum, rather than the apparently reciprocal "shedding" of light and odor, that is fundamental to the establishment of a mutual atmosphere. This is especially clear in the case of the sensitive plant: "Radiance and odour are not its dower," and all it can contribute is "love" (1:75–76). Because a mutual atmosphere binds its parties to one another through longing, rather than need—because it is a mutual seduction rather than an exchange—each side remains less and less what it was the longer that the atmosphere persists. Shelley's use of analogy operates according to this same involutionary logic, for we only "grasp" the evanescence of atmosphere when we sense that the familiar trope of young love does not yet bring us close enough to the logic of mutual atmosphere. The distance implicit in the "like" of analogy thus serves as impetus for a movement that invites us to create new concepts and thoughts for this excessive form of love.[46]

The love that makes mutual atmospheres possible is not a structure of defense against an external world but rather a practice of ecstasis, of going beyond oneself. Shelley would expand on this understanding of love a year later in A Defence of Poetry (1820), in which he described love as "a going out of our nature, and an identification of ourselves with the beautiful which exists in thought, action, or person, not our own" (487). Yet "The Sensitive-Plant" makes clear that love does not depend upon psychological acts of identification, or upon psychology at all, for though the sensitive plant lacks psychological interiority, it nevertheless "desires what it has not—the Beautiful!" (1:77). The love that creates mutual atmospheres is the effort by temporal, organic beings to draw close to that which exceeds the organism, and to those experiences within time that nevertheless seem out of time:

> For love, and beauty, and delight,
> There is no death nor change: their might
> Exceeds our organs—which endure
> No light—being themselves obscure. (Conclusion, lines 21–24)

Love, in short, is not what binds organic beings to one another in relationships of reciprocity; rather, it is what draws organic beings to that which exceeds the organic.

Recognizing the love that enables mutual atmospheres as a form of apprenticeship to cryptogamia, rather than as a variant of human sexuality, also refines our sense of how to explain Romantic plant poetry. At the very least, it suggests that Romantic plant poetry cannot be explained in terms of traditional gendered understandings of (male) activity and (female) passivity. Later critical embarrassment over Romantic plant poems was in a sense a consequence of seeing Romantic plant love through the lens of traditional understandings of human sexuality. When viewed through such a lens, the apprenticeship Romantic plant poems valorized and demanded seems like a form of triple passivity: instead of redeeming poetry—an activity that itself risks being cast as feminine indulgence—by actively choosing and manhandling masculine poetic subjects, the Romantics passively gave themselves over to plants, which were themselves indices of passivity. Yet rather than giving in to embarrassment—giving in to the demand that our relationship to verse correspond to traditional notions of male activity and female passivity—we ought instead to understand these poems as forms of apprenticeship that cannot be understood in terms of traditional gendered relationships. Nor, for the same reason, should Romantic plant poems be understood as projections of, or compensation for, love that would otherwise have been directed into human sexual relations. While it may have been the case that Jean-Jacques Rousseau turned, in his old age, to a love of plants as a way of "distract[ing]" himself from social relations, this should not be taken as a paradigm for Romantic plant love, despite the importance of Rousseau for Romanticism more generally.[47]

APPRENTICESHIP, SIGNS, AND ART

Drawing together my points about tracing, cryptogamia, and mutual atmospheres, we can describe Romantic plant love as a form of apprenticeship, an active, patient practice of becoming a part of something foreign to oneself. As in the case of every apprenticeship, learning to create a mutual atmosphere is a matter of interpreting signs.[48] For the gardener of "The Sensitive-Plant," this means determining by means of signs which insects kill and "gnaw," and which bring benefits.[49] The particular signs will differ significantly from one kind of plant to another; they will even differ between two individual instances of the same plant species. Nor can such an apprenticeship limit itself to interactions with plants, but it will also (and necessarily) include interactions with those signs about plants generated by other humans, whether these signs are spoken, written, or left in the form of living or dead tended plants. A mutual atmosphere is in this sense not a community of two individuals but rather an intersection between at least two populations, and between multiple generations.

Romantic plant poetry emphasizes that art, as a specific kind of sign, can also play a role in this apprenticeship. As I noted above, Romantic plant poetry aims to provide neither particular facts about plants nor an expansive sense of nature's beneficence; instead, it enables a *tracing* of the "serpentine / Upcoiling, and inveterately convolved" paths of plant generativity. Poetry thus exercises that capacity to locate signs that is essential to apprenticeship; or, as Shelley put it in *A Defence of Poetry*, "Poetry enlarges the circumference of the imagination by replenishing it with thoughts of ever new delight, which have the power of attracting and assimilating to their own nature all other thoughts, and which form new intervals and interstices whose void forever craves fresh food" (488). Romantic plant poetry refines an otherwise vague sense of the complete otherness of plant life, yet it also ensures that this refinement of one's sense of cryptogamia comes by way of delight.

My emphasis on cryptogamia leads us away from the understanding of plant generativity—and, as a consequence, away from the understanding of the relationship between plants and art—Paul de Man advanced in his seminal essay on the "Intentional Structure of the Romantic Image." Drawing heavily on Friedrich Hölderlin's comparison of words (*Wörte*) to flowers (*Blumen*) in the poem "Brot und Wein" (Bread and Wine) de Man contends that the Romantic poetic image expresses an impossible wish that words could come into being (*entstehen*) in the manner of flowers. He provides a clear and distinct account of the differing manners in which flowers and words emerge:

How do flowers originate? They rise out of the earth without the assistance of imitation or analogy. They do not follow a model other than themselves which they copy or from which they derive the pattern of their growth. By calling them *natural* objects, we mean that their origin is determined by nothing but their own being . . . They originate out of a being which does not differ from them in essence but contains the totality of their individual manifestations within itself. All particular flowers can at all times establish an immediate identity with an original Flower, of which they are as many particular emanations. (4)

Words, by contrast, are productions of consciousness, and thus originate through a constitutive negation and self-difference:

It is the essence of language to be capable of origination, but of never achieving the absolute identity with itself that exists in the natural object . . . The word is always a free presence to the mind, the means by which the permanence of natural entities can be put into question and thus negated, time and again, in the endlessly widening spiral of the dialectic. (6)

Though one recognizes in de Man's claims echoes of many (mostly German) aspects of Romantic-era reflections on plants—Hegel's emphasis on the lack of vegetable interiority, Goethe's emphasis on the transcendental plant—it is striking how little de Man's discussion depends upon the specific example of flowers (which is itself already a rather abstract example). He treats "flowers" as a synonym for "living plants"—rather than, for example, as dead cuttings taken from living plants—and he suggests that Hölderlin could just as well have written "stones" (*Steinen*) in place of flowers (6). Flowers are, for de Man, simply an instance of the nonhuman, and thus equivalent to all other instances of the nonhuman. Vis-à-vis this fundamentally undifferentiated nature, the function of human words and concepts is to testify to human freedom through negation of that which is given.[50]

Because it leads us away from the Romantics' interest in cryptogamia, de Man's neo-Hegelian understanding of language gives us little purchase on the variety or specificity of Romantic plant poems. One could question whether the poem in which Hölderlin mentions flowers (*Blumen*) is really about plants (*Pflanzen*), but even were we to grant that point, it is certainly not a representative example of the Romantic interest in plants that I have described above. I suspect that this point would not have troubled de Man, for his argument suggests that there is really only *one* Romantic-era poem about plants—namely, an abstract, transcendental poem that clearly opposes the *in-itself* of nature to the *for-itself* of subjectivity—

and thus every existing Romantic plant poem, including Hölderlin's or Shelley's or Wordsworth's, is simply a partial expression of this truth. Yet were we to follow Étienne Geoffroy Saint-Hilaire's teratological approach (see chapter 5), rather than de Man's version of Goethe's interest in the *Urblatt*, we could then see Romantic plant poems as less interested in the *transcendental* aspect of the transcendental Plant than in the monstrous variants that the premise of a transcendental Plant enables and allows one to trace.

De Man's neo-Hegelian understanding of language also obscures the Romantic-era practice of apprenticeships that seek to engender mutual atmospheres. For the Romantics, it was *not* the case that flowers—or, more accurately, plants—"rise out of the earth without . . . assistance." Plants, as well as animals, require the assistance of both indifferent milieux and members of other species that help establish mutual atmospheres, such as the insects that spread pollen or the gardener of "The Sensitive-Plant" who culls out the killing and gnawing insects. Mutual atmospheres, in other words, confuse the distinction between an undifferentiated in-itself of nature and the for-itself of subjectivity that opposes and negates it. In place of that abstract schema, we have a field of living beings, each of which seeks to tighten the plena of the mutual atmospheres to which it belongs. From the perspective of the mutual atmosphere, the function of words in Romantic plant poetry is not to negate the given world—a task for which, as de Man admits, words about plants are no more appropriate than words about any other "natural" entities—but rather to amplify the possibilities of the creation of mutual atmospheres between populations of humans (and animals) and populations of plants.

Romantic plant poetry is in this sense a technology that enables something like the "tuning" that sociologist of science Andrew Pickering suggests characterizes the process of scientific experimentation. Pickering uses the concept of tuning to capture the roles of passivity and activity on the part of both the scientist and the elements of the natural world that are engaged by an experimental device:

> As active, intentional beings, scientists tentatively construct some new machine. They then adopt a passive role, monitoring the performance of the machine to see whatever capture of material agency it might affect. Symmetrically, this period of human passivity is the period in which material agency actively manifests itself. Does the machine perform as intended? Has an intended capture of agency been effected? Typically the answer is no, in which case the response is another reversal of roles: human agency is once more active in the revision of modeling vectors, followed by another bout of human passivity and material performance, and so on . . . [This has] the form of a *dialectic of resistance and*

accommodation, where resistance denotes the failure to achieve the intended capture of agency in practice, and accommodation an active human strategy of response to resistance.[51]

The experimental device allows a scientist to attune herself to, and move in step with, a previously unknown or unverified element or process, and the device does so by enabling a "dialectic of resistance and accommodation" that alternately privileges what Pickering describes as the agency of matter and the agency of the scientist.[52]

Though the technology of the plant poem differs significantly from the scientific experimental device, it also enables a kind of tuning. Plant poems cannot engage the vegetable world in the same way as scientific experimental devices, of course: beyond the fact that a plant poem may be printed in a book that employs paper and inks derived from plants, there is no ontological connection between a specific plant and a poem about plants. As a consequence, plant poems cannot enable the dialectic of resistance and accommodation Pickering describes as characteristic of scientific experimentation, for the poem does not present living plants with a surface that is able to capture and register plant "resistance." Nevertheless, plant poems can function as technologies for amplifying one's awareness of the strangeness of plants. They do this by embedding moments of what Nicola Trott calls "eerie"-ness within the field of pleasure established by the Romantic poem. It is a given for the Romantics that a good poem will produce pleasure; however, plant poems treat this pleasure as a unified field—or, to appropriate the language of Coleridge's reflections on meter, a medicated atmosphere—within which the reader or auditor can *sense* the fugitive nature of plant vitality. The serpentine bends of Wordsworth's Borrowdale yew trees give way to upcoiling and convolution; Clare's dark creeping ivy and red clover each suspend time in different ways; Percy Shelley's plants interpenetrate one another to create mutual atmospheres. Each of these representations of the cryptic nature of plant life "plant-ifies" the poem itself, soliciting from the reader or auditor an attention to the plantlike upcoiling and involution of the poem's rhythms, or the poem's tenuous suspension of time, or its creation of a mutual atmosphere between poem and reader. The stable form of the poem and the pleasure it produces allows auditors and readers to return again and again to these moments of cryptogamia and, through patient tracing of their convolutions, to amplify the sense of vegetable eeriness.

The tuning that plant poems enable thus takes the form of feedback, rather than dialectic. In place of the comedic narrative that Pickering associates with the scientific experimental device—that dialectic of resistance and accommodation that enables a dance of agency and (with luck) a reciprocal relationship between an autonomous scientist and an autonomous material agent—we find instead a narrative structure of errant amplification. And in place of de Man's understanding of poetic language as a "means by which the permanence of natural entities can be put into question and thus negated, time and again, in the endlessly widening spiral of the dialectic," we see words functioning as elements in the creation of a mutual atmosphere between reader and poem, which in turn can serve as the starting point for the creation of mutual atmospheres between readers and "natural entities."

TRANCE-PLANTATION

Understanding Romantic cryptogamia as an attempt to create mutual atmospheres between plants and humans helps us more accurately situate this literary project within a wider field of Romantic-era practices. Ecocritics such as Jonathan Bate have begun this work, but often guided by more contemporary concepts of "sustainability" and "respect for the environment," they have tended to position Romantic-era literary representations of the natural world as a virtuous reaction to the ecological devastation that modern capitalism was already beginning to produce. Romantic literature thus emerges as a prescient awareness of the need to return to traditional, more ecologically friendly agricultural practices.[53] Recognizing the cryptogamic impulse of Romantic-era representations of plants, by contrast, means recognizing the *newness* of this approach to the vegetable world: far from constituting a reaction against modern institutions, Romantic representations of plants were instead part of an explosion of new institutions that sought to engage plants differently than before. These new institutions included the concerted expansion of British and French royal botanical gardens, the cultivation of new tastes for vegetables and vegetable products formerly unavailable in Europe, and the creation of new markets for "exotic" (nonindigenous) plants and flowers. Like Romantic-era literary representations of plants, each of these practices was motivated by a sort of interspecies desire, a longing to create new mutual atmospheres between humans and plants that would change existing patterns of life. Many of these new mutual atmospheres were established, of course, primarily for the purpose of creating surplus value (profit), which begs the question of how one might make evaluative judgments among different mutual atmospheres, and I return to that point below. What I wish to stress at the moment, though, is that the deictic character of Romantic representations of plants—the attempt to trace cryptogamia in such a way that readers could learn to draw themselves closer to the life of plants—was not, fundamentally, a conservative desire to retain what had come before but rather a desire to become otherwise than one had been.

From this perspective, Romantic plant poems were perhaps closest in spirit to the newly expanded royal botanical gardens of Britain and France. These scientific gardens represented a global experiment in plant love, as species were transported across the world and tended by gardeners who had to develop techniques for allowing nonnative plants to thrive in new conditions. Government botanical gardens have been viewed with some suspicion in recent literary criticism, largely because they were dependent upon those same imperial networks of power by means of which France and Britain (among other countries) sought to exert political and economic control at a distance.[54] Yet if we consider these gardens and the networks of which they were a part from the perspective of plants, they also emerge as vectors for a vegetative colonialism moving in the reverse direction, as species formerly limited to other continents suddenly had new European vistas open up.[55] Moreover, while botanical gardens may have indirectly served the purposes of capital, this was not their primary purpose; what motivated the researchers who brought back plants was generally an inchoate, fugitive desire for the new and unknown.

These similarities between Romantic-era botanical gardens and literary representations of the vegetable world emerge especially clearly in John Clare's "Ballad—A Weedling Wild." Clare has recently emerged as a privileged figure of Romantic-era ecological "protest" poetry, a laborer and poet who decried large-scale changes in Britain's agricultural practices and encouraged his readers to develop a sense for the beauty of wilds, swamps, and common lands ("ponder . . . Nature, when she blooms at will").[56] Yet Clare's verse also manifests that same urge for transplantation that motivated botanical gardens: in "Ballad—A Weedling Wild," the narrator's initial urge to "crop" a beautiful wildflower that he finds on a "lonely lea" is transformed into the desire to transform the weed into a garden plant by taking "the root and all" with him.[57] To allow oneself to be solicited by cryptogamia did *not* mean just "letting nature be"—it was not a stepping back from nature—but rather involved an active *falling-for* that encouraged literal transplantations of plants into human lives.

Though the transplantations Clare and the botanical gardeners practiced can certainly be seen as modes of imperialist acquisition and an effort to control the disseminative powers of plants, such a reading misses the ecstatic, self-overcoming impetus of these movements. In order to make that impetus clear, it is helpful to see Romantic cryptogamia as a precursor of what Richard Doyle has described as "transgenic involution."[58] Doyle uses this term to describe processes in which a mutual attraction between two different species ends up functioning as a vector of selection (in the genetic sense). His discussion of what he felicitously terms "cannaboid porn" has particular relevance for our understanding of Romantic cryptogamia. He notes that contemporary marijuana growing—an illegal and decentralized endeavor that nevertheless produces both increasingly high potency crops and genetic diversity—has produced its own haze or atmosphere of high-resolution images of marijuana plants, propagated in both magazines and on the Internet. Doyle stresses that this "pornography of plants" should not be understood solely as an epiphenomena of intoxication (after all, "the Internet has not yet become home to photo galleries of home brew liquor or beer"). Instead, these images of plants are a means—albeit one without an agent in the traditional sense—for producing ever more intimate interpenetrations of pot plants and gardeners. For the tender and tremulous pot gardener, images of pot plants produce a gardening desire—"some of the pictures almost take you inside the bud"—which solicits further hybridizations of plants from these illegal growers.[59] Aesthetic images of pot are part of an "involution" of plants and humans, creating lineages of plants that require a "human assay" for their continued transformation. In the case of pot growing, this assay moves through the lungs and brain, as pot growers seek to judge whether the "high" produced by this plant promises a future, different high that can come into being through further hybridization. In Romantic cryptogamia, the assay was more likely to make use of the eyes and nose, as gardeners sought to judge the future of these petals and odors—could they become even more beautiful? In both cases, though, aesthetic appreciation becomes a vector for processes of plant and human transformation.

EVANESCENCE

In both its poetic and scientific modes, Romantic cryptogamia does not seek stable equilibrium, whether of the kind Serres describes under the rubric of exchange or that valorized by many ecocritics under the rubric of harmony; rather, it seeks to intensify intertwinings and involutions. Romantic plant poems are not prescriptive descriptions of virtuous states of natural equilibrium ("Look how these plants and people are living in harmony—now make this happen in *your* life!"). Rather, they are solicitations to create mutual atmospheres, to give in to beauty by becoming part of an involution that produces real transformations by means of aesthetic experiences and patient practices of selection and tending. One can, of course, also create mutual atmospheres with more familiar forms of life, such as birds or mammals. However, in favoring plants in their literary depictions, Romantic-era authors emphasized the ecstatic logic of the mutual atmosphere: rather than being a quasi-contract with forms of life that are like us, the mutual atmosphere is an entwining embrace with a form of life that is emphatically not our own. Plants live, but not in the way that we do, and cryptogamia is in this sense always a love affair with an alien form of life. Romantic cryptogamia thus remains quite distant from later nineteenth-century invocations of *Blut und Boden* (blood and soil): where the ideology of *Blut and Boden* is based on the premise of a selfsame essence that must be kept pure, Romantic cryptogamia is premised on the necessity of becoming-other.

Because it threatens, and promises, to transform both parties, interspecies love often fails to become actual, and when it does become actual, it is often an evanescent affair. Romantic cryptogamia is an attempt to make these virtual loves actual, to extend what is otherwise fleeting. Shelley suggested in his *Defence of Poetry* that poetry "arrests" otherwise "evanescent visitations of thought and feeling," extending in time those moments in which we feel as though there were an "interpenetration of a diviner nature through our own."[60] Shelley's interest in arresting evanescence should not be understood as a form of Platonism but rather as an attempt to extend mutual atmospheres. "The Sensitive-Plant" describes an evanescent event, the creation and dissolving of the mutual atmosphere of gardener and ornamental plants, in order to arrest and extend the reader's otherwise fugitive cryptogamia. The poem is not about reaching the Idea of Beauty through plants; instead, it is an attempt to prolong the reader's desire for *plant-ing*. To be moved by a Romantic plant poem is thus to amplify and give in to a desire to be part of a becoming that links plants and humans, to help create an atmosphere between humans and plants.

To the extent that they move readers, Romantic plant poems participate in the creation of mutual atmospheres, and in this sense they may function like imperatives, silently commanding readers to go forth and create mutual atmospheres with alien forms of life. However, this imperative is itself silent on the question of harmony (and its modern synonym, "sustainability"). If there is any lesson in Shelley's poem, it is that what one finds beautiful now cannot last forever; only "spawn, weeds, and filth" remain fixed and frozen. The best one can do is to extend an ultimately evanescent atmosphere enough that its beauty has at least a chance to be perceived by others and its evanescence becomes bearable. Shelley's poem is no Grecian urn, and even its atmosphere will eventually give way to other atmospheres.

It is this deep commitment to mutability and evanescence that fundamentally distinguishes Romantic cryptogamia from ecocriticism. Romantic plant poems are grounded in the premise that human things—that is, that which we right now take to be most human—ultimately do not persist in the fact of vegetable vitality. Mutual atmospheres between beautiful plants and humans dissipate, and weeds overgrow and erase all traces of humanity; in the face of cryptogamia, as the sage pedlar of Wordsworth's *The Ruined Cottage* notes, "very soon/Even of the good is no memorial left" (lines 71–2). Recognizing that nature for the Romantic poets was fundamentally indifferent to human concerns, an earlier generation of literary critics argued that the Romantics placed their hopes in a redemptive "imagination" that was able to prune back the creeping void of plant life; because the life of nature exorbitantly exceeds and overwhelms human life, nature must be imaginatively "humanized" if we are to dance with the daffodils. Ecocritics correctly emphasize that, however helpful this earlier line of criticism was in locating the psychological dramaturgy of Romantic poems, there is nevertheless something fundamentally wrong in this agonistic paradigm: the Romantics did not present nature as a foe but as something to be embraced.[61] Yet the ecocritics err by willfully ignoring the importance of evanescence in Romantic plant poems and by seeking to understand the embrace of human and nonhuman in terms of reverence for, and harmony with, nature.[62] This embrace is more akin to the twining of bindweed and peas that Wordsworth described in the version of the story of the ruined cottage that appeared in *The Excursion*: "The cumbrous bind-weed, with its wreaths and bells,/Had twined about her two small rows of peas,/And dragged them to the earth." To see the short-term stability of the pea garden as a form of "harmony" with nature is to ignore its tighter and ultimately deadly embrace with the bindweed. From the perspective of humans, the fixed and frozen atmospheres of the "loathliest weeds" are always pressing in on that of the more tenuous atmospheres that link us to beautiful flowers and fruits.

This is simply another way of making explicit the intersection of ecology and astronomy, an intersection that ensures that no "atmosphere," not even that of the earth as a whole, will last forever.[63] This does not render distinctions between good and bad mutual atmospheres impossible. However, it does mean that establishing "harmony" as a criterion for value ignores what ought to be a central truth of ecology, namely, that harmony is not a natural state into which we can slip, but instead a *judgment* about a state of affairs, made by human beings on the basis of quite human priorities and time scales rather than according to some transcendent value. Learning from Romantic cryptogamia means embracing the fact that the mutual atmospheres that we collectively create necessarily force us to change, to become different. There is no transcendent payoff for such submission to change—both we and our plants will eventually die, like the gardener and her garden—and so all "rewards" have to be understood as immanent to this life; what one "gets" is the continued opportunity for ecstatic self-transformation. Such change does not occur in response to moral imperatives but when an individual allows him or herself to be abducted by a fascination with other forms of life. Being abducted is not (just) a matter of wide-eyed fascination but (also) a patient commitment to the opacity and specific materiality of other forms of life (a fascination and commitment that one is as likely to find among scientists as among organic farmers).

In the absence of any transcendent source of judgment, distinguishing between good and bad mutual atmospheres becomes a matter of assessing the relationships to aesthetic judgments (and thus, commitments) that these atmospheres enable. Good mutual atmospheres are disseminative, in the sense that they facilitate, rather than hinder, abductions via aesthetic judgments; good mutual atmospheres make possible the formation of new and different mutual atmospheres. What is problematic about the weeds for the narrator of Shelley's "The Sensitive-Plant" is that they seem to create a "fixed" atmosphere, one in which, as William Keach notes, "previously vital, expansive processes of dissolving and evaporating are reversed and give way to thickening, condensation and frozen constriction" (33). One can imagine another narrator—a botanist, or perhaps even John Clare—who, more attentive to the particularities of the "weeds," would find beautiful flux and evanescence where Shelley's narrator perceived only stasis and fixity. A truly bad mutual atmosphere, though, would be one in which plant or human life becomes standardized and homogenized to the extent that it is no longer possible to locate variations that could be embraced and tended.

From this perspective, then, many of the goals of contemporary ecological practice would indeed also be virtues for an approach that draws on Romantic cryptogamia. Literal or even metaphorical tree hugging is simply a confused way of embracing the obscurity of plant vitality, but protecting plants in order to encourage genetic diversity *is* a virtue because it makes possible (and likely) the continual creation of new mutual atmospheres, allowing multiple sites of aesthetic and affective engagement. The relative stability of ecosystems—relative, that is, to the time scales of human beings—is also a virtue, for it allows individuals time to make aesthetic discriminations and to engage in those forms of patient practice that proceed from such discriminations. Yet these goals—protection of plants, genetic diversity, relative ecosystem stability—are neither justified as means to achieve harmony nor provide any assurance that humans will stay just as they are. Instead, they promise nothing more (or less) than perpetual transformation.

Biopolitics and Experimental Vitalism

Though the focus of this book has been on Romantic-era experiments in art and science, both my introduction and my various references to the work of Gilles Deleuze, Félix Guattari, Michel Foucault, and other contemporary theorists emphasize that it is also an attempt to address, and intervene in, our own contemporary vital turn. I have sought to intervene in part by introducing the concept of "experimental vitalism," for my hope is that such a concept allows us to historicize, to reveal the Romantic-era origins of, our own contemporary interest in phenomena such as suspended animation, bare life, and non-organic life. Yet the point of historicizing in this way is *not* to shunt contemporary experimental vitalism back into the past; my goal is not to suggest that a purportedly vitalist contemporary author such as Deleuze should be understood as an atavism who belongs more properly to the eighteenth than to the twenty-first century. Rather, I hope to amplify resonances between Romantic-era and contemporary experimental vitalisms in ways that make Romanticism—or, at any rate, a certain strain of Romanticism—itself contemporary, rather than a part of our distant past. The relationship that I establish between Romantic and contemporary experimental vitalism has the form of chiasmus: on the one hand, I draw on contemporary theory in order to reveal alongside or beneath the Romanticism of organic form an experimental Romanticism focused on the vertiginous, the non-organic, and suspensions of animation; on the other hand, my theorization of Romantic literary experimentation is intended to allow us to grasp more clearly than in the case of much contemporary theory the extent to which experimental vitalism depends on practices that confuse distinctions between "art" and "science."

Linking Romantic and contemporary experimental vitalisms through the form of the chiasmus also means, however, that I cannot adopt the historian's stance of distance from my subject matter and thereby inoculate Romantic-era experimental vitalism from our own debates about vitalist language and logic. To put this another way, if the theoretical work of Deleuze, Guattari, and Foucault helps us to understand Romantic-era experimental vitalism, then it is also true that criticisms of contemporary neo-vitalism are likely to apply to Romantic-era experimental vitalism (in both its strictly historical dimension and in that dimension of contemporary resonance that I hope to give it).

There are at least two quite distinct kinds of criticisms leveled at contemporary neo-vitalisms. On the one hand are criticisms of the purportedly *homogenizing* tendencies of contemporary concepts of life, claims that concepts of life are necessarily univocal, and as a consequence obscure important differences among living beings. On the other hand are criticisms of the *ideological function* of concepts of life in our current political economic situation, claims that because our ever-expanding "bioeconomy" depends upon tight feedback loops between speculative concepts of life and speculative debt and finance structures, any discussion of theories of life that avoids political economic critique risks serving as a vector for expansion of the bioeconomy. In what follows, I consider each of these criticisms in more detail, in the hopes of using these as means for further clarifying precisely what is at stake in the concept of experimental vitalism.

THE UNIVOCITY OF LIFE

The potentially homogenizing tendency of concepts of life comes to the fore in criticisms of Roberto Esposito's influential *Bíos: Politics and Philosophy*. Esposito's book is part of a wide-ranging debate on the forms and nature of biopolitics, a debate grounded in Foucault's discussions of populations but which takes his approach in new directions. Esposito's book explicitly engages—and draws its title from—the link that Giorgio Agamben established between the Foucaultian problematic of biopolitics and the Greek political distinction between *zoe* (the "simple fact of living common to all living beings") and *bios* ("the form of life proper to a given kind of living being"—in the case of humans, a political life).[1] Agamben argues that biopolitics operates on populations by establishing institutions—the concept and institutions of the "citizen" and the "refugee," the discourse of human rights, legal distinctions between "coma" and "vegetative state," and so on—that paradoxically function in part by depriving some individuals of *bios* (their "right" to political existence), which in turn allows these individuals to be killed (or prevented from dying) without such actions constituting either murder or an abridgement of rights. Agamben thus sees in biopolitics a "thanatopolitical" logic that received its most explicit expression in the Nazi concentration camps: "the camp," Agamben writes, is "the hidden paradigm of the political space of modernity" (123).

Esposito builds on Agamben's analysis but argues that the thanatopolitical examples that Agamben cites are not *intrinsic* to biopolitics but are rather the result of an "immunitary paradigm." Esposito argues that immunity has both a medical sense ("a condition of natural or induced refractoriness on the part of a living organism when faced with a given disease") and a political-juridical sense ("a temporary or definitive exemption on the part of the subject with regard to concrete obligations or responsibilities that under normal circumstances would bind one to others").[2] The immunitary paradigm links these two senses of immunity in the field of "life," thereby justifying the position that in some instances, the life of a given population can be saved—for example, in the face of a plague—only when some individuals are "exempted" from their normal rights. Esposito argues that while Agamben is correct to see the Nazi camps as the clearest expression, and logical consequence, of this immunitary logic, it is nevertheless possible to forge an *affirmative* biopolitics that does not lead inexorably to a thanatopolitics. The only way to avoid turning biopolitics into a thanatopolitics is by valorizing *all* life: "any thing that lives needs to be thought in the unity of life [which] means that no part of it can be destroyed in favor of another" (194). Esposito's affirmative biopolitics is thus a neo-vitalism not in the sense that he affirms a transcendent principle of life that rules over matter or is contained within matter but rather in the sense that life is affirmed as the ultimate value.

Criticisms of Esposito's affirmative biopolitics have emphasized the practical political and ethical ambiguities of his univocal valorization of life. What, for example, are the implications of this valorization for the debate around abortion in the United States? Does Esposito's position imply that one must adopt a "pro-life" position, and if so, would it also commit one to a decision about *when* "life begins"?[3] The implications for activists interested in altering relations between humans and animals are equally uncertain. Esposito's position may initially seem to provide support to those who advocate for animal rights—his point, after all, is that we must move away from an immunitary logic that sacrifices *these* lives for the sake of *those* lives. Yet his position also seems necessarily to obscure differences between forms of life that are at the center of many forms of animal-human ethics (i.e., animal activism is not oriented toward "the animal," as though all animals were the same, but at specific relations between specific humans and specific animals). Cary Wolfe suggests that Esposito's univocal understanding of life fails to take seriously the "differences between different forms of life—bono-

bos versus sunflowers, let's say."[4] Life, in other words, strangely becomes in Esposito's work a concept that de-differentiates, rather than allowing us to make any
political and ethical distinctions between different kinds of living beings. Wolfe
notes that the practical consequences of such a de-differentiating approach to life,
explored in depth and at length in the Deep Ecology movement of the 1970s and
1980s, are forcefully summarized in Tim Luke's litany of rhetorical questions:

> If *all* forms of life are given equal value, then we face questions such as the
> following: "Will we allow anthrax or cholera microbes to attain self-realization
> in wiping out sheep herds or human kindergartens? Will we continue to deny
> salmonella or botulism micro-organisms their equal rights when we process the
> dead carcasses of animals and plants that we eat?"[5]

At the very least, these practical criticisms demonstrate the difficulties in understanding how to translate Esposito's valorization of life into concrete political,
legal, and ethical situations.

Wolfe suggests that these practical ambiguities of Esposito's position are the
consequence of at least two deep theoretical impasses implicit in Esposito's invocation and valorization of "life." First, insofar as Esposito argues that all "forms
of life are taken to be equal," then this "can only be because" (by which I take
Wolfe to mean that Esposito must be asserting implicitly that) "they, as 'the living,' all equally embody and express a positive, substantive principle of 'Life' not
contained in any one of them" (56). Second, Wolfe argues that Esposito's interest
in the population aspect of Foucault's notion of biopolitics leads him to stress
the role of the human biological *species* in ways that do not seem consistent with
Esposito's own neo-vitalism. Noting that Esposito elsewhere loops his affirmative
biopolitics through an appeal to forge "a new alliance between the life of the
individual and the life of the species," Wolfe astutely observes that the life forces
Esposito emphasizes "clearly don't stop at the water's edge of species."[6] As a consequence, though "Esposito may be right that the body is the immunitary site upon
which biopolitics seizes control over life . . . the cordoning off of 'the body' within
the domain of 'species' simply reinstates the very autoimmunitary, thanatological
movement that his affirmative biopolitics wants to resist" (58). Wolfe charges, in
other words, that Esposito is only able to address the question of which lives to
privilege (e.g., anthrax or human) by returning to the immunitary paradigm that
is ostensibly the object of Esposito's critique.

Though both of these are fair criticisms of some kinds of vitalism, neither quite
strikes the mark in the case of Esposito's approach. The first criticism, that every
invocation of life implicitly appeals to a substantive "principle" of life, seems

especially difficult to pin on Esposito. He explicitly claims that life should not be understood as a *property* of individual entities but rather as a *dynamic relation* between impersonal singularities which, "rather than being imprisoned in the confines of the individual, opens these [i.e., individuals] to an eccentric movement that 'traverses men as well as plants and animals independently of the matter of their individuation and the forms of their personality.'"[7] Wolfe takes Esposito to mean that each singularity "manifests" life, and thus his "neovitalism . . . ends up radically dedifferentiating the field of 'the living' into a molecular wash of singularities that all equally manifest 'life.'"[8] This is not what Esposito, following Deleuze, means when he invokes the term "singularities"; he is not attributing life to these singularities, as though a singularity were a form of matter that had "life" as one of its qualities or properties. Rather, life is understood as a dynamic relationship between singularities. Living *forms*—in this case, living organisms—are one result of these dynamic relationships. However, organic form is not the only result of these relationships, for it is also in reference to these that Deleuze refers when he writes of "non-organic life" (and this non-organic life is in turn something like what John Hunter sought to investigate in his experiments into the life that is evident in fertilized eggs prior to the formation of organs). Neither Esposito nor Deleuze nor Hunter is claiming that these non-organic singularities that make organic form possible possess the "quality" of life but rather that organic life itself depends upon a dynamic, vital movement—a dynamism that Deleuze refers to as non-organic life—between singularities. Since non-organic life, like Hunter's simple life, never appears in the absence of organic life, one cannot seek to "protect" it; for that same reason, however, it cannot be subjected to an immunitary logic. One can only "protect" non-organic life by protecting the conditions for the emergence of *new* forms of life.

We can thus propose that what remains (unfortunately) only implicit in Esposito's account is the intrinsic role of *experimentation* to his conception of life. His valorization of life is not ultimately an effort to cast a net of immunity around all existing life forms but rather a valorization of that which makes possible the emergence of current and new life forms. Life itself experiments, Esposito suggests, and this has important consequences for an affirmative biopolitics. It means, first and foremost, that we cannot seek to derive any stable "norm" for a specific species of living being—for example, human beings—that can then be used to justify immunitary measures intended to protect the "normal" from the "abnormal." Rather, he argues, we must follow Georges Canguilhem, who focuses on normativity, the ability to create new norms. From this perspective, "normal man is normative man, the being capable of establishing new, even organic forms."⁹ Rather than "following nature" by deriving stable norms for human beings—a project that, Esposito claims, will always lead to a thanatopolitics—we must instead follow nature by experimenting with norms.

Redescribing Esposito's neo-vitalism as an experimental vitalism also helps us to address Wolfe's second criticism, namely, that Esposito limits his account to the human species. The non-organic vital movement between singularities that Esposito privileges implies as a necessary consequence, as Wolfe points out, that life "clearly do[es]n't stop at the water's edge of species." Yet Esposito makes precisely this point when he emphasizes how biotechnologies enable us not only to transfer organs from one person, and even species, to another but also to introduce nonliving assemblages into the human (and we might here recall Hunter's artificial breathing apparatus). The human can always be—and, in fact, is constitutively—"interpenetrated" by the nonhuman (to draw on P. B. Shelley's term). This suggests that the goal of Esposito's affirmative biopolitics is something like what I described in chapter 6 as the creation of "mutual atmospheres." Mutual atmospheres are invariably affairs among multiple species and multiple forms of nonliving matter; moreover, they end up vexing clear distinctions between the inside and outside of the species (for one species becomes a condition of possibility for another) and between what is living and what is not.

The concept of the mutual atmosphere also helps us to better understand the place of death in Esposito's neo-vitalism. As Jacques Derrida presciently noted much earlier in his critique of Husserlian phenomenology, although vitalisms are generally understood as claims about the nature of life, they are also thus—and necessarily—claims about death. Derrida charged Husserlian phenomenology (and by extension, all phenomenology), with being a "philosophy of life," "not only because at its center death is recognized as but an empirical and extrinsic signification, a worldly accident, but because the source of sense in general is always determined as the act of *living*, as the act of a living being, as *Lebendigkeit* [vitality]."¹⁰ Derrida's critique of Husserl is a challenge for all vitalisms, experimental or otherwise: what, precisely, is the role of death in a given vitalism? Is death to be understood as an empirical accident that is of no real significance in the face of a more primal life?

As I noted in chapter 6, the concept of a mutual atmosphere enables us to address the fact of death but to do so without allowing mortality (and decisions about death) to be seen within the logic of a thanatopolitical immunitary norm. The thanatopolitical immunitary norm leads to a logic that commands that living individual X be sacrificed to save the "pure" life form Y. The ecstatic logic of the mutual atmosphere, however, denies the possibility altogether of any "pure" life form, while the evanescent nature of the mutual atmosphere demands that what must be valorized is not a (specific) norm but the capacity to continue to experiment with norms.

THE POLITICAL ECONOMY OF LIFE ITSELF

Where critics such as Wolfe emphasize the theoretical and ethical costs of a univocal concept of life, another set of critics focuses on what we might call the political economy of life itself: that is, the ways the recent theoretical valorization of life as self-transcending is entirely consonant with, and in fact an intrinsic part of, changes in the global system of capitalist finance and production. Melinda Cooper has developed one of the most compelling versions of this line of criticism, noting that since the 1970s, efforts in the sciences to rethink the nature and limits of life have been inextricably linked with changes in the nature and reach of capitalist debt and finance structures. So, for example, petrochemical and pharmaceutical companies, faced in the late 1960s and 1970s with falling profit margins from industrial modes of production, turned to—and actively encouraged the growth of—the emerging field of "biotechnology." For a petrochemical giant such as Monsanto, for example,

> The commercial calculus was straightforward—instead of profits from mass-produced chemical fertilizers and herbicides, the agricultural business would displace its claims to invention onto the actual generation of the plant, transforming biological production into a means for creating surplus value. Moreover, it was predicted that biotechnology would expand the geological spaces open to commercial agriculture, making it possible to create plants that would survive on arid land or flourish in the degraded environments created by industrial agriculture . . . In short, the geochemical laws ruling over Fordist industrial production would be replaced by the much more benign, regenerative possibilities of biomolecular production.[11]

Yet creating new modes of production and markets also meant establishing a "new regime of accumulation, one that relies on financial investment to a much greater extent than the Fordist economy had" and in which "the evaluation of future profits becomes the decisive factor in determining price" (23). What emerged was a speculative logic that produces present economic value through the promise to overcome current biological limits in the future. A company—or country—redescribes its present debts as "investments," which enables it to guarantee its debts not through existing collateral (e.g., gold reserves or short-term production capacity) but rather by means of the claim that its bioengineering endeavors will

enable living tissues and organisms to overcome their current limits at some indefinite point in the future (e.g., the future possibility of engineering bacteria to
clean up industrial wastes).

Cooper suggests that we should see recent theoretical valorizations of life in
the sciences and the humanities as part and parcel of the "vitalist" logic of neoliberalism. Neoliberalism is vitalist in the sense that the speculative logic of the
bioeconomy depends on two premises: first, the claim that life itself—that is, life
in the absence of human intervention—constantly exceeds its own limits and,
second, the claim that this self-exceeding force can be tapped and channeled by
humans to enable endless economic growth. Neoliberal theorists also draw an
additional conclusion, that "because life is self-organizing . . . we should reject
all state regulation of markets." Insofar as the "market" names for neoliberals a
constant bubbling over of human ingenuity—or, to put this another way, the
market is how the excess of life is expressed in human terms—market regulation
can do nothing more than hinder the natural excess of life.[12] From this perspective, moreover, neoliberalism seems not only to be a form of vitalism, but even
a form of *experimental* vitalism, since never-ending economic growth depends
upon constant scientific innovation.

Cooper's point is *not* that depictions of life as self-exceeding are "incorrect" or
that theorists should keep their distance from the topic of life. Nor is she suggesting that we can "decode" contemporary theoretical discussions of life as no more
than an ideological screen for transformations of capitalist structures of debt and
finance, for she argues that biotechnology ensures that debt and life truly do cross
through the other. Rather, her point is that "in the absence of any substantive
critique of political economy, any philosophy of *life as such* runs the risk of celebrating *life as it is*. And the danger is only exacerbated in a context such as ours,
where capitalist relations have so intensively invested in the realm of biological
reproduction" (42). In short, we risk privileging the life-forms-to-come over those
present life forms—including our fellow humans—with whom and with which
we are actually living in the present.

Though I have examined the political economy of both the Romantic era
and our present moment elsewhere, such a critique is indeed lacking in the present volume, and my project here thus arguably runs the risk of lending implicit
support to the vitalist dimension of neoliberalism.[13] Elements of my argument
here may even seem to exacerbate the problem that Cooper diagnoses: my valorization of experimentation may come across as an attempt to create excitement
around that key concept of neoliberalism, "innovation"; my interest in those liminal states, such as suspended animation and cryptogamia, that fascinated the

Romantics may seem to participate in the "destandardization" of life especially evident in the life sciences, which now often focus on "the extremes rather then the norms of biological existence" (33); and my emphasis on conceiving media in terms of populations could be taken as a crypto-valorization of the redemptive power of markets. Moreover, my interest in establishing resonances between Romantic and contemporary experimental vitalisms may appear to be an attempt to circumvent precisely that analysis of contemporary political economy that Cooper demands, in the sense that I loop our own interest in vitality through theory and literature from the "dead" past of the late eighteenth and early nineteenth century rather than through accounts of the still vital economic structures of the much more recent past. How, then, does this book avoid unreflectively valorizing those comportments, theoretical loci, and affective engagements essential to neoliberalism?

Though I do not develop a political economic critique here, I nevertheless hope to have provided a set of distinctions and tools with which we can guide contemporary interest in the excessive nature of life away from a purely economic capture. *Experimental Life* is premised on the principle that paying close attention to the historical emergence of concepts such as "suspended animation" or "media" is itself a means of achieving *both* a critical distance from and an activist relation to our own contemporary use of such terms. It is not the case, of course, that every historical account can produce either a critical distance or an activist relationship to the present. Historical accounts that simply point out to us that people in the past have understood a concept otherwise—or, even more problematically, suggest that the Romantics "anticipated" our contemporary approach to a given concept—generally do *not* enable critical distance or activism, for such merely descriptive approaches almost invariably lend themselves to narratives of progress that position the past as mere preparation for the present. My historical account is thus less of an attempt to show how the present builds progressively on the past than it is an effort to change the direction of the present by locating and reviving the virtual dimensions of the past.

Locating and reviving the virtual dimensions of the past means, in part, approaching the central concepts of these chapters as indices of (linked) problems that generated a *variety* of "solutions." The concept of suspended animation, for example, linked John Hunter's cryogenic dreams and therapeutic organs with Keats's and Shelley's poetics of cold pastoral and with Coleridge's reflections on suspensions of will and belief. Yet despite a common problem—can there be a state *between* life and death, and if so, how can such a state be investigated experimentally—these authors produced different solutions, which have been differentially revived and suspended over the subsequent two centuries. Elements of Hunter's dream of using suspended animation as a life-extension technology, for example, were revived in twentieth-century laboratory practices of freezing cells in order to maintain them for later research, as well as in the hopes of human cryonics advocates that sick and dying human beings can be kept in suspended animation until such a point in the future that scientists have discovered therapies for their illnesses. However, we can create a less delirious present and future not by abjecting a concept such as suspended animation but by reviving elements of the other "solutions" to the problem of liminal vital states established in the Romantic era. In place of Hunter's link between suspended animation and life extension, we can turn to Keats's and Shelley's practices of cold pastoral; in place of a conception of a linear history characterized by ever-accelerating technological and economic growth, we can seek to develop narratives of reversals, slowness, and endings.

A significant part of this effort involves reconfiguring the concept of experimentation. This means, in part, de-synonymizing "experiment" and "innovation." Since innovation properly names neither a practice nor a kind of experience, but rather a commercially available outcome of certain kinds of research and business practices, to understand experiment and innovation as synonymous is inevitably to encourage a view of experimentation as simply a neoliberal means for creating new products and services. Shapin and Schaffer's "constructivist" account of the communities, witnesses, and literary technologies of scientific experiments helps us to de-synonymize these terms by bringing to the fore the specific, and competing, political and disciplinary interests that are served by a given regime of scientific experimentation. This critical approach also suggests other avenues along which we can investigate other forms of experimentation, such as the Romantic literary engagement with experimentation that is the subject of this book. This approach encourages us to consider for example, contemporary relationships of artistic experimentation to questions of witnessing, community-formation, as well as the historical continuity (or lack thereof) between different past instances of artistic experimentation.

Yet even as Shapin and Schaffer's approach is able to explain those social dynamics of artistic experimentation that are *like* those of scientific experimentation, it cannot explain the *differences* between these two modes of experimentation, or how these differences could emerge as a consequence of a practice of experimentation that was itself neither properly scientific nor artistic. Since this latter practice of experimentation involved linking, or hybridizing, science and art, I have also turned to accounts of pre-scientific understandings of experiment (e.g., Dear's account of Aristotelian experiments as common experience) and to theoretical accounts that show how modern scientific experiments establish new links among otherwise separated materials, individuals, communities, and institutions. These latter accounts included Rheinbürger's stress on experimental systems over the isolated experiment, Latour's emphasis on actants and Pickering's focus on the "mangle of practice" and its attendant dialectic of accommodation and resistance. Collectively, this work emphasizes the extent to which scientific experimental systems bring into being new configurations of materials, humans, and institutions that cannot be fully parsed into the interests of (solely) human agents.

Though Romantic artistic experimentation, at least as I have presented it here, aspires to create linkages between different parts of the world, its quite different "materiality" may seem to distance it from these accounts of scientific experimentation. That is, insofar as the elements of Romantic artistic experimentation seem less "material," or at least less extensive in scope, than those of the sciences— Romantic poetry, for example, draws together only printed texts, language, and human perceptual capacities, rather than the various metals, fluids, chemical elements, and nonhuman living bodies that scientific experiments also link to printed texts, language, and human perceptual capacities—artistic experimentation may either seem less efficacious or more purely conceptual than scientific experimentation. Yet without denying the differences between artistic and scientific modes of experiments, the issue of materiality does not strike me as the proper axis along which to plot this difference. Drawing on Fleck's and Pickering's notions of tuning and attunement, for example, I highlighted the ways Romantic artistic experimentations exploited—or, to use Serres's terminology, parasited—the work produced within scientific botanical networks by linking them to experiences of local flora. Romantic plant poems point to the (scientific) reality of plants but parasite this reality to create new experiences and practices. From this perspective, it does not make sense to see Romantic literary experimentation as a less material imitation of scientific experimentation within the different realm of art; rather, Romantic literary experimentation functioned as a sur-science that created a vital medium around the sciences, drawing off their methods and results for other ends.

Philosopher of science Isabelle Stengers captures some of this Romantic sense of vital experimentation when she argues that the point of contemporary philosophical experimentation ought to be the production of "friction." By friction, Stengers means accounts that *expand* our sense of the various processes by which something can come to be what it is. She notes that the natural sciences are premised on reducing friction, whether literally or metaphorically. So, for example, in the case of Galileo's experiments with "carefully-polished, round balls, rolling down an equally carefully smoothed, inclined plane . . . the whole aim of the experimental activity of polishing and smoothing is that the auto-biography of the rolling ball would tell nothing about the ball, as such, in order for the speed it gains to reliably testify to what we now call terrestrial attraction (gravity). The intricate adventure that we call friction must not be recorded."[14] Stengers holds that though the natural sciences necessarily depend upon reducing friction—this is the cost of the specialized knowledge that they produce—this cannot be the path of experimentation in philosophy, which must instead aim at "a maximization of friction, recovering what has been obscured by specialized selection" (95).

Artistic experimentation shares with Stengers's image of philosophical experimentation the desire to produce something like friction. Yet in place of Stengers's concept of friction, which is drawn from the physical sciences, I have developed here a series of other terms drawn more from the sciences of life, such as suspension, disorientation, nausea, media, and cryptogamia. There is no either/or contrast to be drawn here, of course, as we can see friction as one among several concepts for "recovering what has been obscured by specialized selection." Yet the contribution of Romantic experimental vitalism is to emphasize that the friction of experimentation always takes place within the milieux and atmospheres of living beings.

Notes

INTRODUCTION: Three Eras of Experimental Vitalism

1. Silver, *Remaking Eden*, esp. 17–26. See also Kirschner, Gerhart, and Mitchison, "Molecular 'Vitalism,'" published in *Cell*, one of the premier scientific journals of biology.

2. See, for example, Margulis and Sagan, *What Is Life?* and *Acquiring Genomes*. Margulis and Sagan distinguish their approach from vitalism per se, but they rely upon a (mis)understanding of vitalism as a mode of animism (the belief that the entire universe is alive [*Genomes* 8]), while their understanding of life as "activity," "process," and "increasing complexity" has considerable overlap with what I call experimental vitalism. For discussions of artificial life, see Levy, *Artificial Life*; Helmreich, *Silicon Second Nature*; and Doyle, *Wetwares*, 19–62.

3. In *A Thousand Plateaus*, Deleuze and Guattari drew on the work of early-twentieth-century embryologist Albert Dalcq to develop the conception of "non-organic life"; for extensions of Deleuze's and Guattari's concept, see De Landa, *Intensive Science and Virtual Philosophy*, and Pearson, *Germinal Life*. For "postvital life," see Doyle, *On Beyond Living* and *Wetwares*. Though philosopher Giorgio Agamben's concept of "bare life" is grounded in political and moral, rather than ontological, categories, he also draws upon medical and biological discussions, arguing that the "over-coma" victim is a pure example of bare life; see Agamben, *Homo Sacer*, 160–61. For "creaturely life," see Santner, *On Creaturely Life*; for an attempt to found an ethics on an ontology of "precarious life," see Butler, *Precarious Life*; for "surplus life," see Cooper, *Life as Surplus*. The ever-increasing bibliography of work on "biopolitics" is too vast to cite here, though for an influential instance, see N. S. Rose, *Politics of Life Itself*, as well as Esposito's attempt to articulate an affirmative biopolitics in *Bíos*.

4. On the "life of images," see W. J. T. Mitchell, *What Do Pictures Want?*

5. On Bergson's influence, see Burwick and Douglass, *Crisis in Modernism*, and Antliff, *Inventing Bergson*.

6. Burke questioned in *Reflections on the Revolution in France* whether it was "true that the French government was such as to be incapable or undeserving of reform, so that it was of absolute necessity that the whole fabric should be at once pulled down and the area cleared for the erection of a theoretic, experimental edifice in its place?" (145), contending that "if commerce and the arts should be lost in an experiment to try how well a state may stand without these old fundamental principles [of the 'spirit of nobility' and religion], what sort of a thing must be a nation of gross, stupid, ferocious, and, at

the same time, poor and sordid barbarians, destitute of religion, honor, or manly pride, possessing nothing at present, and hoping for nothing hereafter?" (90).

7. Dear, *Discipline & Experience*.

8. See Richards, *Romantic Conception of Life*; Ruston, *Shelley and Vitality*; and Gigante, *Life*.

9. In *History and Theory of Vitalism*, Driesch defines vitalism as "the recognition of the 'Autonomy of vital powers'" (6) and contends that the history of vitalism begins with Aristotle. In *Vitalism*, Wheeler follows Driesch, defining vitalism as "all the various doctrines which, from the time of Aristotle, have described living things as actuated by some power or principle additional to those of mechanics and chemistry" (vii), and asserts that insofar as the ancient physician Galen was a vitalist, "all medical men and biologists from A.D. 200 down to A.D. 1628 or longer were vitalists because they were Galenists" (12).

10. Rogers, *Matter of Revolution*, 1. In *Vitalizing Nature in the Enlightenment*, Reill employs a variation of this kind of definition to account for a specifically "Enlightenment" mode of vitalism. According to Reill, Enlightenment-era naturalists such as Georges-Louis Leclerc Buffon also "posit[ed] the existence in living matter of active or self-activating forces" and believed that these forces "had a teleological character" (7).

11. Foucault, *Order of Things*, 160. See also Foucault's claim in his debate with Noam Chomsky that from the late eighteenth century, the concept of "life" has not functioned as "a *scientific concept*" but rather "has been an *epistemological indicator* of which the classifying, delimiting, and other functions had an effect on scientific discussions, and not on what they were talking about" (Chomsky and Foucault, "Human Nature," 110).

12. Benton, "Vitalism," 18.

13. Benton's solution multiplies categories of vitalism, distinguishing between phenomenological vitalists, realist vitalists, teleological vitalists, nomological vitalists, etc.

14. Lenoir, *Strategy of Life*.

15. See, e.g., Shapin and Schaffer, *Leviathan and the Air-Pump*.

16. Jacob, *Logic of Life*, 39.

17. For his own account of his experiments, see Driesch, *History and Theory of Vitalism*. For an account of Driesch's career, see Freyhofer, *Vitalism of Hans Driesch*.

18. I discuss this "commonsense" view of scientific experiments further in chapter 1.

19. Yet Rheinberger reminds us in "Experiment, Difference, and Writing," that "the experimentalist does not deal with single experiments" (309). Quoting Ludwik Fleck, he notes that "every experimental scientist knows just how little a single experiment can prove or convince. To establish proof, an entire *system of experiments* is needed, set up according to an assumption . . . and performed by an expert" (309, emphasis added by Rheinberger; quote from Fleck, *Genesis and Development*, 96).

20. Rheinberger, "Experiment, Difference, and Writing," 324, my emphasis.

21. Hunter, "Experiments on Animals and Vegetables," 446.

22. Canguilhem, *Vital Rationalist*, 339. For a discussion of Canguilhem's description of medicine and the life sciences as "vital needs," see R. Smith, "As Yet Unknown," and Osborne, "What Is a Problem?," in which Osborne notes that "for Canguilhem the business of a vitalist philosophy is not at all to propose, normatively, a vitalist philosophical 'theory' but to keep the door open for the further mutations of vitalism" (9).

23. Drawing on the work of David Hyder, Rheinberger notes that Canguilhem's emphasis on the need to appreciate the link between epistemology and "scientific" ways of

being-in-the-world that are directed toward life had its proximate origin in Husserlian phenomenology, particularly in Husserl's late work on the "crisis of European sciences" ("Gaston Bachelard" 313); see also Hyder, "Foucault, Cavaillès, and Husserl." However, as I document throughout this book, the belief that ontological investigations of life require a reassessment of epistemology is implicit in all of the authors I consider, and explicit in many. Modernist vitalists were especially emphatic about the importance of linking ontology and epistemology: Driesch, for example, argued in *The Problem of Individuality* that his vitalist experiments required the construction of a new logical category of causality (see 44–81), while in *Creative Evolution*, Henri Bergson contended that *"theory of knowledge* and *theory of life* seem to us inseparable," and it was "necessary that these two inquiries, theory of knowledge and theory of life, should join each other, and, by a circular process, push each other unceasingly" (xiii).

24. I discuss Abernethy in chapter 3. The quotation from Thomas Mann is from his diary entry of 1 March 1921, cited in Harrington, *Reenchanted Science*, 59.

25. Cooper, *Life as Surplus*, 42.

26. Examples of political-contextual explanations of Abernethy's vitalism include Marilyn Butler's introduction to Mary Wollstonecraft Shelley, *Frankenstein: The 1818 Text*, xviii–xxi, and Ruston, *Shelley and Vitality*, 24–73. An example of a political-contextual explanation of Uexküll's work is Harrington, *Reenchanted Science*, 34–71.

CHAPTER 1: **Romanticism, Art, and Experiments**

1. Enzensberger, "Aporias," 35.

2. Adorno, *Aesthetic Theory*, 37.

3. In *Experimental Cinema: The Film Reader*, for example, editors Foster and Winston-Dixon do not discuss the term "experimental" at all, while in their introduction to the *Routledge Companion to Experimental Literature*, editors Bray, Gibbons, and McHale suggest that "unfettered improvisation and the rigorous application of rules, accidental composition and hyper-rational design, free invention and obsessively faithful duplication, extreme conceptualism and extreme materiality, multimediality and media-specificity, being 'born digital' and being hand-made—all of these, and many others, are ways of being experimental in literature" (1). While Bray, Gibbons, and McHale seek to corral this diversity somewhat by proposing that "one feature that all literary experiments share is their commitment to raising fundamental questions about the very nature and being of verbal art itself," they acknowledge that such an account also "intimat[es]" "that the history of experimentation in literature might be considered as old as the history of literature itself" (17).

4. The paradigm for this approach was arguably established by Renato Poggioli in his well-known *Theory of the Avant-Garde* (1968), for Poggioli claimed both that experimentalism is "one of the primary characteristics of avant-garde art" (131) and that experimentalism pursued to its end "results in the contradiction or negation of the purely aesthetic end of the work of art" (136). For Poggioli, art is premised on *creation*, yet "experiment precedes creation"; as a consequence, while an experiment can "fus[e] into creation" and still result in art, if an artist seeks to fuse artistic "creation into experiment," then he or she will have left the realm of art (137). Such a position allows the avant-garde to have its cake and eat it too: used judiciously, experiments allow the avant-garde to produce art; however, should we desire something more political from the avant-garde than mere art,

it can produce that as well by simply pushing its tendency to experiment to the extreme. Perhaps not surprisingly, then, Poggioli ends up seeking to distinguish "good" from "bad" experimentalism: "if in the best cases [of avant-garde art] the experiment does become an authentic experience (in the most profound sense of the word), all too often, in the more literal-minded and narrow avant-garde, it remains merely an experiment" (135). The link Poggioli establishes between (true) experiments and experience is important, and it is one to which I return below, but he unfortunately does not clarify this distinction in his account.

5. See, e.g., Bray, Gibbons, and McHale, *Routledge Companion to Experimental Literature*, 2; Bender, "Novel Knowledge."

6. Coleridge and Wordsworth, *Lyrical Ballads*, 47. Robert Southey (anonymously) reviewed the volume's "experimental poems" in the October 1798 *Critical Review*; the anonymous reviewer for the 1799 *New Annual Register* complained of the volume's "unfortunate experiments"; and the anonymous reviewer of the January 1799 *New London Review* hoped that this review would convince the poet of "the failure of these 'Experiments.'" (These reviews are collected in Coleridge and Wordsworth, *Lyrical Ballads*, 148–67.) Nor was explicit invocation of the term "experiment" limited to British Romantic literary authors, for German poets and prose authors such as Novalis (Georg Philipp Friedrich Freiherr von Hardenberg) and Friedrich Schlegel invoked the term in similar ways at roughly the same time; see Daiber, *Experimentalphysik des Geistes*, and Brain, "Romantic Experiment as Fragment." For other accounts that locate the origin of the terminology of experiments to the Romantic era, see Poggioli, *Theory of the Avant-Garde*; Siskin, *Work of Writing*; and Gamper's cottage industry of edited volumes on *Experiment und Literatur* (which, though technically commencing with late sixteenth-century texts, in fact date the concept of "Literatur"—and hence, the explicit connection between concepts of "literature" and "experiment"—to the late eighteenth century): Gamper, Wernli, and Zimmer, *Es ist nun einmal*; Gamper, Wernli, and Zimmer, *Wir sind Experimente*; Gamper, *Experiment und Literatur*; and Bies and Gamper, *Es ist ein Laboratorium*.

7. On Zola's view of experiments, see Zola, "Experimental Novel," and Bender, "Novel Knowledge," esp. 284–86; on experimentalism in the avant-garde, see Poggioli, *Theory of the Avant-Garde*, 130–47; on the link between Stein's psychological research and literary writing, see Kittler, *Discourse Networks*, 225–29, and Cecire, "Sense of the Real," esp. 60–105; on experimental and minimalist music (and the opposition of both to avant-garde music), see Nyman, *Experimental Music*; on Fluxus, see Friedman, *Fluxus Reader*; on Oulipo, see Motte, *Oulipo*.

8. My division of science studies into three "waves" draws inspiration from Hans-Jörg Rheinberger's historiography in *Toward a History of Epistemic Things*, 11–37, as well as (though more loosely) on Stengers, *Invention of Modern Science*, 3–37.

9. Stengers, *Invention of Modern Science*, 49.

10. Kuhn, *Structure of Scientific Revolutions*, 151.

11. Max Planck, cited in ibid., 151.

12. Kuhn's account had an implicit policy lesson for federal funding of science: if "Science exhibited a self-organizing structure," it should be well funded by the government without being asked to justify its research priorities (citation from Mirowski, "Kuhn," 90; see also Stengers, *Invention of Modern Science*, 4–6).

13. Kuhn, *Structure of Scientific Revolutions*, 151.

14. See, e.g., Bloor, *Knowledge and Social Imagery*; Shapin and Schaffer, *Leviathan and the Air-Pump*; and Galison, *How Experiments End*.

15. Gooding, Pinch, and Schaffer argue in *Uses of Experiment* that the "experimenter's task" is "the persuasion of others" (xiv). Because advocates of the sociological approach bracket the question of the truth of scientific accounts of the world, they are often described as "social constructivists," which implies that scientific truth claims about the natural world are neither objective nor discovered but rather constructed. Yet some social constructivists have argued that their position is not necessarily opposed to the claim that science provides true accounts of the world; see, e.g., Bloor's contributions to his debate with Latour: Bloor, "Anti-Latour"; Latour, "For David Bloor"; and Bloor, "Reply to Bruno Latour." In order not to confuse these issues, I second Barbara Herrnstein Smith's point that "constructivism" is less a description of an actual methodology than a rubric—often used derisively by critics—under which many different approaches have often been lumped together. See her *Scandalous Knowledge*, 1–17.

16. Latour, *Pasteurization of France*.

17. Latour's description of science as "politics by other means" (*Pasteurization*, 229) encourages a reading of his work as sociological in focus. However, Latour's understanding of "politics" goes beyond human relationships, for it encompasses all relationships of "force," including those among nonhuman entities (153–226).

18. Rheinberger, "Experiment, Difference, and Writing," 324, my emphasis.

19. The ambiguity of Latour's concept of the "parliament of things," for example, is arguably a consequence of the way in which his writings hybridize sociological and ontogenetic approaches to experimentation. Depending on whether one prioritizes sociology or ontology, Latour's theory appears as either a sociological theory about the role of nonhuman entities and assemblages in mediating human relationships or as an ontogenetic theory that positions relationships among humans as simply one instance of relationships of force, power, and alliances among things in general. Latour has thus been simultaneously appropriated by scholars in the humanities interested in sociological questions (see, e.g., Leask, *Curiosity*); savaged by advocates of more traditional sociological accounts of science (see, e.g., Bloor, "Anti-Latour"); and valorized by philosophers interested in moving away from human-centered perspectives (see, e.g., Harman, *Prince of Networks*).

20. Cage, *Silence*, 7.

21. Adorno, *Aesthetic Theory*, 23.

22. Ibid., 326, 330.

23. Cage, *Silence*, 8.

24. Adorno, "Difficulties," 658–59. Adorno's account seeks to resolve dialectically Poggioli's otherwise paradoxical account of the effects of artistic experimentation, for Poggioli simultaneously suggests that avant-garde experimentalism *expanded* the field of art by increasing the number of artistic formal techniques, subjects, and kinds of art (*Avant-Garde*, 133), yet also effectively *contracted* the field of art by engaging in a scorched-earth policy of evaluation that "liquidated"—that is, cast as uninteresting—all past art (132).

25. Adorno, *Aesthetic Theory*, 24.

26. Adorno, "Difficulties," 654, 659.

27. Quotations from Southey's review, cited in Coleridge and Wordsworth, *Lyrical Ballads*, 149, 150.

28. Coleridge and Wordsworth, *Lyrical Ballads*, 47. As Mayo notes in "Contemporaneity of the *Lyrical Ballads*," the poems in *Lyrical Ballads* did not differ greatly in content or form from other poetry written at the same time. Yet Mayo's attempt to establish an ontological continuity between the poems in *Lyrical Ballads* and other contemporary verse misses completely the difference produced when poets explicitly *label* their poems "experiments" and that label is taken up by critics and readers. The difference between the poems in *Lyrical Ballads* and contemporary poems is thus not located solely, or even primarily, in the realm of formal devices and poetic content but rather in the kinds of communities that emerge (or do not emerge) around the poems (a point to which I return below). Understanding how and why communities emerge is not solely a matter of investigating the properties of objects (in this case, poems) but also requires that we understand that language can be used to produce an "event" simply by naming and dating something (e.g., reviewers' invocations of "Wordsworth's and Coleridge's poetic 'experiments' of 1798"). On the importance of names and dates for the constitution of events, see Deleuze and Guattari, *Thousand Plateaus*, 75–110.

29. See Schaffer, "Self Evidence" and "Consuming Flame"; Golinski, *Science as Public Culture*; Stewart, *Rise of Public Science*; and Lynn, *Popular Science and Public Opinion*.

30. Lynn, *Popular Science and Public Opinion*, 126.

31. Nollet, *Leçons*, 1:xxx–xxxi.

32. Bewell, *Wordsworth and the Enlightenment*, 17.

33. Siskin, *Work of Writing*, 46.

34. Ibid., 21. Siskin contends that the "self-critical turns [that lyric] enabled conformed productively to the need within the Baconian experimental method to alternate (think of the repetition of the chorus in ballads, the strophic sequences of the ode, the apostrophic turns of sonnets) between episodes of empirical practice (its lining up of finely detailed images) *and* the formulation of general propositions (its epodic marshaling of those details into large-scale claims about temporality and humanity). The very abruptness of these alternations between the empirical and the general left an aftereffect: what is usually experienced as the lyric's supposedly defining traits of personal intensity" (138).

35. Dear, *Discipline & Experience*, 4. For neo-Aristotelians, "experience" had to be expressed in universal propositions (e.g., "heavy bodies always seek to fall") rather than as a specific proposition (e.g., "On this day at this time, this heavy body fell"). For a discussion of the shifting relationships between concepts of experience and experiment in the sixteenth and seventeenth centuries, see Daston and Park, *Wonders*, 215–53.

36. Dear, *Discipline & Experience*, 11.

37. See, e.g., Shapin and Schaffer, *Leviathan and the Air-Pump*.

38. Sprat, *History of the Royal Society*, 113.

39. See Wordsworth's 1800 note to "The Thorn" in Wordsworth and Coleridge, *Lyrical Ballads*, 211–12. On the distinction between preternatural and supernatural causes, see Daston and Park, *Wonders*, 159–72.

40. Shapin, "Pump and Circumstance"; Shapin and Schaffer, *Leviathan and the Air-Pump*.

41. Wordsworth later sought to blunt the experimental impetus of the 1798 volume of *Lyrical Ballads*, both by appropriating to his name the 1800 and 1802 volumes, rather than allowing the work to remain anonymous, and by contending in 1800 and 1802 that the volume *as a whole* was *an* experiment (rather than, as in the original formulation, positioning the volume as containing multiple experiments). Where the multiple experiments of the first volume pointed toward an decentered collective able to take up the challenge of future experiments, Wordsworth's repositioning of the volume as his singular experiment made it easier to place the text within the biography of an author ("first came the locodescription of *Descriptive Sketches* and *An Evening Walk*; then the 'experiment' of *Lyrical Ballads*; followed by . . . ").

42. Coleridge and Wordsworth, *Lyrical Ballads*, 47.

43. My approach thus differs from Bender's claim in "Novel Knowledge" that the early-eighteenth-century British novel should be understood as experimental because it "participate[d] in the aspirations and uncertainties about knowledge, experience, and experiment pervasive during the scientific revolution of which it was a part" (288). Precisely because early-eighteenth-century British novels were *not* explicitly described by their authors or others *as* "experiments," Bender must assert a relatively diffuse continuity of "aspirations and uncertainties" between novels and science in order to make his case that the two were related. This difficulty underscores the important link among events, naming, and dating that I describe above (n. 28).

44. Rheinberger, "Experiment, Difference, and Writing," 324, my emphasis. Rheinberger's work has inspired many recent German-language reflections on the relationship between literature and experiments: e.g., Solhdju's *Selbstexperimente* and many essays in Gamper's edited collections (*Es ist einmal*; *Wir sind Experimente*; and *Experiment und Literatur*, as well as Bies and Gamper, *Es ist ein Laboratorium*). Yet because much of this work focuses primarily on the epistemological, rather than the ontogenetic, dimensions of Rheinberger's approach—by, for example, understanding both science and art as modes of "research" by means of which "knowledge" is produced—it tends to miss what I describe as the eccentric relationship of artistic to scientific experiments.

45. Though Rheinberger uses the word "reproduction," he has in mind not so much situations in which one science (say, biochemistry) uses experimental systems to maintain its "essence" in the face of other competitor sciences (say, molecular biology), but rather situations in which an experimental system serves as the means by which different sciences are hybridized and new sciences created. In the case study that forms the subject of *Toward a History of Epistemic Things*, Rheinberger traces the way in which the experimental system of protein synthesis emerged in one field (cancer biochemistry) but then took root in a completely different field (the new science of molecular biology).

46. Citation from Francis Jeffrey's review of Robert Southey's *Thalaba the Destroyer*, reproduced in Coleridge and Wordsworth, *Lyrical Ballads*, 409–10, 413. Jeffrey's concern about the Lake Schoolers was so great that he did not manage to get around to his negative assessments of Southey's experiments, such as Southey's "English Sapphics" and the ostensible subject of the review, *Thalaba the Destroyer*, until midway through the essay.

47. Wordworth, *Lyrical Ballads, with Pastoral and Other Poems*, xxxviii–xxxix.

48. In the remainder of this chapter, I use "Art" and "art-in-general" as synonyms. For several different, though ultimately not disjunctive, accounts of the transformation of "the arts" into "Art" in the late eighteenth and early nineteenth centuries, see Shiner,

Invention of Art; Rancière, *Politics of Aesthetics*; de Duve, *Kant after Duchamp*; and Bourdieu, *Rules of Art*. As de Duve notes, the transformation of the arts into Art also eventually enabled the fundamentally modern aspiration of becoming an all-purpose "artist" (rather than, for example, "a poet" or "a painter") (375).

49. Lessing, *Laocoön*. However, as David Wellbery emphasizes, the apparent modernity of Lessing's book is purchased only at the cost of ignoring the hierarchy between time-spaced and space-based arts Lessing subtly worked to establish. See Wellbery, *Lessing's Laocoon*.

50. See Bloom, *Anxiety of Influence*.

51. Rancière, *Politics of Aesthetics*, 23.

52. See Fleck, *Genesis and Development*, and Kuhn, *Structure of Scientific Revolutions*. For Kuhn's criticisms of attempts to see a past paradigm as a special case of a later paradigm, see *Structure of Scientific Revolutions*, 101–2.

53. Deleuze and Guattari, *What Is Philosophy?*, 124–25. For a related account of modern science as a project of making time "irreversible," see Latour, *Pasteurization of France*, 49–48, and *Pandora's Hope*, 145–73.

54. As Kahn notes (critically) in *Noise, Water, Meat*, several generations of composers have sought to ensure the persistence of a Cageian approach to music by rallying around the concept of "experimental music" and thereby distinguishing themselves from "avant-garde" music. Yet it is still difficult to imagine perceiving anything like "progress" in this series of experiments.

55. See, e.g., Greenberg, "Modernist Painting."

56. Adorno, *Aesthetic Theory*, 1.

CHAPTER 2: Suspended Animation and the Poetics of Trance

1. From Coleridge, *Biographia Literaria*, part 2, 6. In notes for his 1808 *Lectures on the Principles of Poetry*, Coleridge argued that theater too depended upon a state of "temporary Half-Faith, which the spectator encourages in himself & supports by a voluntary contribution on his own part," this latter by means of a willing "suspension of the Act of Comparison" (*Lectures* 134–35).

2. Quoted in Tomko, "Coleridge's 'Suspension of Disbelief,'" 241. Tomko also surveys different twentieth-century approaches to the interpretation of Coleridge's phrase.

3. Coleridge, *Lectures*, 124.

4. Mary Shelley, "Roger Dodsworth," 43. For an account of this hoax, see Robinson, "Roger Dodsworth Hoax," 20–28.

5. Culler, "Why Lyric?" 205.

6. Suspended animation remains both a literary trope and vexing object of scientific research, still balanced between scientific reality and literary dream. For discussion of the dreams and technologies currently associated with suspended animation, see Doyle, *Wetwares*, 63–87, and the chapter entitled "Freezing Time" in Mitchell, Burgess, and Thurtle, *Biofutures*.

7. See Wang, "Romantic Sobriety."

8. The Royal Humane Society still exists, though now as a "charity that assesses acts of bravery in the saving of human life and makes awards" (www.royalhumanesociety.org.uk/, accessed 17 December 2006). For historical accounts of the society, see Coke, *Saved from a Watery Grave*; and Carolyn D. Williams, "Luxury of Doing Good."

9. Anon., *Transactions*, x.

10. For accounts of various therapies and technologies, see Anon., *Transactions*; Hunter, "Proposals," 412–25; and Ogden, *Plain Directions*.

11. Anon., *Transactions*, 280n.

12. Blagden, "Experiments and Observations" and "Further Experiments."

13. Blagden, "Experiments and Observations," 118.

14. Ibid., 122. Though Blagden and Hunter implicitly positioned heat as a kind of fluid or substance that could be created or destroyed, both were primarily interested in the ability of living beings to regulate bodily temperature, rather than the question of the origin of animal heat; from this perspective, Everett Mendelsohn's contention in *Heat and Life* that Hunter's "interpretation of his results added nothing new and left the discussion of animal heat in as confused a state as before" (101–2) is based on a misunderstanding of the focus of Hunter's experiments.

15. Hunter primarily used the term "simple life" in his unpublished *Lectures on the Principles of Surgery* (e.g., 21, 36, 38, 107), though see also his *Treatise on the Blood*, 78. Since Hunter provided no term for the form of life tied to action, "practical life" is my own term (drawn from the Greek *pratteo*, to act).

16. Hunter, *Lectures*, 223.

17. Hunter, "Of the Heat," 15, 20.

18. Hunter, *Observations* (1786), 115n; for a slightly different wording, see also Hunter, *Observations* (1792), 130.

19. Hunter, *Observations* (1786), 155n. In the 1792 edition of *Observations*, Hunter was more hesitant, describing the stimulus as "preceding" death and suggesting that in this "contracted state, possibly, absolute death is produced" (130).

20. Schelling, *First Outline*, 5. Žižek emphasizes Schelling's commitment to the principle that "there is something that precedes the Beginning itself—a rotary motion whose vicious cycle is broken . . . by the Beginning proper" (Schelling and Žižek, *Abyss of Freedom*, 13); see also Krell, *Contagion*, 90–100, and Rajan, "First Outline." Brown discusses the importance of suspension for German Romantic-era philosophy in *Gothic Text*, 92–104.

21. For Hunter's most extensive discussion of his work with fowl eggs, see "Of the Progress." Though it is arguably anachronistic to use the term "model organism" to describe eighteenth- and nineteenth-century scientific interest in poultry eggs, there are nevertheless significant parallels between Romantic-era interest in eggs and more recent use of model organisms such as *Drosophila melanogaster* (fruit fly) or *Mus musculus* (common mouse). For Romantic-era researchers, the advantages of poultry eggs were their ready availability, their relatively large size, and the fact that development occurred outside the parent animal, which made it possible to observe and document developmental processes. For these reasons, Hunter noted, "It would almost appear that this mode of propagation [of fowl] was intended for investigation" (205).

22. Hunter, "Of the Heat," 28; for Hunter's description of the different elements of the egg, see "Of the Progress," 200–202. I draw the term "non-organic life" from Deleuze, *Cinema 1*, 50–51. For Deleuze's more detailed discussion of the role of non-organic life in embryological development, see *Difference and Repetition*, 249–54.

23. Hunter, *Lectures*, 223; see also his "Of the Progress," 202, and "Experiments and Observations," 119–20. For an example of Hunter's claim that simple life could not be

reduced to organization, see "Experiments and Observations," in which he noted that he "had long suspected, that the principle of life was not wholly confined to animals, or animal substance endowed with visible organization and spontaneous motion; but I conceived, that the same principle existed in animal substances, devoid of apparent organization and motion, where the power of preservation simply was required" (119).

24. Hunter, "Of the Heat," 30.

25. Hunter, "Of the Progress," 207.

26. Though James F. Palmer, the nineteenth-century editor of Hunter's *Works*, identified Hunter's distinction between simple and active life with Xavier Bichat's distinction between "organic life" and "animal life" (Palmer in Hunter, *Works* 1:242 n.1), the relevant comparison is rather Jean-Baptiste Lamarck's distinction between "active" and "suspended" life. Bichat defines "organic life" as the capacity of a living being to "transfor[m] into its proper substance the particles of other bodies, and afterward rejec[t] them when they are become heterogeneous to its nature," while "animal life" is the capacity of the living to "fee[l]," "perceiv[e]," and "reflec[t] on its sensations" (*Physiological Researches*, 13). Organic life characterizes "all organized beings, whether animal or vegetable . . . because organic texture is the sole condition necessary to its existence"; animal life "is exclusively the property of the animal" (14). Yet for Hunter, simple life could *precede* the emergence of organs (e.g., fertilized eggs) and could persist even when organs and organic functions were suspended (e.g., suspended animation). Hunter's distinction is much closer to Jean-Baptiste Lamarck's claim that "life . . . is an order and state of things which permit of organic movements; and these movements constituting active life result from the action of a stimulating cause" (*Zoological*, 202; *Philosophie*, 390). For Lamarck, life had two modalities, "active" (*active*) or "suspended" (*suspendue*); the latter mode was exemplified by the ability of desiccated polyps, rotifers, infusorians, mosses, and algae to return to "active life"; by hibernation in mammals; and by cases of people who were revived after having apparently drowned (*Zoological*, 203–4; *Philosophie*, 381–93).

27. See Hunter, "Proposals," 417–19, 422, and "Case of Paralysis."

28. As Sir William Lawrence noted in the context of an homage to Hunter, "The eighteenth century will be ever memorable for the advancement, not only of general civilization, but of all branches of knowledge" (*Introduction*, 53–54).

29. The *Transactions*'s authors suggested that the fact that a society devoted to victims of suspended animation first appeared in Amsterdam was due to the "great abundance of canals" there (Anon., *Transactions*, 2); for discussion of canal victims in Britain, see 85, 105, 230. On the importance of canals to eighteenth-century development of modern structures of communication, see Mattelart, *Invention of Communication*, esp. 29.

30. On the invention of "history" in the late eighteenth and early nineteenth centuries, see White, *Metahistory*; Fabian, *Time and the Other*; and Chandler, *England in 1819*.

31. Koyré, *Metaphysics and Measurement*, 8–9, 124–26. Koyré describes the modern (i.e., post-Brunonian) account of this "experiment" (most often performed in thought rather than in fact): "if we imagine two men, one of them on the top of the mast of a ship passing under a bridge, and the other on that bridge, we may imagine, further, that at a certain moment, the hands of both of them will be in the selfsame place. If, at that moment, each of them shall let a stone fall, the stone of the man on the bridge will fall down (and in the water), but the stone of the man on the mast will follow the movement

of the ship, and (describing, relatively to the bridge, a peculiar curve) fall at the foot of the mast" (9).

32. Suspended animation was in this sense the objective, vitalist, and experimental correlate of the unhinging of time that Immanuel Kant effected in the *Critique of Pure Reason*. As Gilles Deleuze notes, "As long as time remains on its hinges, it is subordinate to movement: it is the measure of movement"; with Kant, however, "Time is no longer related to the movement which it measures, but movement is related to the time which conditions it: this is the first great Kantian reversal in the *Critique of Pure Reason*" (*Kant's Critical Philosophy*, vii).

33. Hunter, *Treatise*, 78; Whiter, *Dissertation*, 14.

34. The bibliography of work on the Romantics' deep knowledge of contemporary scientific concepts and developments is large, though see esp. Levere, *Poetry Realized in Nature*; Sperry, "Keats and the Chemistry of Poetic Creation"; Goellnicht, *Poet-Physician*; De Almeida, *Romantic Medicine*; Richardson, *British Romanticism*; Ruston, *Shelley and Vitality*; Holmes, *Age of Wonder*; and Gigante, *Life*.

35. Mary Shelley, *Frankenstein*, 50.

36. Keats, "To Autumn," lines 26, 25, 12.

37. Keats, "Ode on a Grecian Urn." For a brief discussion of the near-suspension of animation in Wordsworth's "Tintern Abbey," see Robert Mitchell, "Suspended Animation," 119–20, n.12.

38. Curran, *Poetic Form*, 66. For a more general history of the shifting expectations associated with the form of the ode, see Maddison, *Apollo and the Nine*, 286–401.

39. Grant Scott, *Sculpted Word*, and W. J. T. Mitchell, *Picture Theory*, 173.

40. Levinson, *Keats's Life of Allegory*, 32, 178.

41. Extending Levinson's argument, Keats's poem seems to seek a middle ground between the unique ancient sculpture it describes and the replicable commodity-form upon which its dissemination depends. Keats's ode solicits readers to return again and again to the poem, just as one might visit a museum object repeatedly. Yet like the commodity-form, the ode seeks less to impart an experience of pastness than to encourage the free play between sensation and imagination that Campbell outlines in his account of consumerism in *Romantic Ethic*. For a short but compelling reading of Keats's ode in relationship to the suspensions of the commodity form, see Collings, "Suspended Satisfaction."

42. Murray Krieger, *Play and Place*, 105.

43. Rajan, "Keats, Poetry," 344. J. Hillis Miller's "linguistic moment" also emphasizes "moments of suspension within the texts of poems" that constitute "a form of parabasis, a breaking of the illusion that language is a transparent medium of meaning" (*Linguistic Moment*, xiv).

44. Shelley and Shelley, *Six Weeks' Tour*, 167.

45. "Torpor" is perhaps most familiar to readers of Romantic-era literature in the psychological sense in which Wordsworth used the term in his preface to the 1802 *Lyrical Ballads*. There he claimed that "a multitude of causes . . . was now acting with a combined force to blunt the discriminating powers of the mind . . . reduc[ing] it to a state of almost savage torpor" (xv). However, the concept of torpor was also central to medical discourse on suspended animation: in the *Transactions* of the Royal Humane Society, Anthony Fothergill compared the state of people apparently drowned with "the TORPID

and apparent LIFELESS STATE which the marmot and other dormant animals undergo in their cells" (Anon., *Transactions* 290), while in *Zoonomia*, Erasmus Darwin described "snails, which have recovered life and motion on being put into water after having experienced many years of torpidity, or apparent death, in the cabinets of the curious" (2:470).

46. This productive difference between living beings and their environments is also evident in Shelley's "Ode to the West Wind" (1820), in which the narrator seeks to channel this wind—"be thou me, impetuous one!" (line 62)—which serves as a vector for simple life by charioting to "their dark wintry bed / The wingèd seeds, where they lie cold and low, / Each like a corpse within its grave, until / Thine azure sister of the Spring shall blow / Her clarion o'er the dreaming earth" (lines 6–10).

47. Keach, *Shelley's Style*, 195.

48. Hollander, *Vision and Resonance*.

49. Keach, *Shelley's Style*, 257 n.18.

50. Wimsatt, *Verbal Icon*, 157, 163–64.

51. In his "Closing Statement," Jakobson draws heavily on Wimsatt's schema (esp. 367–68).

52. The premise of an adjudicating consciousness undergirds many recent theories of rhyme and meter, including the account of the "metrical code" Finch describes in *Ghost of Meter*—and which itself draws on Hollander's account of the "metrical contract"—and accounts of the "embodied" dimensions of poetry recitation, such as Abrams's and Richardson's suggestions that feelings of pleasure result from the "fit" between the physiology of enunciating, and the meaning of, lines of verse (e.g., Keats's description of "Joy's grape" bursting "against [a] palate fine" in "Ode to Melancholy," "gains much of its feeling of rightness from its physical enacture in the mouth of the reader"). Where Wimsatt emphasizes judgments concerning relationships between the fact of repetition and the semantic meaning of words, Abrams and Richardson stress judgments that relate the physical act of speaking lines of verse to physical actions described in the verse. Richardson, *British Romanticism*, 139, paraphrasing Abrams, "Keats's Poems."

53. Pound, "Retrospect," 3; Olson, "Projective Verse."

54. I expand on this latter conception of medium/milieu in chapter 5.

55. For a discussion of suspension in Milton's poetry, see Frontain, "Typological Identity."

56. Milton, "Sonnet XIX," line 14. For a brief but helpful discussion of Milton's sonnet in the context of a discussion of rhyme in Wordsworth's poetry, see Fry, *Poetry of What We Are*, 111.

57. Keats, *Letters*, 1:185. I return to Keats's claim about sensation below.

58. Coleridge, *Lectures*, 124.

59. There is arguably something a bit unfair about my use of Coleridge as synecdoche for a critical strain of Romantic thought about suspended animation, since some of his poetry (e.g., "Kubla Kahn") seemed to aim at the same kind of trance I have described as the goal of Keats's and Percy Bysshe Shelley's cold pastoral. My next chapter addresses this potential injustice by focusing on the complexities of Coleridge's thought.

60. Buck-Morss, "Aesthetics and Anaesthetics," 18, 22, 12.

61. Percy Bysshe Shelley, *Queen Mab*, 80 (lines 147–49). This image in *Queen Mab* was a poetic reworking of a line of thought that Shelley first developed in correspondence with Elizabeth Hitchener. In his 2 January 1812 letter to Hitchener, for example,

he argued that an understanding of life as "infinite" was equivalent to the position that "every thing is animation" (*Letters* 1:156).

62. Shelley's doctrine of passive resistance, exemplified in a poem such as "The Mask of Anarchy" (1819), has often been interpreted from a purely psychological view: passive resistance is effective because it "shames" aggressors (see, e.g., Frosch, "Passive Resistance in Shelley"). It is more productive to approach Shelley's understanding of passive resistance from the perspective of suspended animation: by suspending one's own action, one enables greater capacities of sensation—that is, lines of linkage with systemic potentials—for others. Such a perspective avoids the dubious causal claims of the psychological approach—it is not at all clear why passive resistance ought "automatically" to produce shame in aggressors—while retaining the emphasis on system to which Shelley was committed.

63. For a useful, though schematic, account of this tradition, see Hamlyn, *Sensation and Perception*. In ancient and medieval philosophy, sensation generally referred to the capacity of sense organs to retain the form or "species" of an object. While modern philosophy since at least Locke has generally understood sensation as the reception of sense "impressions" or data about—rather than the form or species of—objects, the basic schema of sensation as the passive receipt of something already formed is assumed by philosophies otherwise as different as those of John Locke, David Hume, Étienne Bonnot de Condillac, and Immanuel Kant.

64. Coleridge, *Aids to Reflection*, 233.

65. In *Aids to Reflection*, Coleridge wrote that the "proper functions of the understanding" are *abstraction* ("generalizing the notices received from the senses in order to the construction of names") and *identification* ("referring particular notices . . . to their proper names; and *vice versa*, names to their correspondent class or kind of notices") (232).

66. This second, ontological approach must be distinguished from another Romantic-era alternative approach to sensation—namely, sensation understood as a form of knowing—described by Jackson in *Science and Sensation in Romantic Poetry*. As Jackson notes, "sensation" was a contested term in the Romantic era, with some authors using it to refer to the brute impression of reality on the senses, while others, such as Wordsworth, regarded sensation "as much as a category of cognition as of physical response" (10). Yet interpreting sensation as itself a mode of cognition does not challenge the epistemological paradigm and its need for a middleman that mediates between reality and cognition; rather, it assumes this paradigm while at the same time leaving unanswered the question of where the hand-off between reality and consciousness occurs.

67. For an acute discussion that emphasizes the extent of Haller's departure from the work of his predecessors (and especially Haller's mentor, Herman Boerhaave), see Vila, *Enlightenment and Pathology*, 16–28.

68. Some physiologists sought to reduce one side of this distinction to the other: in *Recherches sur l'histoire de la médecine* (1767), for example, Théophile de Bordeu argued that sensibility "seems much easier to understand than irritability, and it [sensibility] can serve quite well as the basis for explaining all of the phenomena of life, whether it be in the state of health or in illness" (*Oeuvres* 2:668, cited in Vila, *Enlightenment and Pathology*, 38), while Erasmus Darwin, by contrast, conjectured that each living being began as a "filament," which possessed only irritability, with sensibility emerging only later, as

a dilution of irritability: "sensibility may be conceived to be an extension of the effect of irritability over the rest of the system" (*Zoonomia* 1:558). For the relative eclipse of irritability by sensibility in nonmedical discussions, see Vila, *Enlightenment and Pathology*.

69. Davy, cited in Hoover, "Some Early Experiments," 14.

70. Deleuze and Guattari, *What Is Philosophy?* 183.

71. I draw the language of "coupling" and "registers" from Gumbrecht, "Rhythm and Meaning," 178–79.

72. In *Essay on Man*, Pope contended that he composed in rhyme, rather than prose, because "principles, maxims, or precepts so written, both strike the reader more strongly at first, and are more easily retained by him afterwards" (270).

73. Gumbrecht, "Rhythm and Meaning," 182.

74. The first quote is from Chew, "Nineteenth Century," 1250; the second from Leavis, *Revaluation*, 263; the third from Walter Jackson Bate, *John Keats*, 584; and the fourth from Nemoianu, "Dialectics of Movement," 208. See also Bush's claim that "To Autumn" presents "less a resolution of the perplexities of life and poetic ambition than an escape into the luxury of pure—if now sober—sensation" ("Keats" 242). In "Keats's 'To Autumn,'" Lindenberger surveys early- and mid-twentieth-century criticism of the poem, while Nemoianu summarizes several interpretations of the poem as "stagnant."

75. In *Fate of Reading*, for example, Hartman describes Keats's ability to avoid introducing a subjective voice into the poem as "true impersonality" (146), while in "Keats's 'To Autumn,'" Lindenberger notes that the poem "seems as impersonal as any poem written during the Romantic period" (123). For a discussion of the importance of impersonality for Romantic poetics, see Khalip, *Anonymous Life*.

76. Culler, *Pursuit of Signs*, 135–54.

77. The suspension of meaning differs from its absence. Twittering is an articulated, rhythmic form of speech, the "semantic" content of which is located outside of human meaning; it thus differs from both noise (which manifests itself as both inarticulate and without meaning) and from instances of languages that we do not know but nevertheless recognize *as* human languages (which manifest themselves as both articulate and possessed of meaning, even if I do not personally have access to that meaning).

78. Keats, *Letters*, 1:185. Keats used "sensation" in a much more limited, derogatory, sense when he despaired that the reading public desired "either pleasant or unpleasant sensation"; "What they want is a sensation of sort" (*Letters*, 2:189).

79. For the strong association between poetry and drugs in Keats, see De Almeida, *Romantic Medicine*, 135–215. My use of the term "addiction" draws on Clark, "Heidegger's Craving."

80. Coleridge and Wordsworth, *Lyrical Ballads*, 46, 171.

81. "The imagination is enlarged by a sympathy with pains and passions so mighty, that they distend in their conception the capacity of that by which they are conceived" (Percy Bysshe Shelley, *Shelley's Prose*, 285).

82. Siskin, *Work of Writing*, 20.

83. See Winter, "Ethereal Epidemic."

84. For discussion of the history of cryogenic technologies, see Scurlock, *Cryogenics*.

85. See Bergson, *Matter and Memory* and *Creative Evolution*. For a history of the importance of bracketing or suspension in phenomenology, see Taminiaux, *Metamorphoses of Phenomenological Reduction*. In *Fundamental Concepts*, Heidegger was es-

pecially attentive to the nuances of the concept of suspension, employing a distinction between *Hingehaltenheit* and *Hineingehaltenheit*—translated as "being held in limbo" and "being held out into"—as explanation for the way in which *Dasein* is able to escape its immersion in the everyday (xix).

86. Deleuze, *Difference and Repetition*, 73.

87. In *Difference and Repetition*, Deleuze writes, "Embryology shows that the division of an egg into parts [i.e., proto-organs] is secondary in relation to more significant morphogenetic movements: the augmentation of free surfaces, stretching of cellular layers, invagination by folding, regional displacement by groups. A whole kinematics of the egg appears, which implies a dynamic. Moreover, this dynamic expresses something Ideal. Transport is Dionysian, divine and delirious, before it is local transfer" (214). For Deleuze, the egg is an acute expression of the conditions of possibility of all organization: "The entire world is an egg. The double differenciation of species and parts always presupposes spatio-temporal dynamisms" (216).

88. See especially Deleuze and Guattari, *What Is Philosophy?*, 163–99.

89. See Deleuze, *Pure Immanence*, 29; see also Agamben, *Potentialities*, 228–30. For Charles Dickens's description of Roger "Rogue" Riderwood's fall into suspended animation, see Dickens, *Our Mutual Friend*, 443–46.

CHAPTER 3: Life, Orientation, and Abandoned Experiments

1. Though I stress here the distinction between the abandoned experiment and the Romantic fragment, the latter encouraged a new kind of scientific experimentation; see Brain, "Romantic Experiment as Fragment."

2. Griggs, "Samuel Taylor Coleridge and Opium," 359. In a letter of 18 November 1802 to his wife, Coleridge wrote that he was "fully convinced, & so is [his friend] T. Wedgewood, that to a person, with such a Stomach & Bowels as mine, if any stimulus is needful, Opium in the small quantities, I now take it, is incomparably better in every respect than Beer, Wine, Spirits, or any *fermented* Liquor—nay, far less pernicious than even Tea" (*Collected Letters*, 2:884).

3. Coleridge, *Collected Letters*, 3:489.

4. For accounts of Coleridge's various attempts to develop a coherent account of the will, see Christensen, *Coleridge's Blessed Machine of Language*; Brice, *Coleridge and Scepticism*; and Hedley, *Coleridge, Philosophy and Religion*.

5. What is clear is the intensity of Coleridge's efforts to find a way to theorize his incapacity. See, for example, his complicated notebook account, probably written between July and August of 1803, in which he suggested that "habits" may be the result of desire becoming caught, or bound, by that which initially served as the object of desire. Describing habit as "the Desire of a Desire," Coleridge suggested a distinction between "desire" and its "desirelets": "May not the Desirelet, a, so correspond to the Desire, A, that the latter being excited may revert wholly or in great part to its exciting cause, a, instead of sallying out of itself toward an external Object, B?" (*Notebooks* 1:1421). For a discussion of this notebook entry, see Youngquist, "Rehabilitating Coleridge," esp. 890–91.

6. This quote is drawn from Coleridge's letter of 10 April 1816 to Lord Byron, in which he described his "daily habit of taking enormous doses of Laudanum which I believed necessary to my Life, tho' I groaned under it as the worst and most degrading of Slaveries—in plain words, as a specific madness which leaving the intellect uninjured

and exciting the moral feelings to a cruel sensibility, entirely suspended the moral Will" (*Collected Letters* 4:626).

7. Coleridge, *Collected Letters*, 4:630.

8. Ibid., 3:489. As Youngquist astutely notes, even Coleridge's later, apparently more limited, claim that he could not *speak* a lie, even if he could act one, put Coleridge's belief in the autonomy of the will at risk: "although Coleridge cannot tell a lie, he can act one. His 'specific madness' splits speech and agency, saying and doing, leaving him subject to a habit that compels behavior he cannot control" (885–86).

9. On the origin and composition of *Sibylline Leaves* and *Biographia Literaria*, see the editors' introduction to Coleridge, *Biographia Literaria*, xlv–lxv.

10. Coleridge, *Collected Letters*, 4:674.

11. Coleridge and Wordsworth, *Lyrical Ballads*, 47.

12. On Wordsworth and Coleridge's discussions about the second edition of *Lyrical Ballads*, see Gamer and Porter's introduction to Coleridge and Wordsworth, *Lyrical Ballads*, 28–34.

13. In the second volume, Wordsworth wrote, "The First Volume of these Poems has already been submitted to general perusal. It was published, as an experiment which, I hoped, might be of some use to ascertain, how far, by fitting to metrical arrangement a selection of the real language of men in a state of vivid sensation, that sort of pleasure and that quantity of pleasure may be imparted, which a Poet may rationally endeavor to impart" (ibid., 171).

14. I draw the phrase "economy of experiment" from Andrews, "American Experiments."

15. Biagioli, *Galileo Courtier*, 36; see more generally 36–60. See also Findlen, *Possessing Nature*, 293–392, and Johns, *Nature of the Book*, 15–26.

16. See Shapin, "House of Experiment."

17. See Johns, *Nature of the Book*, 482; and more generally 44–542.

18. Boyle, *Works*, 2:63. On Boyle's view of science as a form of gift-exchange with God, see Shapin, *Social History of Truth*, 126–92.

19. *Pace* the model, developed by Hagstrom in *The Scientific Community*, of twentieth-century science as reliant on a form of gift exchange, Latour and Woolgar argue that scientific practices of information-gathering and attribution—that is, reading and citing the work of other scientists—should not be explained as *responses* to unwritten "norms" of gift exchange among scientists but rather as a *"demand* for credible information" that allows individual scientists to maximize the amount of scientific work they can produce (201–8). Though Latour and Woolgar's description is likely also applicable to eighteenth-century science, their model ignores what we might call the persistence of the gift-function in the sciences. As my analysis above suggests, the gift-function operates differently in different kinds of science—the site of the gift is different in sixteenth-century patronage science than in eighteenth-century institutional science—but the function itself nevertheless persists across different modes of science.

20. On the transformation of the patronage system, see Folkenflik, "Patronage and the Poet-Hero"; Korshin, "Eighteenth-Century Literary Patronage"; and Mark Rose, *Authors and Owners*, 1–17.

21. For a compelling, nuanced account of Wordsworth's "Calvert legacy," the poet's interest in benevolence, and his consequent "discovery of debt," see Liu, *Wordsworth,*

311–58. On Coleridge's Wedgwood annuity, see Griggs, "Coleridge and the Wedgwood Annuity."

22. Coleridge, *Biographia Literaria*, part 2, 119. Significantly, Coleridge inserted into his criticism of Wordsworth's experiments a long digression on the rules—or, in Coleridge's terminology, the "code"—that he thought ought to guide periodical critics in their judgment of works, and this code is remarkably similar to those rules of propriety that the Royal Society sought to enforce in its communications: for example, Coleridge contends that critics must comment only on the work, not on the author; to transgress this boundary is to engage in *"personal* injury" and *"personal* insults" rather than "criticism" (part 2, 109, my emphasis). Having earlier failed to consider fully the role of agonism in the reception of the experiments of *Lyrical Ballads*, Coleridge subsequently sought in *Biographia Literaria* to establish rules that would determine how literary works were to be judged by their public.

23. For the background of the text, see the editors' introduction to Coleridge, "Essay on Scrofula," 454–57.

24. See, e.g., Coleridge's letter of 14 December 1802 to his brother James, in which he wrote that he had no doubt that "there is a taint of Scrofula in my constitution . . . Where you find a man indolent in body & indisposed to definite action, but with lively Feelings, vivid ideal Images, & a power & habit of continuous Thinking, you may always, I believe, suspect a somewhat of Scrofula—With me it is something more than a suspicion—I had several glandular Swellings at School—& within the last four years a Lump has formed on my left cheek, just on the edge of my whisker—" (*Collected Letters* 2:987). For a helpful (though unsympathetic) discussion of the role of scrofula in Coleridge's correspondence and his composition of "Dejection: An Ode," see Wallen, "Coleridge's Scrofulous Dejection."

25. Coleridge, *Collected Letters*, 4:688.

26. Ottley, "Life of John Hunter," in Hunter, *Works*, 1:139–46.

27. Though Abernethy had begun as an experimentalist, publishing several volumes in the 1790s in which he discussed his research on topics such as "the composition and analysis of animal matter," he moved away from an experimental methodology when discussing Hunter's theory of life. See, for example, "An Essay on the Composition and Analysis of Animal Matter" in Abernethy, *Surgical and Physiological Essays*, 77–113.

28. Abernethy, *Physiological Lectures*, 16, 34.

29. Ibid., 40. Both Saumarez and Abernethy were assisted in their representations of Hunter's "real" beliefs about life by the fact that Hunter's notes were not published until the mid-nineteenth century. Saumarez and Abernethy could thus imply that their representations of Hunter were grounded in "esoteric" sources only available to an inner circle. Yet student notes from Hunter's lectures, published in the 1840s, do not support their "transcendent" interpretation of Hunter's theory of life, for there the surgeon stressed that "we find something like the life of mechanics" in investigations of inanimate matter, and he suggested that living matter itself may have emerged as a sort of heightening of the life of mechanics; "the animal and vegetable or organic portion of the world," Hunter wrote, appears to be a sort of "monstrosity in the combination of matter" (*Works* 1:213–14). For a compelling reading of Hunter's theory of monstrosity as indicative of a more general Romantic-era shift in conceptions of monstrosity, see Gigante, "Monster in the Rainbow."

30. See Sir William Lawrence, *Comparative Anatomy and Physiology*.

31. For Lawrence's account of Abernethy's attack, see Sir William Lawrence, *Lectures*, 1.

32. Coleridge, "Theory of Life," 486, 488.

33. It was thus of little consequence to Coleridge whether one defined life by taking "some one particular function of Life common to all living objects—nutrition, for instance" (490) or by selecting "some property characteristic of all living bodies," such as the "power of resisting putrefaction" (494) or by enumerating "the sum of all the functions by which death is resisted" (489). For Coleridge, all of these theories of life were problematic insofar as each proceeded from the same axiomatic and unacknowledged orientation toward the properties of things; each assumed, in other words, that life could be understood as a property that distinguished one set of things from another.

34. Coleridge, "Theory of Life," 537. I describe this as *quasi*-evolutionary because there is an apparent tension between evolutionary and hierarchical tendencies in Coleridge's account. When he demarcates the four "stages," this clearly cannot describe a temporal progression, since the third state—peat and coral—require the fourth (plants and animals). On the other hand, his reference to the "progress of Nature" (537) implies evolutionary development. For a dated but useful discussion of the tensions between these tendencies in Coleridge's work more generally, see Potter, "Coleridge and the Idea of Evolution."

35. In their notes to Coleridge's "Theory of Life," Jackson and Jackson contend that "the concept individuation was [Heinrich] Steffens's distinctive contribution to *Naturphilosophie*" (510), and they refer readers to Steffens's *Beyträge zur innern Naturgeschichte der Erde* (Contributions to an Inner Natural History of the Earth, 1810) and *Grundzüge der philosophischen Naturwissenschaft* (Fundamentals of Philosophical Natural Science, 1806). Without questioning Jackson and Jackson's basic claim, I nevertheless note that Steffens's use of the term *Individualisiren* (individualizing) tends to emphasize the *result* of a process of individuation, while Coleridge's term "individuation" emphasizes the process itself. For an overview of medieval theories of individuation, see Gracia, *Problem of Individuation*.

36. Coleridge, "Theory of Life," 518, 535.

37. The premise of Coleridge's life-manuals depart significantly from the view outlined in that other great Romantic-era text about mental orientation, Kant's "What Is Orientation in Thinking?," first translated into English in 1798 for *Essays and Treatises on Moral, Political, and Various Philosophical Subjects*. In his short essay, Kant sought to account for the ability "to orientate oneself in *thought*, i.e. [to orient oneself], *logically*" (Kant, *Political Writings*, 239). He suggested that logical orientation, which he understood as the precondition for moral orientation, originated in the empirical sense of geographical orientation that allows us to comport ourselves in physical space, an ability that was itself dependent upon the capacity to "feel a difference within my own *subject*" (238). Yet for Kant the empirical origin of orientation was a singular event that, once accomplished, was of little further consequence for subsequent "logical" orientation. Coleridge, by contrast, implied that orientation never ceased to have an embodied dimension, for one required constant returns to the sense of "life" in order to determine the sense of what "makes sense."

38. It is thus no coincidence that the creature kills Elizabeth, the sister-to-be-wife

"given" to Victor by his parents when he was four years old. Mary Shelley's additions to the 1831 edition of *Frankenstein* made this dimension of gift-giving much more explicit: Victor's mother describes Elizabeth as "a pretty present for my Victor" (323), and Victor describes Elizabeth as his "promised gift" (323), one that he later "fondly prized before every other gift of fortune" (337). For a modern example of the sense of unendurable debt that "gifts of life" can produce, see Fox and Swazey's discussion in *Spare Parts* of the ambivalent feelings of modern organ transplant recipients toward their donors.

39. Though writing his sections of the essay on scrofula cost Coleridge time and effort, his goal was to provide Gillman with neither generic intellectual labor (i.e., saving Gillman time by writing what Gillman might in principle have written himself) nor a specialized form of intellectual labor (e.g., supplementing Gillman's practical knowledge with his own theoretical specialization). Instead, Coleridge wrote and gave to Gillman the form of orientation.

40. Judging by Coleridge's 10 November 1816 letter to Gillman, Gillman seems to have recognized this, for Coleridge acknowledged the "formal" nature of his contributions to the essay, admitting that "the same Truths may be taught in a great variety of Symbols." Though Coleridge stressed that he had sought to find a terminology that would allow the two authors to express a more universal truth to medical readers—to "connect Physics with Physiology, the connection of matter with organization, and of organization with Life" (*Collected Letters* 4:690)—the implication that they could have presented this truth without any reference to medicine or physiology was likely a bit unsettling for Gillman.

41. Coleridge died in 1834 and Gillman in 1839. Coleridge's parts of the essay surfaced in 1848, when they were found among a bundle of Gillman's effects, which had been given to Dr. Seth Watson by Gillman's son. The resurfaced essay sections threatened to provoke a rivalry between Gillman's heirs, who felt Gillman had written them, and Coleridge's followers, who were certain (and correct) that Coleridge had. On the textual history of these manuscripts, see the editors' comments on Coleridge, "Essay on Scrofula," 454–56.

42. Hooke, *Micrographia*, 54. In *The Structure of Scientific Revolutions*, Kuhn drew on the concept of the crucial experiment to explain his seminal notion of a scientific "paradigm," suggesting that "'crucial experiments'—those able to discriminate particularly sharply between . . . two paradigms—have been recognized and attested before the new paradigm was even invented" (153).

43. Galison, *How Experiments End*, 3.

44. See the editor's introduction to Coleridge, *Statesman's Manual*, xxix–xlvii.

45. Ibid., 7, 9, 8.

46. Mary Shelley, *Frankenstein*, 226.

47. Coleridge, *Aids to Reflection*, 9.

48. Coleridge, *Friend*, Part 1, 14. On Coleridge's "generic experiments" in this period, see Malachuk, "Coleridge's Republicanism." Coleridge's development of the concept of experimentative faith may have been a consequence of the fact that not all of his reviewers appreciated the demands placed on the reader by his earlier life-manuals. In his review in *The Examiner* of *The Statesman's Manual*, for example, Hazlitt described a reading experience closer to the disorientating nausea of seasickness than the ecstatic inspiration of polar voyages: "The effect is monstrously like the qualms produced by the

heaving of a ship becalmed at sea; the motion is so tedious, improgressive, and sickening" (571).

49. Coleridge, *Aids to Reflection*, 9.

50. Youngquist, "Rehabilitating Coleridge," 901.

51. Ibid., 893. Youngquist quotes from Nietzsche, *Birth of Tragedy*, 37. The example by means of which Youngquist links Coleridge's opium use to his early verse is, naturally enough, "Kubla Khan." Youngquist argues that the poem's framing—i.e., Coleridge's claim that the poem was composed under the influence of laudanum—"forces us to approach his 'vision in a dream' as an occurrence that opium made possible; whether it really happened that way or not is beside the point" (896).

52. Youngquist, "Rehabilitating Coleridge," 901. The quoted passage is from Coleridge's 26 June 1814 letter to Josiah Wade, in Coleridge, *Collected Letters*, 3:511.

53. Youngquist, "Rehabilitating Coleridge," 901. For an account of Coleridge's success at obtaining "unprescribed" opium even after 1816, see Griggs, "Samuel Taylor Coleridge and Opium."

CHAPTER 4: **Nausea, Digestion, and the Collapsurgence of System**

1. See, e.g., Bataille, *Story of the Eye*; Burroughs, *Ticket That Exploded*; Pynchon, *Gravity's Rainbow*.

2. For discussion of some of these works of art in the context of the categories of disgust and abjection, see Bois and Krauss, *Formless*, and Menninghaus, *Disgust*. On *Disembodied Cuisine*, see Robert Mitchell, *Bioart and the Vitality of Media*.

3. See, e.g., Bakhtin, *Rabelais and His World*.

4. On the role of the concept of "system" in the Romantic era, see Siskin, "1798," and Robert Mitchell, *Sympathy and the State*.

5. Pfau, *Romantic Moods*, 21 (quoted text in italics in the original). I am adopting Pfau's account of paranoia as one of the dominant "moods" (*Stimmungen*) of the Romantic era.

6. Golinski, *Science as Public Culture*; Schaffer, "Consuming Flame"; Stewart, *Rise of Public Science*; Lynn, *Popular Science*.

7. Todd, *Imagining Monsters*.

8. The English word "nausea" stems from the Greek *naus* (ship), since, according to eighteenth-century encyclopedias, "people, at the beginning of their voyages, are usually liable to this disorder." See, e.g., "Nausea" in Chambers, *Cyclopaedia*.

9. These include Wilson, "Gut Feminism," which develops a feminism capable of a productive, rather than oppositional, relationship to the sciences, and *Affect and Artificial Intelligence*, which argues that early research in artificial intelligence was as much an effort to understand the role of affect in developmental processes as an attempt to "simulate" human intelligence; as well as Doyle's analysis, in *Wetwares*, of the function of science for advocates of human cryonics, and his advocacy of modes of first-person science, developed in *Darwin's Pharmacy*. See also Massumi's *Parables for the Virtual*, which promotes the Jamesian-Whiteheadian-Deleuzian goal of a "radical empiricism" capable of developing a science of *all* aspects of experience, not simply those aspects amenable to quantification.

10. McKendrick, "Consumer Revolution of Eighteenth-Century England," provides

a concise account of the importance of urban concentration and wage increases for the eighteenth-century British "consumer revolution" (9–33).

11. Addison, *Spectator*, 18 May 1711, quoted in Mackie, *Commerce of Everyday Life*, 205.

12. As Cohen notes in *Networked Wilderness*, digestion—or, more specifically, indigestion—often played an important role in the contact zones between early European settlers and native populations in the Americas (76–91).

13. Mintz's *Sweetness and Power* provides the classic account of the means by which sugar had, by "no later than 1800 . . . become a necessity—albeit a costly and rare one—in the diet of every English person" (6). Schivelbusch, in *Tastes of Paradise*, and Walvin, in *Fruits of Empire*, describe the introduction of foreign spices and fruits, respectively, to European polities.

14. Nor were the border crossings of these digestible substances solely a matter of national boundaries, for occasionally medicinal liquids and solids previously available only through the mediation of physicians became available in nonmedical establishments. Thomas Trotter complained in *View of the Nervous Temperament* that "*aërated soda water*"—a drug "manufactured by the apothecary"—was now available "in coffeehouses," which was for Trotter a disturbing example of "a medicine converted into a tavern beverage!" (318–19).

15. Eighteenth-century physicians agreed that the goal of digestion was the production of "chyle," a fluid that "resembles a natural Emulsion, made of soft, oily, insipid, watery, and mucilaginous Particles" (Hoffman 5). See entries for "chyle," "chylification," "chylosis," and "chyme" in Chambers, *Cyclopaedia*; "chyle" and "chylification" in Quincy, *Lexicon Physico-Medicum*; and "chylificatio" in Barrow, *Dictionarium Medicum Universale*. "Chyle" stems from the Greek *cheo* (to draw forth).

16. Chylopoietic discourse was also a site of contestation between physicians and non-physicians over the authority to speak about the physiological effects of commodities. See, for example, the anonymously authored *Remarks on Mr. Mason's Treatise upon Tea*, in which Mason is criticized for his lack of medical authority.

17. Mason, *Effects of Tea Consider'd*, 19 (the latter quote is Mason's gloss of Thomas Short's *Dissertation upon Tea*); Anon., *Best and Easiest Method*, 60; Thomas Trotter, *View of the Nervous Temperament*, 75, 70.

18. On the Galenic understanding of digestion, see Winslow and Bellinger, "Hippocratic and Galenic Concepts of Metabolism"; and Leicester, *Development of Biochemical Concepts*, 28–36. For the role of Galenic principles in seventeenth- and eighteenth-century models of digestion, see Leicester, *Development of Biochemical Concepts*, 92–129, and Estes, "Medical Properties of Food."

19. "Nutrition" in Quincy, *Lexicon Physico-Medicum*. Advocates of the "mill" model of the stomach often stressed the purported compressive power of the stomach muscles. In his entry for "Digestion," for example, Quincy attributed the power of the stomach to break down aliment primarily to "the continual Motions of [the stomach's] Sides, whose absolute Power is demonstrated to be equal to the Pressure of 117088 Pound Weight; To which, if be added the absolute Force of the Diaphragm, and Muscles of the *Abdomen*, which likewise conduce to Digestion, the Sum will amount to 250734 pound Weight." Quincy disqualified chemical interpretations, arguing that any "ferment in the Stomach

... cannot contribute any Thing to the Digestion of the Food, any further than by [non-chemically] softening it, whereby it is capable of being further divided."

20. In his influential *Dissertations*, Spallanzani suggested that by the mid-eighteenth century, five different theories of digestion were present: the "trituration" theory, which held that food was ground into particles; the "solvent" theory, which ascribed digestion to a chemical process of decomposition; the "fermentation" theory, which represented digestion as akin to the processes that occurred in bread and beer; the "putrefaction" theory, which held that the stomach accelerated the pace of natural decomposition; and what we might call the "combination" theory, which Spallanzani attributed to Boerhaave, that digestion "depends upon all these causes operating in conjunction" (1:i). For additional eighteenth-century descriptions of digestion, see Hoffman, *Treatise*, 38–41; and "Chylification" in Chambers, *Cyclopaedia*.

21. In *Dissertations*, Spallanzani argued that trituration (grinding) contributed to digestion in animals with muscular stomachs, for it helped to crack the shells of hard grains, but played little, if any, role in animals with membranous stomachs. Through an extensive series of experiments—many involving the introduction of perforated tubes, filled with food substances, into the stomachs of animals—Spallanzani demonstrated that digestion always depended upon the chemical action of "gastric juice." For Hunter's work on digestion, see his *Observations* (1786), 147–88. For a historical account of theories of digestion, see Leicester, *Development of Biochemical Concepts*.

22. In *English Malady*, Cheyne considered "Nervous Illness of all Kinds" but limited himself to the effects of these diseases on the aristocratic body, while in *Treatise on Tobacco, Tea, Coffee, and Chocolate*, Paulli suggested that tea was beneficial only for Chinese bodies and produced ill health in European bodies.

23. Thomas Trotter, *View of the Nervous Temperament*, 37–53.

24. "Nutrition" in Quincy, *Lexicon Physico-Medicum*.

25. In *Drink and the Victorians*, Harrison contends that Trotter was the "first scientific investigator of drunkenness" (92), while in *Nerves and Narratives*, Logan notes that Trotter's text was the "first book on mental medicine ever printed in the United States" (16).

26. For discussion of Trotter and his *View of the Nervous Temperament*, see Porter's introduction to Thomas Trotter, *Drunkenness and Its Effects*, ix–xliii; Porter, "Addicted to Modernity"; Logan, *Nerves and Narratives*, 15–44; George Rousseau, "Coleridge's Dreaming Gut," 115–16; and Jonsson, "Physiology of Hypochondria," 20–21, 24–25.

27. Thomas Trotter, *View of the Nervous Temperament*, xvii; see also viii. I draw the term "statistical panic" from Woodward, "Statistical Panic."

28. Thomas Trotter, *View of the Nervous Temperament*, xv, xvi. While this description appears as a quote, its source is Trotter's own *Medicina Nautica*, an account of naval health written during the military conflict with France. Trotter's description of nervous illness was thus generated in a context in which an "inaptitude for work" and "sympathetic selfishness" were directly related to national security, for a "nervous" navy man represented a threat to the safety of his ship and, by extension, to British security.

29. Thomas Trotter, *View of the Nervous Temperament*, 46, 40. Even those who provided the capital for manufacture were at risk, and created risk for others, for "it is that puddle of corruption, the Stock Exchange . . . that has filled the nation with degenerate fears, apprehensions, and hypochondriackism" (147).

30. Ibid., 201. This understanding of the stomach as physiological center of the body was not peculiar to Trotter. In his "Proposals for the Recovery of People Apparently Drowned," for example, Hunter claimed that "the stomach sympathizes with every part of an animal, and . . . every part sympathizes with the stomach" (414). On medical understandings of sympathy in the eighteenth and early nineteenth centuries, see Christopher Lawrence, "Nervous System"; Leys, *From Sympathy to Reflex*; and Vila, *Enlightenment and Pathology*.

31. For more on Trotter's account of the body's nervous capacity to store impressions, see Logan, *Nerves and Narratives*. Logan suggests that those afflicted with nervous illness suffer from their own narratives, for in these individuals, the past is physiologically stored in the nerves in an unhealthy way. While I agree with the gist of Logan's analysis, it seems more useful to understand the effects of the environment on bodies in Trotter's account in terms of feedback rather than "narrative" (though a concept of recursive narration might mediate between these two positions).

32. See Thomas Trotter, *View of the Nervous Temperament*, 97, 148.

33. As Logan notes in *Nerves and Narratives*, Trotter's account of the physician's capacities is notably "Romantic," for the true physician possessed a kind of sensibility and genius reminiscent of Wordsworth's description of the poet (38–40).

34. On the relationship of the gothic and paranoia to *Caleb Williams*, see Punter, *Literature of Terror*, 1:123. The second quote is Siskin's description of the novel in "1798," 211.

35. Godwin, *Caleb Williams*, 263, 249. For readings that situate the novel in the context of the development of governmental domestic surveillance mechanisms in the 1790s, see Ousby, "My Servant Caleb," and Thompson, "Surveillance."

36. For a complementary reading of *Caleb Williams* as a novel both about, and instantiating, the logic of paranoia, see Pfau, *Romantic Moods*, 123–45.

37. After Falkland describes Caleb's fate, Caleb is overcome by a feeling that "every atom of my frame seemed to have a several existence, and to crawl within me" (151). After being imprisoned, Caleb is again overcome by nausea: "No language can do justice to the indignant and soul-sickening loathing that these ideas excited" (190). When he discovers a broadsheet purporting to recount his exploits in Laura's house, Caleb feels "a sudden torpor and sickness that pervaded every fibre of my frame" (312), and when he is informed that he may never leave England, this information occasions "an instantaneous revolution in both my intellectual and animal system" (324).

38. Cabanis, *Rapports*, 1:124, my translation.

39. The term "digest" stems from the Latin *dis* (apart) and *gerere* (to carry).

40. Godwin, *Enquiry*, 280. For examples of Caleb's disappointments, see Godwin, *Caleb Williams*, 177, 182, 220, 319. The power of the print system over Caleb is obvious within the novel. Even without Gines's interference, broadsheets describing Caleb circulate everywhere. Moreover, Caleb's attempts to master print fail miserably: he becomes an author, but this only serves to announce his location within the print system, for he writes barely disguised autobiographies that reveal his location to Gines. See also "*Caleb Williams* and Print Culture," in which Sullivan notes that "Caleb's contributions [to the printed "histories" of criminals] . . . lead to his detection by Gines" (324).

41. Habermas, *Philosophical Discourse of Modernity*. Habermas notes the "performative contradiction" of Adorno's and Horkheimer's critique of "instrumental reason," which relies on a "suspicion of ideology" that has "becom[e] total": this suspicion is thus

"not only turned against the irrational function of bourgeois ideals, but against the rational potential of bourgeois culture itself . . . But the goal remains that of producing an effect of unmasking . . . [T]his description of the self-destruction of the critical capacity is paradoxical, because in the moment of description it still has to make use of the critique which has been declared dead" (119).

42. Patton and Mann in Coleridge, "Lecture on the Slave-Trade," 232.

43. It is worth noting that Coleridge's Unitarian auditors were denied the means of embracing such public responsibility, for their religious affiliation meant they could neither vote nor run for office.

44. Hartley, *Observations*, 8. Christensen's *Coleridge's Blessed Machine of Language* remains one of the most perceptive accounts of Coleridge's struggles with Hartley's philosophy, and my chapter as a whole draws inspiration from Christensen's acute observation of the paradoxes at the heart of Hartley's philosophy.

45. Coleridge to Robert Southey, 11 December 1794, in Coleridge, *Collected Letters*, 1:137.

46. Coleridge to Thomas Poole, 24 September 1796, in Coleridge, *Collected Letters*, 1:236.

47. Coleridge, *Conciones ad Populum*, 47; see also 19.

48. On Coleridge's lifelong interest in the philosophy of language, see Christensen, *Coleridge's Blessed Machine of Language*; McKusick, *Coleridge's Philosophy of Language*; and Fulford, *Coleridge's Figurative Language*.

49. McKusick, *Coleridge's Philosophy of Language*, 61.

50. Coleridge, "Lecture on the Slave-Trade," 248. In the version of the lecture Coleridge published in the 25 March 1796 *Watchman*, the passage reads, "A part of that food among most of you, is sweetened with Brother's Blood" (138–39).

51. Morton, "Blood Sugar," 87–88.

52. Coleridge, "Lecture on the Slave-Trade," 248. For the nearly identical published version, see Coleridge, *Watchman*, 139.

53. Morton, "Blood Sugar," 92.

54. Coleridge, "Lecture on the Slave-Trade," 241; Coleridge drew his description from Clarkson, *Essay*, 118, unnumbered note.

55. The classic locus for this understanding of performative speech is Austin, *How to Do Things with Words*; see as well the widely influential use of this terminology in Derrida, *Limited Inc.*, and Butler, *Bodies That Matter*.

56. The distinction between "states of affairs" and "events" is outlined in Deleuze and Guattari, *Thousand Plateaus*, 75–110; and Deleuze, *Logic of Sense*. Deleuze and Guattari relate this distinction to the category of performative speech in *Thousand Plateaus*, 77–78, 82.

57. Deleuze and Guattari, *Thousand Plateaus*, 87.

58. Hedley, *Coleridge, Philosophy and Religion*, 56.

59. Coleridge, "Lecture on the Slave-Trade," 235.

60. As Deleuze notes in *Logic of Sense*, "To eat and to be eaten—this is the operational model of bodies, the type of their mixture in depth . . . To speak, though, is the movement of the surface" (23).

61. Coleridge, *Watchman*, 131. For a discussion of the role of digestion in Coleridge's later thought, see George Rousseau, "Coleridge's Dreaming Gut."

62. Coleridge, *Biographia Literaria*, Part 1, 154–55.

63. Nietzsche, *Portable Nietzsche*, 271.

64. For an acute reading of Nietzsche's snake image and its connection with his interest in rumination, see Menninghaus, *Disgust*, 170–71 (though see as well my qualification below in n. 71).

65. Menninghaus notes in *Disgust* that "beginning in the seventeenth century . . . and more fully in the eighteenth, disgust—as represented or reflected in texts—attains to a 'life of its own,' becoming worthy of consideration for the sake of its own (anti-) aesthetic and moral qualities. Even the words *dégoût*, disgust, and *Ekel* first come into general usage in the sixteenth or seventeenth century and make a more than isolated entry into theoretical texts for the first time in the eighteenth century" (3–4).

66. Schlegel, "Anmerkungen über Ekel," 111, cited in Menninghaus, *Disgust*, 25. Kant's more well-known formulation of this principle appears in Kant, *Critique of Judgment*, 180.

67. Mendelssohn, "Rhapsodie," 140, cited in Menninghaus, *Disgust*, 26.

68. As Menninghaus puts it in *Disgust*, for eighteenth-century aesthetics, "*Ekel* is both lower and upper limit, adversary and innate tendency of the beautiful" (30).

69. Disgust is generally described as a cultural phenomenon—that is, an affect that signals deep-seated structures of "taste"—and distinguished from nausea (the latter understood as a natural phenomenon, e.g. the sign of an ill body). For key discussions of disgust, see Kolnai, *On Disgust*; William Ian Miller, *Anatomy of Disgust*; and Menninghaus, *Disgust*. For accounts of nausea as something other than a purely natural phenomena, see Derrida, "Economimesis"; and Gigante, "Keats's Nausea" (and the related book chapter, "Endgame of Taste," in *Cultures of Taste*).

70. Douglas's famous formula appears in *Purity and Danger*, 35, and plays an important role in both Miller's *Anatomy of Disgust* and Menninghaus's *Disgust*.

71. One cause of the confusion between the concepts of disgust and the nausea of collapsurgence is that while disgust was most thoroughly theorized in the German philosophical tradition, German also lacks the distinction between "disgust" and "nausea" available in both English and French. Thus, despite my admiration for Menninghaus's reading of Nietzsche's *Ekel*-images, his overall stress on disgust, rather than nausea, encourages him to underestimate the importance of the nautical image of exploration that immediately follows Nietzsche's snake image—an image that suggests that in this case, *Ekel* denotes for Nietzsche not "disgust" but rather the nausea associated with voyages into new seas.

72. For cannibalism, see "A Modest Proposal"; for fecal imagery, see in *Gulliver's Travels* the narrator's account of being assaulted with human feces by Yahoos (in Swift, *Writings*, 194), the effort of one Laputan scientist "to reduce human Excrement to its original Food, by separating the several Parts, removing the Tincture which it receives from the Gall, making the Odour exhale, and scumming off the Saliva" (153), and the claim by another Laputan professor that one could "discove[r] Plots and Conspiracies against the Government" by analysis of "the Colour, the Odour, the Taste, the Consistence [*sic*], the Crudeness, or Maturity of Digestion" (163).

73. Poggioli, *Theory of the Avant-Garde*; Bürger, *Theory of the Avant-Garde*; Greenberg, "Avant-Garde and Kitsch."

74. Though Gigante technically focuses on taste and disgust, rather than nausea,

in both "Keats's Nausea" and "Endgame of Taste" she implicitly distinguishes between nausea and disgust by emphasizing the lineage that connects the Romantic-era with the work of Sartre and Beckett.

75. Ironically, the discovery that about 95% of the body's serotonin—the "target" of antidepressant drugs such as Prozac—is located in the gut rather than the brain, makes the premise of a "thinking gut" *more* believable now than in Trotter's time; see Wilson, "Gut Feminism," 85.

76. For a trenchant account of the function of "risk" in contemporary society, see Beck, *Risk Society*.

77. Bersani, "Pynchon, Paranoia, and Literature," 99.

78. Kittler, "Media and Drugs," 106.

79. Bersani, "Pynchon, Paranoia, and Literature," 116, 106–7.

CHAPTER 5: The Media of Life

1. For a compelling account of this convergence, see Thacker, *Biomedia*. I document the divergence of cultural and biological senses of media in the late nineteenth and early twentieth century in Robert Mitchell, *Bioart and the Vitality of Media*. On hybridizations of communication theory and biology in mid-twentieth-century understandings of genetics as information, see Keller, *Refiguring Life*, 79–118.

2. On the rise of historicism, see Bann, *Romanticism and the Rise of History*, and Chandler, *England in 1819*; for the emergence of "deep time," see Spears, "Evolution in Context"; for a penetrating discussion of the importance of Kant's rethinking of time, see Deleuze, *Difference and Repetition*, 85–91, and *Kant's Critical Philosophy*, esp. vii–viii.

3. Relationships between the terms "medium" and "milieu" in English, French, and German are complicated and, to my knowledge, have not been fully articulated. As Canguilhem notes in "Living and Its Milieu," in eighteenth-century French natural philosophy, the English word "medium" was often translated by the French term *milieu*, and that latter term subsequently moved from French physics into French biology (7). Yet Canguilhem does not address that English authors in the nineteenth century retained the term "medium" while importing "milieu" from French, and German authors also used both terms (*Medium, Milieu*). For additional accounts of nineteenth-century use of the term "milieu," see Rabinow, *French Modern*, 126–67; Tresch, *Romantic Machine*, 4–5, 223–48, 270–73; and (though less reliable) Spitzer, *Essays*, 179–316. For accounts of "medium" in nineteenth- and twentieth-century German philosophy and social theory, see the entry for "Medien/medial" in Barck and Fontius, *Ästhetische Grundbegriffe*. For the term in English, see Raymond Williams's entries for "Media" and "Mediation" in *Keywords*, 203–7, and Guillory, "Genesis of the Media Concept" (a condensed version of which appears as "Enlightening Mediation"). Unfortunately, none of these accounts mentions—or seems aware of—the biological sense of "medium." Eliassen and Jacobsen, "Where Were the Media," gesture toward biological conceptions of media but are ultimately interested only in theorizing eighteenth-century "communication structures" (67) and "information flows" (68). Siskin and Warner, "This Is Enlightenment," seek to outline the conditions for a history of "mediation" but define this term so broadly—as "everything that intervenes, enables, supplements, or is simply in between" (5)—that it is not clear how such a history would be written (nor what the goal of writing it would be). Siskin and Warner's "variables" for such a history—"infrastructure, genres and formats,

associational practices, and protocols" (12)—echo those theorized much more precisely in Kittler's important essay "Towards an Ontology of Media" (to which I return below).

4. I draw the term "material space" from Diderot and D'Alembert's *Encyclopédie*, in which it is noted that "medium" is a term from mechanical philosophy identical in meaning to "fluid or milieu," though in French, "the latter term [*milieu*] is more frequently employed" (*"ce dernier est beaucoup plus usité"*). They define "milieu" as "signif[ying]" a material space [*un espace materiel*] through which a body passes as it moves, or, more generally, a material space in which a body is situated, whether it moves itself or not" (my translation). For Bacon and Newton on media, see Bacon, *Philosophical Works*, 3:277, 122, 55, and Newton, *Optical Lectures*, 5ff.

5. Godwin, *Enquiry*, 363, 771.

6. Adams, *Lectures*, 2:116.

7. Fulton, *Canal Navigation*, 4; Raithby, *Peace*.

8. Adam Smith, *Wealth of Nations*, 2:940, V.iii.81.

9. Godwin, *Enquiry*, 674.

10. Adams, *Lectures*, 2:3.

11. On the shift from "natural history" to "biology" in the early nineteenth century, see Foucault, *Order of Things*, 125–65, 226–32.

12. Saumarez, *New System of Physiology*, 1:23; Treviranus, *Biologie*, 4:120.

13. Lamarck, *Zoological Philosophy*, 79; French original in Lamarck, *Philosophie zoologique*, 156. Hereafter, citations refer to the English translation, with page numbers for the French edition indicated by PZ.

14. This understanding of medium was thus quite different from the classical claim, revived in the eighteenth century by authors such as Montesquieu, Rousseau, Buffon, and Herder, that the "environment" affected individuals (e.g., by influencing the physical appearance, desires, and morals of groupings of people). Where the environment was understood as a way of explaining, by means of an occasional cause or influence, individual or local differences, the concept of medium was intended to account for the condition of possibility of life itself as well as dynamic change over time. On eighteenth-century notions of environment, see Spitzer, *Essays*, 179–316, and Bewell, *Wordsworth and the Enlightenment*, 237–45. On the distinction between an objective concept of enviroment and Jacob von Uexküll's "subjective" concept of *Umwelt* (also translated as "environment"), see Pollmann, "Invisible Worlds, Visible."

15. Jenner, *Inquiry*, 29. As Jenner explained later in his treatise, the "human subject" could also serve as a medium through which infectious matter could be passed (36).

16. Thelwall, *Animal Vitality*, 20.

17. Saumarez, *New System of Physiology*, 1:23.

18. Sir William Lawrence, *Comparative Anatomy and Physiology*, 128. This did not mean, of course, that every difference between a living body and its medium benefited the organism. Physician Thomas Beddoes warned that remaining in an overly warm medium was injurious to human health: *"Continued warmth renders the living system less capable of being called into strong, healthy, or pleasurable action,"* for "every muscle, steeped in a heated medium, loses its contractility" (*Hygëia*, 2: Essay 5, 53–54). See also my discussion in chapter 2 of Hunter's experiments with animal and vegetable heat.

19. Burkhardt's *Spirit of System* provides a useful intellectual biography of Lamarck.

20. In his preface to *Zoological Philosophy*, Lamarck contends that animal "organisa-

tion gradually increases in complexity" (*l'organisation se compose et meme se complique graduellement, dans sa composition*) (1; PZ, iii).

21. Lamarck acknowledged in *Zoological Philosophy* that one could attribute such creative power to "the Supreme Author of all things," but he insisted that zoological philosophy also explained this diversity. He thus suggested that his account did not question the power of the Supreme Author but simply revealed that "nature" was the means by which the Supreme Author acted (40–41; PZ, 67).

22. Canguilhem contends, "Lamarck always speaks of milieus in the plural, and by this he specifically means fluids like water, air, and light. When Lamarck wants to designate the whole set of outside actions that are exercised on a living thing, in other words what we call today the 'milieu,' he never says 'milieu,' but always 'influential circumstances.' As a result, circumstance is a genus within which climate, place, and milieu are species" ("Living and Its Milieu," 9). Lamarck's account of the fish's *milieu dense* establishes that he does write about a singular milieu, but Canguilhem's larger point is nevertheless valid.

23. Canguilhem, "Living and Its Milieu," 12; translation modified.

24. Lamarck, *Zoological Philosophy*, 190 (PZ, 375–76).

25. On the theoretical difficulties presented by the term "complexity" in nineteenth- and twentieth-century discussions of evolution, see McShea, "Complexity and Evolution."

26. Lamarck argued in *Zoological Philosophy* that the concept of species was not grounded in nature but rather in the temporal limits of human experience: though animal forms seemed stable to humans, this was due to the fact that human observations collectively "only extend back a few thousand years; which is a time infinitely great with reference to himself, but very small with reference to the time occupied by the great changes occurring on the surface of the globe" (43; PZ, 72).

27. Coleridge, *Biographia Literaria*, Part 1, 162–63.

28. Saumarez, *New System of Physiology*, 1:ix.

29. Saumarez contended that the "rational power . . . known by the various appellations of SOUL and MIND, constitutes the paramount presiding principle, of which the brain is the immediate recipient . . . [T]he brain itself is evolved and organized, perfected and preserved by the energy of the living principle" (ibid., 1:151).

30. Saumarez suggested in *A New System of Physiology* that even biological conception is not simply the transmission of form, for semen transmits not only form (i.e., the paternal character) but also "energy and specific power," which are dependent upon the sensible qualities of this medium. In similar fashion, Saumarez contended that "without [the] specific action of the brain, the mind would be in a dormant state, like the specific power of the foetal testes," and the mental development of the mind depended upon the organic development of the brain (1:157).

31. Like many eighteenth-century dualistic accounts, Saumarez's new system of physiology begged the question of how two different modes of being, such as mind and brain, could come into contact with one another. However, this was generally treated as a pseudo-problem by dualist authors, for, they contended, the answer to this question could be known only to God.

32. Duster, "Lessons from History."

33. Herrnstein and Murray, *Bell Curve*.

34. See, e.g., Krieger et al., "Racism, Sexism, and Social Class."

35. See, e.g., Lukács, *Young Hegel.*

36. On the relationship of Kant's critical philosophy to nineteenth-century German biology, see Lenoir, *Strategy of Life.*

37. My discussion of the role of the concept of "medium" in Hegel's work has benefited from Benjamin's reflections, developed in *The Concept of Criticism*, on the concept of medium (and "reflection-medium") in German Romanticism.

38. Hegel, *Phenomenology of Spirit*, 1, 46; German original in Hegel, *Werke*, 3:11, 68. Hereafter, citations refer to the English translations of Hegel's works, with page and volume numbers for the German edition appended and indicated by W.

39. This is also the case even when one speaks of something passing from one medium to another (e.g., a sound that passes from a fluid to a solid medium). In such cases, one can speak of the "law of refraction" of each medium precisely because one already assumes a basic homogeneity between the two media, such that something—in this case, a sound—can pass between the two.

40. Though Kant was clearly the object of Hegel's critique in the introduction to *Phenomenology*, Kant also had established the basic schema for Hegel's argument, contending in *Critique of Pure Reason* that time—understood not as an external objective reality but as the form of inner sense—was "the medium [*Medium*] of all synthetic judgments" (192, B94); original in Kant, *Gesammelte Schriften*, 3:B94.

41. In *Science of Logic*, Hegel contends that "all that we have to do to ensure that the beginning will remain immanent to the science of this knowledge is to consider, or rather, setting aside every reflection, simply to take up, *what is there before us*" (47; W, 5:68). He had established at the end of *Phenomenology of Spirit* that what is "there before is" is "pure knowledge," which, "thus *withdrawn* into this *unity*, has sublated every reference to an other and to mediation; it is without distinctions and as thus distinctionless it ceases to be knowledge; what we have before us is only *simple immediacy*" (47; W, 5:68).

42. Hegel, *Science of Logic*, 46 (W, 5:66). Guillory cites the first clause of this passage in "Genesis of the Media Concept," but he curiously draws from it the claim that for Hegel, "the principle of mediation denies the possibility of an immediate (*unmittelbar*) relation between subject and object, or the immediacy of any knowledge whatsoever" (343). Though such a claim may seem to capture the spirit, if not the letter, of Hegel's text, it ultimately suggests a significant misunderstanding of the relationship between immediacy and mediation in Hegel's philosophy. As Hegel emphasizes in the second clause of the passage, the point is not to privilege mediation over immediacy but rather to see the opposition between immediacy and mediation as a false problem.

43. Though Hegel tended to distinguish between the different arts in terms of their elements (*Elemente*) or materials (*Material*)—e.g., "there is an essential distinction between the different arts according to the elements [*Elemente*] in which they are expressed" (*Aesthetics*, 1:254; W, 13:329)—he is one of the first nineteenth-century authors to discuss art (in general) as a "medium" of self-expression.

44. As in the case of Lamarck, Hegel's criterion of perfectibility is quasi-quantitative. For Lamarck, the "proof" that the world is more complex now than in the past is the fact that the organic structures of relatively recent beings (e.g., humans) are more "canalized" than those of earlier organisms. For Hegel, the proof that we have reached the final stage

of complexity—the state of purely fluid thought—is that philosophy can connect fluidly all other (purportedly rigid) forms of thinking, such as art and religion or the individual sciences, whereas those other forms of thought are each bound to their own domains.

45. Hegel, *Hegel's Philosophy of Nature*, 277 (W, 9:342).

46. This too had its Aristotelian origins. Aristotle argued in Book III of the *Physics* that motion is "the fulfillment of what exists potentially, in so far as it exists potentially"; thus, when everything that exists potentially has become actual, movement is at an end (201a10–11).

47. For a compelling account of Hegel's difficulties in describing the post-vital condition of post-historical Absolute Spirit, see Rajan, "(In)Digestible Material," esp. 230–31.

48. On the ways national botanical gardens served as hubs of networks intended to link state centers to the rest of world, see Fulford, Lee, and Kitson, *Literature, Science, and Exploration*, 33–45; for Lamarck's relationship to the Jardin du Roi more specifically, see Burkhardt, *Spirit of System*, 27–30. On the appreciation of Lamarck's philosophy by British liberal and radicals intent on *changing* the nature of the British state, see Desmond, *Politics of Evolution*, esp. 1–100, 276–334.

49. On the consolidation of the Hegelian school and its links with the Prussian state, see Toews, *Hegelianism*, esp. 85–88.

50. Kittler, *Discourse Networks*, 370.

51. My reading of Schelling, and Schelling's relation to Hegel, is indebted to Rajan's extensive work on this topic, and especially "(In)Digestible Material" and "Encyclopedia and the University of Theory."

52. Schelling, *First Outline*, 5, 6; the German original appears in *Erster Entwurf*, 2:5, 6.

53. As Rajan notes, "In *The First Outline* fluidity is the medium in which the freedom of unconditional knowledge becomes possible, as a releasing of fixed forms from what Schelling curiously calls figure (*Figur*)" ("Outline" 316).

54. Schelling, *System of Transcendental Idealism* (1800), 232; the German original appears in *System des Transcendentalen Idealismus*, 477.

55. Schelling, *Philosophy of Art*, 84.

56. In *System of Transcendental Idealism*, Schelling explicitly describes the will as a "medium": "it is only through the medium of willing [*das Medium des Wollens*] that the intelligence becomes an object to itself" (156; *System des Transcendentalen Idealismus*, 325). This description of *willing* as a medium captures in condensed form the complexity of Schelling's understanding of media, for in his system, a medium is neither merely a means of communication nor a simple conduit of teleological growth; rather, it both requires and enables the perpetual possibility of de novo origination.

57. Schelling, *First Outline*, 35; German original in *Werke*, 2:42. On the evolutionary biological implications of *The First Outline* and *The World-Soul* (*Die Weltseele*, 1798), see Richards, *Romantic Conception of Life*, 294–306.

58. Gilles Deleuze is especially attentive to the importance of "exhaustion" within Schelling's work, noting, e.g., that Schelling's theory of potencies should be understood as a sort of "differential calculus": that is, a "method of an exhaustion and evolution of powers" (*Difference* 191).

59. For an account of Geoffroy's work that emphasizes Cuvier's increasing renown and success, see Appel, *Cuvier-Geoffroy Debate*.

60. The basic conflict between Geoffroy's and Cuvier's approaches to biological media and animal forms came to a public head in 1830 in the form of a debate between the two zoologists—a debate from which Cuvier seemed, at least in the short term, to emerge victorious. See Appel, *Cuvier-Geoffroy Debate*. However, as Desmond notes in *Politics of Evolution*, it was Geoffroy's philosophical anatomy (and, to a lesser extent, Lamarck's philosophical zoology) that was much more attractive to British radicals and reformers (see esp. 1–100, 193–235).

61. Deleuze and Guattari, *Thousand Plateaus*, 46.

62. Deleuze and Guattari emphasize that both Lamarck's and Geoffroy's concepts of organic variation differed significantly from that of Charles Darwin, who understood the relationship of living beings to their media more explicitly in terms of populations (rather than simply "species") and who substituted for "degrees of development" concepts of "speeds, rates, coefficients, and differential relations" (ibid., 48). I take up the issues of both population and speed below.

63. Geoffroy, "Des différns états de pesanteur des oeufs"; see also Appel, *Cuvier-Geoffroy Debate*, 128–29.

64. For a discussion of the questionable nature of Geoffroy's success in creating monsters, see Appel, *Cuvier-Geoffroy Debate*, 129. In "Monster in the Rainbow," Gigante documents the extent to which this Romantic-era understanding of monstrosity as excessive life was not peculiar to Geoffroy but also underwrote the experimental approach of John Hunter and the poetry of John Keats.

65. Mary Shelley, *Frankenstein*, 79, 80. The latter, non-organic option seems more likely, for Victor stresses that his construction of a living being by "bestowing animation upon lifeless matter" was, strictly speaking, only the first phase of a more grand project, and he would only *later* be able to reanimate dead humans: "I thought, that if I could bestow animation upon lifeless matter, I might in process of time (though I now found it impossible) renew life where death had apparently devoted the body to corruption" (82).

66. For readings of the novel in terms of psychological doublings and projection, see Kaplan, "Fantasy of Paternity"; Levine, "Ambiguous Heritage"; and Collings, "Monster and the Maternal Thing."

67. Mary Shelley, *Frankenstein*, 49–50. On the role of magnetic flows in Romantic-era science generally, and in *Frankenstein* in particular, see Fulford, Lee, and Kitson, *Literature, Science and Exploration*, 149–76.

68. Walton writes to his sister Margaret that "I have no friend . . . I desire the company of a man who could sympathize with me" (53). For Walton's account of his desire for "glory," see 51.

69. Mary Shelley, *Frankenstein*, 128–29, 131. On the role of the sun in Romantic-era literary and scientific thought, see Underwood, *Work of the Sun*.

70. Mary Shelley, *Frankenstein*, 138. On the role of sympathy in *Frankenstein*, see Marshall, *Surprising Effects of Sympathy*, 178–227, and Daffron, "Male Bonding."

71. We might even see the Romantic-era conception of vital media as a "legitimated" version of Paracelsus's earlier, and less scientifically respectable, alchemical discussions of the role of generative matrices within "elements" such as water. See, e.g., Paracelsus's description of the generation of minerals from "primal matter" in Paracelsus and Waite, *Hermetic and Alchemical Writings*, 1:89–113.

72. It is from this perspective that we can explain Victor's otherwise inexplicable

initial reaction to his creature: the fact that Victor had anticipated an experience of the beautiful but instead experienced "horror and disgust" results from the unanticipated conversion of means into end announced by the creature's self-animation (85). For a different, but complementary, account that emphasizes the difficulties one encounters in trying to understand Victor's response solely in aesthetic terms, see Gigante, "Facing the Ugly."

73. Mary Shelley, *Frankenstein*, 226. On Victor's automatism, see Levine, "Ambiguous Heritage," esp. 6, 10.

74. For Burke's description of culture as a kind of "second nature," see Burke, *Reflections*, 215. From the perspective on media I outline here, Burke's *Reflections* acknowledges that cultural media *can* be grasped by (i.e., provide handholds for action for) human beings, but Burke argues that such manipulations are always counterproductive, for they initiate changes that cannot be controlled. As a consequence, he suggests, cultural media ought to be treated *as though* they were natural media.

75. The quoted passages are from the narrator's description of Clerval's imperial aims, added by Shelley to the 1832 edition of the novel; see appendix F in Mary Shelley, *Frankenstein*, 344.

76. From this perspective, one of the key contributions of deconstruction was its illumination of the ways a medium that normally counts as communicational, such as writing, can also function as something akin to a natural medium (i.e., as a medium that grasps, and cannot be grasped by, an individual).

77. For a comparison between Wordsworth and Hegel, see Abrams, *Natural Supernaturalism*, 236–37; for Hegel and Faust, see Kittler, *Discourse Networks*, esp. 155–73; for Hegel and the *Bildungsroman*, see Moretti, *Way of the World*, esp. 55, 60, 85–86.

78. Foucault, *Madness and Civilization*, 219.

79. Walter Scott, "Remarks on Frankenstein"; originally published in *Blackwood's Edinburgh Magazine*, the review is partially reprinted in appendix D in Mary Shelley, *Frankenstein*, 300.

80. See, e.g., Scott's review in *Blackwood's Edinburgh Magazine*, as well as the anonymous review in *Edinburgh Magazine and Literary Miscellany* 2 (1818): 249–53 (partially reprinted in appendix D in Mary Shelley, *Frankenstein*, 306–8); the anonymous review in *Literary Panorama and National Register* 8 (1818): 411–41; and (arguably) the anonymous review in *Belle Assemblée; or, Bell's Court and Fashionable Magazine* 17 (March 1818): 139–42. All of these reviews are available in Romantic Circles, "Mary Wollstonecraft Shelley."

81. Citation from Mary Shelley, *Frankenstein*, 306, 308.

82. Ibid., 306. Though my reading of *Frankenstein* differs significantly from the practice of "distant reading" outlined in Franco Moretti's *Graphs, Maps, Trees*, his interest in thinking novels in terms of species, populations, and forces of selection highlights a related way of understanding the milieux of literary production.

83. Mary Shelley, *Frankenstein*, 104. It is worth calling attention in this context to Auerbach's important reflections in *Mimesis* on the concept of "milieu," especially to his claim that the pathos we feel for a character such as Julien Sorel in *The Red and the Black* is a consequence of Stendhal's tendency to treat a character as having "been thrown almost by chance into the milieu in which he lives; it is a resistance with which he can

deal more or less successfully, not really a culture-medium with which he is organically connected" (464–65).

84. See, e.g., Anderson's claim in *Imagined Communities* that novels and newspapers enabled the emergence of a sense of nationalism by serving as media that enabled geographically dispersed readers to orient themselves toward one another, and Armstrong's contention in *Desire and Domestic Fiction* that eighteenth-century novels helped to *constitute* the affective and epistemological structures necessary for a middle class that was not yet fully in existence, rather than simply "reflecting" preexisting structures. The fact that most people in the nineteenth and twentieth centuries first encountered the story of *Frankenstein* in other media, such as theater, film, or television, simply highlights its status as content *and* medium. (As St. Clair documents in *The Reading Nation*, the small initial print runs of *Frankenstein*, combined with its copyright term, meant that "for most of the 1850s, 1860s, and 1870s, *Frankenstein* was out of print"; since it had also "disappeared from the circulating libraries" [364], for "most of the nineteenth century, it was not the reading of the text of the book, but seeing adaptations of the story on the stage which kept *Frankenstein* alive in the culture" [367].) *Frankenstein* has continued to serve as a point of orientation in debates about changes that seem to bear simultaneously on social and vital matters, including the proper medical uses of human body parts, the dangers and virtues of nuclear power, and, most recently, the promise and peril of genetically engineered "Frankenfruits" and animals. See Levine and Knoepflmacher, *Endurance of Frankenstein*; Baldick, *In Frankenstein's Shadow*; and Dijck, *Imagenation*, esp. 53, 115–16.

85. Foucault, *Security, Territory, Population*, 67, 74.

86. The creature says that he read as "true histor[ies]" all the volumes that initially had "fallen into [his] hands": "*Paradise Lost*, a volume of *Plutarch's Lives*, and . . . *Sorrows of Werter*" (152). On the roles of books, reading, and education in *Frankenstein*, see Richardson, "Darkness Visible," and McLane, *Romanticism and the Human Sciences*, 84–108.

87. After reading *Paradise Lost*, the creature concludes not only that he, like Adam, "was created apparently united by no link to any other being in existence," but that he lacked even Adam's connection to, and conversation with, a generative figure (i.e., God) (Mary Shelley, *Frankenstein* 154).

88. More specifically, the creature would have had to read at least some of the books that had fallen into his hands *as* novels rather than as histories. (I owe this reflection on the relationship between epics and novels in *Frankenstein* to Zurawski, "Mis-Education of Frankenstein"). Though I have profited from McLane's emphasis in *Romanticism and the Human Sciences* on the importance of the concept of population to *Frankenstein*, I ultimately disagree with her claim that the novel seeks to show, through the example of the creature's reading, the failure of an understanding of "Literature" as something that leads to *Bildung* (i.e., individual cultural development). Our disagreement stems in part from our different understandings of the term "population": where McLane treats "population" as simply a synonym for a reproductive "multitude" (138), I emphasize, in addition, the importance of variation. We also interpret the creature's mode of reading differently, for I see as especially significant that he *mis*-reads both poetry and the novel (*Paradise Lost* and *Werther*): i.e., he reads *all* of his books as though they were histories

(accounts of events that really happened). What the creature lacks, in other words, is the very concept of fiction, a concept that is essential to both *Bildung* and the humanities. This in turn suggests that the novel is less a critique of the premises of *Bildung* or the humanities than of what we might describe as exclusively "realist" reading practices (which tend to be privileged by those who critique the humanities, rather than by those who assert the importance of *Bildung*).

89. Kittler contends in "Towards an Ontology of Media" that since ontologies of form and matter are predicated on the assumption that we never actually encounter form in the absence of matter or matter in the absence of form, "the togetherness or concrescence of these two categories in one and the same present thing suppresses all distance, absence, and nihilation from its entelechy" (25). Calls for more attention to the "materiality of media" are often premised on precisely this distinction between form and matter.

90. On the nebular hypothesis in eighteenth- and nineteenth-century astronomy, see Schaffer, "Nebular Hypothesis"; and Secord, *Victorian Sensation*, 9–10, 57–59, 90–91.

91. Schaffer, "Nebular Hypothesis," 134.

92. Rajan has done invaluable work in showing that despite Hegel's desire to create a univocal project, he in fact "produc[ed] encyclopedias of particular subjects that sometimes dissent from the *Encyclopedia*" ("Encyclopedia and the University of Theory," 341), which further emphasizes the impossibility of understanding the *variety* of Romantic-era theories of media from the perspective of Hegel's (or Kittler's) account.

93. As Kittler writes in *Discourse Networks*, "When the one Mother gave way to a plurality of women, when the alphabetization-made-flesh [i.e., Poetry] gave way to technological media . . . Poetry also disintegrated . . . from the magic of letters to a histrionics of media" (178). Guillory makes a similar point in "Genesis of the Media Concept": "The emergence of the media concept in the later nineteenth century was a response to the proliferation of new technical media—such as the telegraph and phonograph—that could not be assimilated to the older system of the arts" (321).

CHAPTER 6: Cryptogamia

1. Kroeber, *Ecological Literary Criticism*, 2.

2. Nichols, "Loves of Plants," paragraph 13.

3. Percy Bysshe Shelley, "Sensitive-Plant," 218, line 98.

4. I have benefited from the discussions of Romantic-era literary interest in the Linnaean category of cryptogams in Trott, "Wordsworth's Loves of the Plants," especially 161–62; and Goldstein, "Sweet Science," 1–31, and "Obsolescent Life."

5. See, e.g., Maniquis, "Puzzling Mimosa" and Nichols, "Loves of Plants." Late-eighteenth- and early-nineteenth-century discussions of the possibility of plant sensation and perception include the entries for "Plant" in Chambers, *Cyclopaedia*; Nicholson, *British Encyclopedia*; *Encyclopaedia Britannica*; and Darwin, *Botanic Garden*, 1:170.

6. Delaporte, *Nature's Second Kingdom*, 13; see also Jacob, *Logic of Life*, 40–42.

7. See Delaporte, *Nature's Second Kingdom*, 18–24. For a helpful but whiggish account of the problems with eighteenth-century "analogical" approaches to plants and animals, see Ritterbush, *Overtures to Biology*. Bewell provides a compelling account of the political and literary importance of analogies between plants and animals in "Jacobin Plants"; see also Browne, "Botany for Gentlemen."

8. For different, though consonant, accounts of increasing suspicion around 1800 of analogical methods of investigating plants and animals, see Ritterbush, *Overtures to Biology*, 158–210; Delaporte, *Nature's Second Kingdom*, 187–91; and Foucault, *Order of Things*, 263–79.

9. See, e.g., Priestley, *Experiments and Observations*.

10. Ingenhousz, *Experiments upon Vegetables*. For a brief biography of Ingenhousz and a contextualization of his research, see Ingenhousz and Reed, *Jan Ingenhousz, Plant Physiologist*.

11. On the emergence of the concept of photosynthesis, see Ingenhousz and Reed, *Jan Ingenhousz, Plant Physiologist*; Rabinowitch, *Photosynthesis and Related Processes*, 1:12–28; and Zallen, "'Light' Organism"; though see also Schaffer's discussion in "Scientific Discoveries" of the difficulties endemic to attempts to date the emergence of the concept of photosynthesis.

12. See, e.g., Aristotle, *On the Soul*, lines 413a23–413b24.

13. "Plant" in Chambers, *Cyclopaedia*.

14. "Plant" in Nicholson, *British Encyclopedia*.

15. On the familiarity of Romantic authors such as Wordsworth, Coleridge, and Goethe with contemporary plant research, see Trott, "Wordsworth's Loves," and Goldstein, "Equivocal Life."

16. My emphasis on the development of a sense for the unknown draws inspiration from Robyn Smith, "As Yet Unknown."

17. Fleck, *Genesis and Development*, 99–100. See also my discussion of mood or *Stimmung* in chapter 4.

18. Wordsworth, "Yew-Trees," lines 10–11.

19. Percy Bysshe Shelley, "The Flower that Smiles Today," lines 1–2.

20. Clare, "To the Ivy," lines 1–8.

21. Wordsworth, *Ruined Cottage*, lines 314–17; "Yew-Trees," lines 17–18. For a contemporary neo-Romantic biological account of the strangeness of plants that has inspired my own thinking, see Hallé, *In Praise of Plants*.

22. "Plant" in *Encyclopaedia Britannica*.

23. In his reading of *The Ruined Cottage*, Alan Bewell stresses Wordsworth's emphasis on the "excessive growth" of the gorse on the English common across which the narrator stumbles, arguing that it "suggests descriptions of colonial wildernesses, now transposed to a British setting" (*Romanticism and Colonial Disease*, 53).

24. Percy Bysshe Shelley, "Mont Blanc," line 90.

25. On the history and symbolism of plants in European literature and social practices, see Knight, *Flower Poetics*, 1–25; and the more popular account in Heilmeyer, *Language of Flowers*.

26. Clare, "To a Red Clover Blossom," lines 12–14.

27. In his *Philosophy of Nature*, Hegel wrote, that "plant-life therefore begins where the vital principle gathers itself into a point [*Punkt*] and this point sustains and produces itself, repels itself, and produces new points" and that the "plant . . . lacks the inwardness [*die Innerlichkeit*] which would be free from the relationship to the outer world" (303, 308; German original in *Werke*, 9:371, 9:377). Goethe also stressed the extent to which plants, as "imperfect" (*unvollkommener*) living beings, lacked individualized parts; see "On Morphology" (citation from 24; German original in *Werke*, 13:56–57). Though

both Goethe and Hegel emphasized the deindividualized character of plant life, Goethe tended to align plant and animal modes of vitality, suggesting that there was no fundamental distinction between the two. The hierarchical character of Hegel's system, by contrast, encouraged him to stress fundamental distinctions between plant and animal life. On the vexed role of Goethe's botany in Hegel's system, see Kelley, "Restless Romantic Plants."

28. Clare makes his point even more emphatically in "To the Cowslip," asserting that the plant to which the poem is addressed is "The self-same flower, the very same / As those which I used to find . . . / . . . Full twenty summers by" — with the consequence that "I'm no more akin to thee . . . / . . . for Time has had a hand with me, / And left an alter'd thing" (lines 3–4, 8–9, 10–11). See also Clare's presentation in "Eternity of Nature" of the daisy that "strikes its little root / Into the lap of time" and is thus able to reemerge perpetually in "the self-same state" (lines 4–5, 15).

29. Fulford, "Cowper, Wordsworth, Clare," 55.

30. Hartman suggests that the poem "intimat[es] a magical or superstitious persistence of the yew's life" and contends that "Yew-Trees" is in "many ways the most ghostly poetry ever written" (*Unremarkable Wordsworth*, 132, 150). For a related discussion of the "eerie" nature of some of Wordsworth's plant images, see Trott, "Wordsworth's Loves," 157.

31. In "Cowper Green," Clare suggests that "when [Nature] blooms at will," the "furze has leave to wreathe / Its dark prickles over the heath . . . / . . . hawthorns spread / Foliag'd houses o'er one's head," while nettles stick out above the other grasses (they are "keen to view") (lines 44, 47–50, 63). In *Wordsworth*, Liu writes, "We know that Robert [a character in *the Ruined Cottage*] inhabits an unenclosed common covered with 'bursting gorse' (11.18–25), for example, and that sheep cross his threshold and garden in his absence (11.388–94, 461). These details depict the area around Racedown, where the Wordsworths walked up gorse-covered Pilsdon Pen and themselves had trouble keeping livestock out of the garden" (330).

32. The parasite "wants to give his voice for matter, (hot) air for solid, superstructure for infrastructure"; he "obtains energy and pays for it in information" (Serres, *Parasite*, 35–36).

33. Serres, it must be said, does not recognize this fundamental distinction between representations of plants and animals, for he equates the cow and the tree in the quote above. Yet his intuition about fables nevertheless seems to hold, for it is in fact rarely the plant but rather generally the animal that is humanized in fables. While there are exceptions to this rule, such as the talking trees of the biblical fable in Judges 9:8–15 or La Fontaine's famous fable of "The Oak and the Reed," these exceptions simply emphasize the dominance of animal life in fables. Moreover, eighteenth-century plant fables, such as John Langhorne's rather saccharine *The Fables of Flora*, invariably depend on the analogical premise that plants are simply variants of animals.

34. Hegel writes in *Philosophy of Nature* that "air and water are perpetually acting on the plant; it does not take sips of water" (308; German original in *Werke*, 9:377).

35. While Erasmus Darwin's botanical-treatise-*cum*-poetic-epic *The Loves of the Plants* was well known during the Romantic era, it is its founding premise — that modes of plant reproduction are simply variations of animal desire and sexuality — and not simply its Augustan meter that explains why this text *feels* so different from other Romantic

engagements with plants (and why it is only with what amounts to special pleading that Darwin's book has recently been included in the Romantic "canon"). Rather than allowing plant love to alter our own conception of what love can and might be, Darwin instead contained plant love within the frame of animal (and more specifically human) relationships. On Darwin's approach to plants, see Bewell, "Jacobin Plants"; Browne, "Botany for Gentlemen"; and Trott, "Wordsworth's Loves of the Plants."

36. Ingenhousz, *Experiments upon Vegetables*, 9–10.

37. Ibid., 48. As Bewell documents in *Romanticism and Colonial Disease*, Romantic-era interest in the beneficial and detrimental effects of plants on local atmospheres was often parsed through hopes and fears about colonial landscapes (a point to which I return below).

38. Ingenhousz tended to refer to atmosphere as one unified entity — "the atmosphere" — but his analyses pointed to "the atmosphere" as a collective phenomenon, composed of numerous local atmospheres linked to one another.

39. My description of the animal cry as establishing a territory draws on both Serres, *Parasite*, 94; and Deleuze and Guattari, *Thousand Plateaus*, 310–50.

40. In *Biographia Literaria* (1817), Coleridge makes a similar, though more complicated point, arguing that the pleasure produced by poetry depends upon the capacity of meter to establish something like an undetected "medicated atmosphere." Coleridge contends that the function of meter is to "increase the vivacity and susceptibility both of the general feelings and of the attention," an "effect it produces by the continued excitement of surprise, and by the quick reciprocations of curiosity still gratified and still re-excited, which are too slight indeed to be at any one moment objects of distinct consciousness, yet become considerable in their influence. As a medicated atmosphere, or as wine during animated conversation; they act powerfully, though themselves unnoticed" (part 2, 66).

41. Pollan has popularized this sense of double domestication in *Botany of Desire*: as the advertising copy on the back cover notes, "just as we've benefited from these plants [i.e., apples, tulips, marijuana, and potatoes], we have also done well by them. So who is really domesticating whom?"

42. Canguilhem, "Living and Its Milieu," 12.

43. "Environment" appears in neither Chambers, *Cyclopaedia*, nor Nicholson, *British Encyclopedia*. The *Oxford English Dictionary* (2nd. ed.) suggests that the term begins to appear by the 1830s, and one suspects that it was imported into English via translations of Lamarck's *Philosophie zoologique*, in which the phrase *les milieux environnans* appeared frequently. See chapter 5 for a more complete discussion of the relationship between the terms "milieu" and "medium."

44. Despite its many virtues, Morton's *Ecology without Nature* systematically confuses these terms. Building on Simpson's critique of Romantic "local detail," Morton develops a sharply critical account of recent ecocriticism that attempts to evoke a sense of local "place" (or "environment" or "ambience"), arguing that such efforts inevitably depend on a kind of bad faith that cannot in the end hold onto anything solid. I am not interested, however, in the invocation of place through local detail but in the quite different endeavor of using detail to facilitate a tracing of the obscurity of plant life.

45. On the logic of ornaments as applied to gardens, see Nadarajan, "Ornamental Biotechnology."

46. Shelley's analogy here emphasizes that the love that enables mutual atmospheres has little, if anything, to do with Erasmus Darwin's anthropomorphized loves of plants. Because Darwin uses analogy to transfer human sexual relations onto plants, Darwin's "loves" are between plants, rather than between humans and plants, and his understanding of plant love draws on traditional, gendered images of passivity and activity. Shelley's Romantic plant poem, by contrast, uses analogy to move *beyond* human sexual relations: the creation of a mutual atmosphere is only *like* the relationship between young lovers.

47. Jean-Jacques Rousseau, *Reveries*, 115. For a reading of Rousseau's botanical pursuits as compensation for absent human sexuality, see Bewell, "Jacobin Plants," 134–35. It is worth noting that though Rousseau depicts his interest in botany in the period after his "sixty-fifth birthday" (105) as a kind of compensation for social relations or a distraction "from [his] misfortunes" (115), this was in fact a *return* to the botanical pursuits of his earlier life, during which period he still enjoyed the company of others, including that of his sexual companion Thérèse Levasseur (84ff.). We can say, at any rate, that for the Romantics, the "raptures and ecstasies" (84) of cryptogamia were not intrinsically asocial.

48. I draw here on the account of the relationship between apprenticeship and signs outlined in Deleuze, *Proust and Signs*.

49. Percy Bysshe Shelley, "Sensitive-Plant," 2:41–56. Plants are also involved in their own apprenticeships, for each individual plant responds differently to the "same" care. This is thus, from both sides, an apprenticeship without masters.

50. De Man's approach to language is indebted to the neo-Hegelian "anthropology" that emerged in the 1930s through the seminars of Alexandre Kojève, among others. For an especially clear articulation of this anthropological version of Hegel and its implications for an understanding of the telos of language, see Blanchot, *Work of Fire*, 300–344. As Borch-Jacobsen notes in *Lacan*, this anthropological understanding of Hegel is difficult to square with Hegel's own absolute idealism (1–20).

51. Pickering, *Mangle of Practice*, 21–22. Pickering draws both the metaphor of tuning and his emphasis on passivity and activity from Ludwik Fleck (22 n.35).

52. In order to capture this sense of moving-in-step-with, Pickering also describes the process of tuning as a "dance of agency" (ibid., 21).

53. Jonathan Bate, *Romantic Ecology*, 22–28.

54. See, for example, Fulford, Lee, and Kitson, *Literature, Science, and Exploration*, 33–45.

55. Bewell is especially attentive to these reverse dynamics in *Romanticism and Colonial Disease*. More generally, my point is intended to provide a partial corrective to Crosby's *Ecological Imperialism*, which emphasizes only the one-way movement of "weeds" from Europe to other parts of the world (see, e.g., 145–70). See also Monique Allewaert's compelling discussion of "plantation spaces" and "ecologies" in "Swamp Sublime."

56. Clare, "Cowper Green," line 44. For a discussion of transplantation, see Clare's "Song—A Beautiful Flower," in which the narrator "transplanted its charms to [his] bosom,/And deep has the root gather'd there" (lines 15–16).

57. Clare, "Ballad—A Weedling Wild," lines 4, 1, 24. Clare detailed his unhappiness with his apprenticeship as a "kitchen" gardener in Clare, Robinson, and Powell, *John Clare by Himself*, 12. Clare's description of his apprenticeship, in combination with his ballad, emphasizes that gardens depend upon two quite different activities—procure-

ment of wild species, followed by domestication of those species—and only the former seems to have interested him.

58. Doyle, "Transgenic Involution," 70.

59. Ibid.

60. Percy Bysshe Shelley, *Defence of Poetry*, 294. My discussion of evanescence draws on Keach, *Shelley's Style*, 118–53.

61. For this critique of Hartman, see Jonathan Bate, *Romantic Ecology*, 7–8.

62. The refrain of "harmony" is especially pronounced in ibid., e.g. 19, 22, 40, but it is implicitly present in much ecocriticism.

63. As ecologist Robert O'Neill notes in "Is It Time to Bury the Ecosystem Concept?": "Ultimately, of course, the ecosystem is unstable. It is only a matter of time until a disturbance of sufficient intensity and spatial extent overwhelms the ecosystem's ability to respond. Examples include broad-scale desertification and rare asteroid collisions. Over a sufficiently long period of time, the cumulative probability of a catastrophic event approaches 1.0 [i.e., 100%]" (3278).

CONCLUSION: Biopolitics and Experimental Vitalism

1. Agamben, *Homo Sacer*, 1.

2. Esposito, *Bíos*, 45.

3. For a discussion of this line of criticism, see Deutscher, "The Precarious, the Immune, and the Thanatopolitical." It is worth noting that in *Bíos*, Esposito explicitly emphasizes the importance of *birth*—"the moment when the umbilical cord is cut and the newborn cleaned of amniotic fluid"—rather than conception, for he is interested in the moment in which the child is clearly and explicitly separated from the mother (176).

4. Wolfe, *Before the Law*, 103. Deutscher suggests that a similar critique applies to other contemporary theoretical valuations of life, including Judith Butler's valorization of all "precarious" or "grievable" life; see Deutscher, "The Precarious, the Immune, and the Thanatopolitical."

5. Wolfe, *Before the Law*, 59; internal quote from Luke, "Dreams of Deep Ecology," 51.

6. The first quote, cited by Wolfe, is from Esposito, "Person and Human Life," 218; the second is from Wolfe, *Before the Law*, 58.

7. Esposito, *Bíos*, 194; internal quote from Deleuze, *Logic of Sense*, 107.

8. Wolfe, *Before the Law*, 59.

9. Esposito, *Bíos*, 191; citing Canguilhem, *Normal and the Pathological*, 139.

10. Derrida, *Speech and Phenomena*, 10.

11. Cooper, *Life as Surplus*, 23.

12. Ibid., 42.

13. On structures of Romantic-era finance, see Robert Mitchell, *Sympathy and the State*. On the financial structure of contemporary bioeconomics, see, e.g., Waldby and Mitchell, *Tissue Economies*, and Robert Mitchell, "U.S. Biobanking Strategies."

14. Stengers, "Constructivist Reading," 95.

Bibliography

Abernethy, John. *Physiological Lectures, Exhibiting a General View of Mr. Hunter's Physiology, and of His Researches in Comparative Anatomy.* 2nd ed. London: Longman, Hurst, Rees, Orme, and Brown, 1822.

———. *Surgical and Physiological Essays.* London: Printed for James Evans, 1793.

Abrams, M. H. "Keats's Poems: The Material Dimensions." In *The Persistence of Poetry: Bicentennial Essays on Keats,* edited by Robert M. Ryan and Ronald A. Sharp, 36–53. Amherst: University of Massachusetts Press, 1998.

———. *Natural Supernaturalism: Tradition and Revolution in Romantic Literature.* New York: Norton, 1971.

Adams, George. *Lectures on Natural and Experimental Philosophy, Considered in Its Present State of Improvement.* 5 vols. London: Printed by R. Hindmarsh, 1794.

Adorno, Theodor W. *Aesthetic Theory.* Edited by Gretel Adorno and Rolf Tiedemann. Minneapolis: University of Minnesota Press, 1997.

———. "Difficulties." In *Essays on Music,* 644–79.

———. *Essays on Music.* Translation with introduction, selection, commentary, and notes by Richard Leppert; new translations by Susan H. Gillespie. Berkeley: University of California Press, 2002.

Agamben, Giorgio. *Homo Sacer: Sovereign Power and Bare Life.* Translated by Daniel Heller-Roazen. Stanford, CA: Stanford University Press, 1998.

———. *Potentialities: Collected Essays in Philosophy.* Translated by Daniel Heller-Roazen. Stanford, CA: Stanford University Press, 1999.

Allewaert, Monique. "Swamp Sublime: Ecologies of Resistance in the American Plantation Zone." *PMLA* 123.2 (2008): 340–57.

Anderson, Benedict. *Imagined Communities: Reflections on the Origin and Spread of Nationalism.* Rev. and extended ed. New York: Verso, 1991.

Andrews, Lindsey. "American Experiments: The Science and Aesthetics of Clinical Practice in American Literature, 1890–1965." Ph.D. diss., Duke University, 2013.

Anonymous. *The Best and Easiest Method of Preserving Uninterrupted Health to Extreme Old Age.* London: Printed by order of his executors, and sold by R. Baldwin, 1748.

———. *Remarks on Mr Mason's Treatise upon Tea,* by "J.N. [Surgeon]." London: Printed for J. Roberts, 1745.

———. *Transactions of the Royal Humane Society: From 1774 to 1784, with an Appendix of Miscellaneous Observations on Suspended Animation.* London, 1795.

Antliff, Mark. *Inventing Bergson: Cultural Politics and the Parisian Avant-Garde*. Princeton, NJ: Princeton University Press, 1993.

Appel, Toby A. *The Cuvier-Geoffroy Debate: French Biology in the Decades before Darwin*. New York: Oxford University Press, 1987.

Aristotle. *The Complete Works of Aristotle*. Edited by Jonathan Barnes. 2 vols. Princeton, NJ: Princeton University Press, 1984.

———. *On the Soul*. In *Complete Works*, 641–92.

———. *Physics*. In *Complete Works*, 315–446.

———. *The Physics*. Edited by Philip Henry Wicksteed and Francis Macdonald Cornford. Rev. ed. 2 vols. Cambridge, MA: Harvard University Press, 1934.

Armstrong, Nancy. *Desire and Domestic Fiction: A Political History of the Novel*. New York: Oxford University Press, 1987.

Auerbach, Erich. *Mimesis: The Representation of Reality in Western Literature*. Translated by W. R. Trask. Princeton, NJ: Princeton University Press, 1953.

Austin, J. L. *How to Do Things with Words*. 2nd ed. Cambridge, MA: Harvard University Press, 1975.

Bacon, Francis. *The Philosophical Works of Francis Bacon*. Edited by Peter Shaw. 3 vols. London: Printed for J. J. and P. Knapton et al., 1733.

Bakhtin, Mikhail. *Rabelais and His World*. Translated by Hélène Iswolsky. Bloomington: Indiana University Press, 1984.

Baldick, Chris. *In Frankenstein's Shadow: Myth, Monstrosity, and Nineteenth-Century Writing*. Oxford: Oxford University Press, 1987.

Bann, Stephen. *Romanticism and the Rise of History*. New York: Twayne, 1994.

Barck, Karlheinz, and Martin Fontius. *Ästhetische Grundbegriffe: Historisches Wörterbuch in sieben Bänden*. 7 vols. Stuttgart: Metzler, 2000.

Barrow, John. *Dictionarium Medicum Universale: Or, a New Medicinal Dictionary*. London: Printed for T. Longman, C. Hitch, and A. Millar, 1749.

Bataille, George. *Erotism: Death and Sensuality*. Translated by Mary Dalwood. San Francisco: City Lights Books, 1986.

———. *The Story of the Eye, by Lord Auch*. Translated by Joachim Neugroschel. San Francisco: City Lights Books, 1977.

Bate, Jonathan. *Romantic Ecology: Wordsworth and the Environmental Tradition*. New York: Routledge, 1991.

Bate, Walter Jackson. *John Keats*. Cambridge, MA: Harvard University Press, 1963.

Beck, Ulrich. *Risk Society: Towards a New Modernity*. Translated by Mark Ritter. London: Sage Publications, 1992.

Beddoes, Thomas. *Hygëia, or, Essays, Moral and Medical, on the Causes Affecting the Personal State of Our Middling and Affluent Classes*. Bristol: Printed by J. Mills for R. Phillips, 1802–3. Facsimile edition edited and with an introduction by Robert Mitchell. Bristol, UK: Thoemmes Press, 2003.

Bender, John. "Novel Knowledge: Judgment, Experience, and Experiment." In Siskin and Warner, *This Is Enlightenment*, 284–300.

Benjamin, Walter. *The Concept of Criticism in German Romanticism*. In *Selected Writings, 1913–1926*, edited by Marcus Bullock and Michael W. Jennings, 1:116–200. Cambridge, MA: Belknap Press of Harvard University Press, 2004.

Benton, E. "Vitalism in Nineteenth Century Thought." *Studies in the History and Philosophy of Science* 5 (1974): 17–48.

Bergson, Henri. *Creative Evolution.* Translated by Arthur Mitchell. Mineola, NY: Dover Publications, 1998.

———. *Matter and Memory.* Translated by Nancy Margaret Paul and W. Scott Palmer. New York: Zone Books, 1988.

Bersani, Leo. "Pynchon, Paranoia, and Literature." *Representations* 25 (1989): 99–188.

Bewell, Alan. "'Jacobin Plants': Botany as Social Theory in the 1790s." *Wordsworth Circle* 20.3 (1989): 132–39.

———. *Romanticism and Colonial Disease.* Baltimore: Johns Hopkins University Press, 1999.

———. *Wordsworth and the Enlightenment: Nature, Man, and Society in the Experimental Poetry.* New Haven, CT: Yale University Press, 1989.

Biagioli, Mario. *Galileo Courtier: The Practice of Science in the Culture of Absolutism.* Chicago: University of Chicago Press, 1993.

Bichat, Xavier. *Physiological Researches on Life and Death.* In *Physiological Researches on Life and Death,* by Xavier Bichat; *Outlines of Phrenology,* by Johann Gaspar Spurzheim; *Phrenology Examined,* by Pierre Flourens. Edited by Daniel N. Robinson. Washington, DC: University Publications of America, 1978.

Bies, Michael, and Michael Gamper, eds. *"Es ist ein Laboratorium, ein Laboratorium für Worte": Experiment und Literatur III: 1890–2010.* Göttingen: Wallstein Verlag, 2011.

Blagden, Charles. "Experiments and Observations in an Heated Room." *Philosophical Transactions of the Royal Society of London* 65 (1775): 111–23.

———. "Further Experiments and Observations in an Heated Room." *Philosophical Transactions of the Royal Society of London* 65 (1775): 484–94.

Blake, William. *The Complete Poetry and Prose of William Blake.* Edited by David V. Erdman and Harold Bloom. Rev. ed. Berkeley: University of California Press, 1988.

Blanchot, Maurice. *The Work of Fire.* Translated by Charlotte Mandell. Stanford, CA: Stanford University Press, 1995.

Bloom, Harold. *The Anxiety of Influence: A Theory of Poetry.* 2nd ed. New York: Oxford University Press, 1997.

Bloor, David. "Anti-Latour." *Studies in the History and Philosophy of Science* 30:1 (1999): 81–112.

———. *Knowledge and Social Imagery.* London: Routledge, 1976.

———. "Reply to Bruno Latour." *Studies in the History and Philosophy of Science* 30:1 (1999): 131–36.

Bois, Yve Alain, and Rosalind E. Krauss. *Formless: A User's Guide.* New York: Zone Books, 1997.

Borch-Jacobsen, Mikkel. *Lacan: The Absolute Master.* Translated by Douglas Brick. Stanford, CA: Stanford University Press, 1991.

Bourdieu, Pierre. *The Rules of Art: Genesis and Structure of the Literary Field.* Translated by Susan Emanuel. Stanford, CA: Stanford University Press, 1996.

Boyle, Robert. *The Works of the Honourable Robert Boyle, in Six Volumes, to Which Is Prefixed the Life of the Author.* 6 vols. London: Printed for J. and F. Rivington et al., 1772.

Brain, Robert Michael. "The Romantic Experiment as Fragment." In *Hans Christian Ørsted and the Romantic Legacy in Science: Ideas, Disciplines, Practices*, edited by R. M. Brain, R. S. Cohen, and O. Knudsen, 217–33. Dordrecht: Springer, 2007.

Bray, Joe, Alison Gibbons, and Brian McHale. *The Routledge Companion to Experimental Literature*. New York: Routledge, 2012.

Brice, Benjamin. *Coleridge and Scepticism*. Oxford: Oxford University Press, 2007.

Brown, Marshall. *The Gothic Text*. Stanford, CA: Stanford University Press, 2005.

Browne, Janet. "Botany for Gentlemen: Erasmus Darwin and 'The Loves of the Plants.'" *Isis* 80.4 (1989): 593–621.

Buck-Morss, Susan. "Aesthetics and Anaesthetics: Walter Benjamin's Artwork Essay Reconsidered." *October* 62 (1992): 3–41.

Bürger, Peter. *Theory of the Avant-Garde*. Translated by Michael Shaw. Foreword by Jochen Schulte-Sasse. Minneapolis: University of Minnesota Press, 1984.

Burke, Edmund. *Reflections on the Revolution in France*. Edited by Thomas H. D. Mahoney. Indianapolis: Bobbs-Merrill, 1955.

Burkhardt, Richard W. *The Spirit of System: Lamarck and Evolutionary Biology*. Cambridge, MA: Harvard University Press, 1995.

Burroughs, Wiliam S. *The Ticket That Exploded*. New York: Grove Press, 1962.

Burwick, Frederick, and Paul Douglass. *The Crisis in Modernism: Bergson and the Vitalist Controversy*. New York: Cambridge University Press, 1992.

Bush, Douglas. "Keats and His Ideas." In *The Major English Romantic Poets: A Symposium in Reappraisal*, edited by C. Thorpe, 231–45. Carbondale: Southern Illinois University Press, 1957.

Butler, Judith. *Bodies That Matter: On the Discursive Limits Of "Sex."* New York: Routledge, 1993.

———. *Precarious Life: The Powers of Mourning and Violence*. New York: Verso, 2004.

Cabanis, P. J. G. *Rapports du physique et du moral de l'homme*. 2 vols. Paris: Chez Béchet jeune, 1824.

Cage, John. *Silence: Lectures and Writings*. Middletown, CT: Wesleyan University Press, 1961.

Campbell, Colin. *The Romantic Ethic and the Spirit of Modern Consumerism*. New York: B. Blackwell, 1987.

Canguilhem, Georges. "The Living and Its Milieu." *Grey Room* 3 (2001): 7–31.

———. *The Normal and the Pathological*. Translated by Carolyn R. Fawcett. New York: Zone Books, 1991.

———. *A Vital Rationalist: Selected Writings from Georges Canguilhem*. Edited by François Delaporte. New York: Zone Books, 1994.

Cecire, Natalia. "A Sense of the Real: Experimental Writing and the Sciences, 1879–1946." Ph.D. diss., University of California, Berkeley, 2010.

Chambers, Ephraim. *Cyclopaedia; or, an Universal Dictionary of Arts and Sciences*. 4 vols. London: Rivington et al., 1788.

Chandler, James K. *England in 1819: The Politics of Literary Culture and the Case of Romantic Historicism*. Chicago: University of Chicago Press, 1998.

———. *Wordsworth's Second Nature: A Study of the Poetry and Politics*. Chicago: University of Chicago Press, 1984.

Chew, Samuel C. *The Nineteenth Century and After (1789–1939)*. In *A Literary History of England*, edited by Albert C. Baugh, 1111–605. New York: Appleton-Century-Crofts, 1948.

Cheyne, George. *The English Malady; or, a Treatise of Nervous Diseases of All Kinds.* London, 1733. Facsimile edition edited by Eric T. Carlson. Delmar, NY: Scholars' Facsimiles and Reprints, 1976.

Chomsky, Noam, and Michel Foucault. "Human Nature: Justice versus Power." In *Foucault and His Interlocuters*, edited by Arnold I. Davidson, 107–45. Chicago: University of Chicago Press, 1998.

Christensen, Jerome. *Coleridge's Blessed Machine of Language.* Ithaca, NY: Cornell University Press, 1981.

Clare, John. "Ballad—A Weedling Wild." In *Village Minstrel*, 1:96–97.

——. "Cowper Green." In *Village Minstrel*, 1:109–120.

——. "The Eternity of Nature." In *The Rural Muse*, 34–37. London: Whittaker, 1935.

——. "Song—A beautiful flower." In *Village Minstrel*, 2:19–20.

——. "To a Red Clover Blossom." In *Village Minstrel*, 2:178.

——. "To the Cowslip." In *The Shepherd's Calendar; with Village Stories, and Other Poems*, 207–9. London: Published for John Taylor, 1827.

——. "To the Ivy." In *Village Minstrel*, 2:165.

——. *The Village Minstrel, and Other Poems.* 2 vols. London: Printed for Taylor and Hessey, 1821.

Clare, John, Eric Robinson, and David Powell. *John Clare by Himself.* Manchester, UK: Carcanet Press, 1996.

Clark, David L. "Heidegger's Craving: Being-on-Schelling." *Diacritics* 27.3 (1997): 8–33.

Clarkson, Thomas. *An Essay on the Impolicy of the African Slave Trade, in Two Parts.* London: Printed and sold by J. Phillips, 1788.

Cohen, Matt. *The Networked Wilderness: Communicating in Early New England.* Minneapolis: University of Minnesota Press, 2010.

Coke, Diana. *Saved from a Watery Grave: The Story of the Royal Humane Society's Receiving House in Hyde Park.* London: Royal Humane Society, 2000.

Coleridge, Samuel Taylor. *Aids to Reflection.* Edited by John Beer. In *Collected Works*, vol. 9.

——. *Biographia Literaria; or, Biographical Sketches of My Literary Life and Opinions.* Edited by James Engell and W. Jackson Bate. In *Collected Works*, vol. 7.

——. *Christabel; Kubla Khan, a Vision; the Pains of Sleep.* London: John Murray, 1816.

——. *Collected Letters.* Edited by Earl Leslie Griggs. 6 vols. Oxford: Clarendon Press, 1956.

——. *The Collected Works of Samuel Taylor Coleridge.* 23 vols. Princeton, NJ: Princeton University Press, 1969–2002.

——. *Conciones ad Populum.* Edited by Lewis Patton and Peter Mann. In *Collected Works*, 1:21–74.

——. "An Essay on Scrofula." Edited by H. J. Jackson and J. R. de J. Jackson. In *Collected Works*, 12:454–79.

——. *The Friend.* Edited by Barbara E. Rooke. In *Collected Works*, vol. 4.

——. "Lecture on the Slave-Trade." Edited by Lewis Patton and Peter Mann. In *Collected Works*, 1:235–51.

——. *Lectures 1808–1819 on Literature.* Edited by R. A. Foakes. In *Collected Works*, vol. 5, part 1.

———. "A Moral and Political Lecture." Edited by Lewis Patton and Peter Mann. In *Collected Works*, 1:235–51.

———. *The Notebooks of Samuel Taylor Coleridge.* Edited by Kathleen Coburn. 4 vols. Princeton, NJ: Princeton University Press, 1957.

———. *The Statesman's Manual.* Edited by R. J. White. In *Collected Works*, 6:1–114.

———. "Theory of Life." Edited by H. J. Jackson and J. R. de J. Jackson. In *Collected Works*, 12:479–557.

———. *The Watchman.* Edited by Lewis Patton. In *Collected Works*, vol. 2.

Coleridge, Samuel Taylor, and William Wordsworth. *Lyrical Ballads, 1798 and 1800.* Edited by Michael Gamer and Dahlia Porter. Peterborough, ON: Broadview Press, 2008.

Collings, David. "The Monster and the Maternal Thing: Mary Shelley's Critique of Ideology." In Mary Wollstonecraft Shelley, *Frankenstein: Complete, Authoritative Text with Biographical, Historical, and Cultural Contexts, Critical History, and Essays from Contemporary Critical Perspectives,* 2nd ed., edited by Johanna M. Smith, 280–95. New York: Bedford / St. Martin's, 2000.

———. "Suspended Satisfaction: 'Ode on a Grecian Urn' and the Construction of Art." In *"Ode on a Grecian Urn": Hypercanonicity & Pedagogy,* Romantic Circles Praxis Series (October 2003), www.rc.umd.edu/praxis/grecianurn/contributorsessays/grecian urncollings.html.

Cooper, Melinda. *Life as Surplus: Biotechnology and Capitalism in the Neoliberal Era.* Seattle: University of Washington Press, 2008.

Crosby, Alfred W. *Ecological Imperialism: The Biological Expansion of Europe, 900–1900.* 2nd ed. Cambridge: Cambridge University Press, 2004.

Culler, Jonathan. *The Pursuit of Signs: Semiotics, Literature, Deconstruction.* Ithaca, NY: Cornell University Press, 1981.

———. "Why Lyric?" *PMLA* 123.1 (2008): 201–6.

Curran, Stuart. *Poetic Form and British Romanticism.* New York: Oxford University Press, 1986.

Daffron, Eric. "Male Bonding: Sympathy and Shelley's *Frankenstein.*" *Nineteenth-Century Contexts* 21 (1999): 415–36.

Daiber, Jürgen. *Experimentalphysik des Geistes: Novalis und das romantische Experiment.* Göttingen: Vandenhoeck und Ruprecht, 2001.

Darnton, Robert. *Mesmerism and the End of the Enlightenment in France.* Cambridge, MA: Harvard University Press, 1968.

Darwin, Erasmus. *The Botanic Garden, a Poem. In Two Parts.* 4th ed. 2 vols. London: Printed for J. Johnson, St. Paul's Church-Yard, 1799.

———. *Zoonomia; or, the Laws of Organic Life.* 2nd ed. 2 vols. London: Printed for J. Johnson, 1796.

Daston, Lorraine, and Katharine Park. *Wonders and the Order of Nature, 1150–1750.* Cambridge, MA: Zone Books, 1998.

De Almeida, Hermione. *Romantic Medicine and John Keats.* New York: Oxford University Press, 1991.

Dear, Peter. *Discipline & Experience: The Mathematical Way in the Scientific Revolution.* Chicago: University of Chicago Press, 1995.

De Bordeu, Théophile. *Oeuvres complètes.* Paris: Caille et Ravier, 1818.

De Duve, Thierry. *Kant after Duchamp*. Cambridge, MA: MIT Press, 1996.

De Landa, Manuel. *Intensive Science and Virtual Philosophy*. New York: Continuum, 2002.

Delaporte, François. *Nature's Second Kingdom: Explorations of Vegetality in the Eighteenth Century*. Translated by Arthur Goldhammer. Cambridge, MA: MIT Press, 1982.

Deleuze, Gilles. *Cinema 1: The Movement-Image*. Translated by Hugh Tomlinson and Barbara Habberjam. Minneapolis: University of Minnesota Press, 1986.

———. *Difference and Repetition*. Translated by Paul Patton. New York: Columbia University Press, 1995.

———. *Kant's Critical Philosophy: The Doctrine of the Faculties*. Translated by Hugh Tomlinson and Barbara Habberjam. Minneapolis: University of Minnesota Press, 1984.

———. *The Logic of Sense*. Translated by Mark Lester, with Charles Stivale. Edited by Constantin V. Boundas. New York: Columbia University Press, 1990.

———. *Proust and Signs*. Translated by Richard Howard. Minneapolis: University of Minnesota Press, 2000.

———. *Pure Immanence: Essays on a Life*. Translated by Anne Boyman. New York: Zone Books, 2001.

Deleuze, Gilles, and Félix Guattari. *A Thousand Plateaus: Capitalism and Schizophrenia*. Translated by Brian Massumi. Minneapolis: University of Minnesota Press, 1987.

———. *What Is Philosophy?* Translated by Hugh Tomlinson and Graham Burchell. New York: Columbia University Press, 1994.

De Man, Paul. "Intentional Structure of the Romantic Image." In *The Rhetoric of Romanticism*, 1–17. New York: Columbia University Press, 1984.

Derrida, Jacques. "Economimesis." *Diacritics* 11.2 (1981): 2–25.

———. *Limited Inc*. Evanston, IL: Northwestern University Press, 1988.

———. *Speech and Phenomena, and Other Essays on Husserl's Theory of Signs*. Translated by David B. Allison. Evanston, IL: Northwestern University Press, 1973.

Desmond, Adrian J. *The Politics of Evolution: Morphology, Medicine, and Reform in Radical London*. Chicago: University of Chicago Press, 1989.

Deutscher, Penelope. "The Precarious, the Immune, and the Thanatopolitical: Butler, Esposito, and Agamben on Reproductive Biopolitics." Unpublished manuscript, 2012.

Dick, Philip K. *Ubik*. New York: Vintage Books, 1991.

Dickens, Charles. *Our Mutual Friend*. Edited by Michael Cotsell. New York: Oxford University Press, 1998.

Diderot, Denis, and Jean Le Rond D'Alembert. *Encyclopédie ou dictionnaire raisonné des sciences, des arts et des métiers, par une société de gens de letters*. [1751–1765]. Compact ed. 5 vols. Elmsford, NY: Pergamon Press, 1969.

Dijck, José van. *Imagenation: Popular Images of Genetics*. New York: New York University Press, 1998.

Douglas, Mary. *Purity and Danger: An Analysis of Concepts of Pollution and Taboo*. New York: Praeger, 1966.

Doyle, Richard. *Darwin's Pharmacy: Sex, Plants, and the Evolution of the Noösphere*. Seattle: University of Washington Press, 2011.

———. *On Beyond Living: Rhetorical Transformations of the Life Sciences*. Stanford, CA: Stanford University Press, 1997.

——. "The Transgenic Involution." In *Signs of Life: Bio Art and Beyond*, edited by Eduardo Kac, 69–82. Cambridge, MA: MIT Press, 2007.

——. *Wetwares: Experiments in Postvital Living*. Minneapolis: University of Minnesota Press, 2003.

Driesch, Hans. *The History and Theory of Vitalism*. Translated by C. K. Ogden. London: Macmillan, 1914.

——. *The Problem of Individuality*. London: Macmillan, 1914.

Duster, Troy. "Lessons from History: Why Race and Ethnicity Have Played a Major Role in Biomedical Research." *Journal of Law, Medicine & Ethics* 34.3 (2006): 487–96.

Eliassen, Knut Ove, and Yngve Sandhei Jacobsen. "Where Were the Media before the Media? Mediating the World at the Time of Condillac and Linnaeus." In Siskin and Warner, *This Is Enlightenment*, 64–86.

Encyclopaedia Britannica; or, a Dictionary of Arts, Sciences, and Miscellaneous Literature; Enlarged and Improved. 6th ed. 20 vols. Edinburgh: Printed for Archibald Constable and Company, 1823.

Enzensberger, Hans Magnus. "The Aporias of the Avant-Garde." In *The Consciousness Industry: On Literature, Politics and the Media*, edited by Michael Roloff, 16–41. New York: Seabury Press, 1974.

Esposito, Roberto. *Bíos: Biopolitics and Philosophy*. Minneapolis: University of Minnesota Press, 2008.

——. "The Person and Human Life." In *Theory after Theory*, edited by Jane Elliot and Derek Attridge, 205–19. New York: Routledge, 2001.

Estes, J. Worth. "The Medical Properties of Food in the Eighteenth Century." *Journal of the History of Medicine and Allied Sciences* 51 (1996): 127–54.

Fabian, Johannes. *Time and the Other: How Anthropology Makes Its Object*. New York: Columbia University Press, 1983.

Finch, Annie. *The Ghost of Meter: Culture and Prosody in American Free Verse*. Ann Arbor: University of Michigan Press, 1993.

Findlen, Paula. *Possessing Nature: Museums, Collecting, and Scientific Culture in Early Modern Italy*. Berkeley: University of California Press, 1994.

Fleck, Ludwik. *Genesis and Development of a Scientific Fact*. Translated by Fred Bradley. Chicago: University of Chicago Press, 1979.

Folkenflik, Robert. "Patronage and the Poet-Hero." *Huntington Library Quarterly* 48 (1985): 363–79.

Forth, Christopher E., and Ana Carden-Coyne, eds. *Cultures of the Abdomen: Diet, Digestion, and Fat in the Modern World*. New York: Palgrave Macmillan, 2005.

Foster, Gwendolyn Audrey, and Wheeler Winston-Dixon. *Experimental Cinema: The Film Reader*. New York: Routledge, 2002.

Foucault, Michel. *Madness and Civilization: A History of Insanity in the Age of Reason*. Translated by Richard Howard. New York: Vintage Books, 1973.

——. *The Order of Things: An Archaeology of the Human Sciences*. Translation unattributed. New York: Vintage Books, 1973.

——. *Security, Territory, Population: Lectures at the College De France, 1977–78*. Translated by Graham Burchell. Edited by Michel Senellart. New York: Palgrave Macmillan, 2007.

Fox, Renée C., and Judith P. Swazey. *Spare Parts: Organ Replacement in American Society*. New York: Oxford University Press, 1992.

Freyhofer, Horst H. *The Vitalism of Hans Driesch: The Success and Decline of a Scientific Theory*. Frankfurt am Main: P. Lang, 1982.

Friedman, Ken. *The Fluxus Reader*. New York: Academy Editions, 1999.

Frontain, Raymond-Jean. "Typological Identity and Syntactic Suspension in Milton's Sonnet 10." *ANQ* 11.2 (1998): 14–22.

Frosch, Thomas. "Passive Resistance in Shelley: A Psychological View." *Journal of English and Germanic Philology* 98:3 (1999): 373–95.

Fry, Paul H. *Wordsworth and the Poetry of What We Are*. New Haven, CT: Yale University Press, 2008.

Fulford, Tim. *Coleridge's Figurative Language*. Basingstoke, UK: Macmillan, 1991.

———. "Cowper, Wordsworth, Clare: The Politics of Trees." *John Clare Society Journal* 14 (1995): 47–59.

Fulford, Tim, and Peter J. Kitson, eds. *Romanticism and Colonialism: Writing and Empire, 1780–1830*. Cambridge: Cambridge University Press, 1998.

Fulford, Tim, Debbie Lee, and Peter J. Kitson. *Literature, Science, and Exploration in the Romantic Era: Bodies of Knowledge*. Cambridge: Cambridge University Press, 2004.

Fulton, Robert. *A Treatise on the Improvement of Canal Navigation*. London: Published by I. and J. Taylor, 1796.

Galison, Peter. *How Experiments End*. Chicago: University of Chicago Press, 1987.

Gamper, Michael, ed. *Experiment und Literatur: Themen, Methoden, Theorien*. Göttingen: Wallstein Verlag, 2010.

Gamper, Michael, Martina Wernli, and Jörg Zimmer, eds. *"Es ist nun einmal zum Versuch gekommen": Experiment und Literatur I: 1580–1790*. Göttingen: Wallstein Verlag, 2009.

———. *"Wir sind Experimente: wollen wir es auch sein!": Experiment und Literatur II: 1790–1890*. Göttingen: Wallstein Verlag, 2010.

Geoffroy Saint-Hilaire, Étienne. "Des différns états de pesanteur des oeufs, au commencement et à la fin de l'incubation." *Journal complémentaire des sciences médicales* 7 (1820): 271–78.

Gigante, Denise. "The Endgame of Taste: Keats, Sartre, Beckett." In Morton, *Cultures of Taste / Theories of Appetite*, 183–201.

———. "Facing the Ugly: The Case of Frankenstein." *ELH* 67.2 (2000): 565–87.

———. "Keats's Nausea." *Studies in Romanticism* 40.4 (2001): 481–510.

———. *Life: Organic Form and Romanticism*. New Haven, CT: Yale University Press, 2009.

———. "The Monster in the Rainbow: Keats and the Science of Life." *PMLA* 117 (2002): 433–48.

Godwin, William. *Enquiry Concerning Political Justice, and Its Influence on Modern Morals and Happiness*. Edited by Isaac Kramnick. 3rd ed. Baltimore: Penguin, 1976.

———. *Things as They Are; or, the Adventures of Caleb Williams*. Edited by Maurice Hindle. New York: Penguin Books, 1988.

Goellnicht, Donald C. *The Poet-Physician: Keats and Medical Science*. Pittsburgh: University of Pittsburgh Press, 1984.

Goethe, Johann Wolfgang von. "On Morphology." In *Goethe's Botanical Writings*, translated by Bertha Mueller, 21–145. Honolulu: University of Hawaii Press, 1952.

———. *Werke: Hamburger Ausgabe in 14 Bänden*. Edited by Erich Trunz. 14 vols. Munich: Deutscher Tashenbuch Verlag, 1981.

Goldstein, Amanda Jo. "Obsolescent Life: Goethe's Journals on Morphology." *European Romantic Review* 22.3 (2001): 405–14.

———. "Sweet Science: Romantic Materialism and the New Sciences of Life." Ph.D. diss. University of California, Berkeley, 2011.

Golinski, Jan. *Science as Public Culture: Chemistry and Enlightenment in Britain, 1760–1820*. Cambridge: Cambridge University Press, 1992.

Gooding, David, T. J. Pinch, and Simon Schaffer. *The Uses of Experiment: Studies in the Natural Sciences*. Cambridge: Cambridge University Press, 1989.

Gracia, Jorge J. E. *Introduction to the Problem of Individuation in the Early Middle Ages*. Washington, DC: Catholic University of America Press, 1984.

Greenberg, Clement. "Avant Garde and Kitsch." In *Art and Culture: Critical Essays*, 3–21. Boston: Beacon Press, 1961.

———. "Modernist Painting." In *Clement Greenberg: The Collected Essays and Criticism*, edited by John O'Brian, 4:85–93. Chicago: University of Chicago Press, 1995.

Griggs, Earl Leslie. "Coleridge and the Wedgwood Annuity." *Review of English Studies* 6.21 (1930): 63–72.

———. "Samuel Taylor Coleridge and Opium." *Huntington Library Quarterly* 17.4 (1954): 357–78.

Guillory, John. "Enlightening Mediation." In Siskin and Warner, *This Is Enlightenment*, 37–63.

———. "Genesis of the Media Concept." *Critical Inquiry* 36 (2010): 321–62.

Gumbrecht, Hans Ulrich. "Rhythm and Meaning." In *Materialities of Communication*, edited by Hans Ulrich Gumbrecht and K. Ludwig Pfeiffer, 170–82. Stanford, CA: Stanford University Press, 1994.

Habermas, Jürgen. *The Philosophical Discourse of Modernity: Twelve Lectures*. Translated by Frederick Lawrence. Cambridge, MA: MIT Press, 1987.

Hagstrom, Warren O. *The Scientific Community*. Carbondale: Southern Illinois University Press, 1975.

Hallé, Francis. *In Praise of Plants*. Translated by David Lee. Portland, OR: Timber Press, 2002.

Hamlyn, D. W. *Sensation and Perception: A History of the Philosophy of Perception*. New York: Humanities Press, 1961.

Harman, Graham. *Prince of Networks: Bruno Latour and Metaphysics*. Melbourne: re.press, 2009.

Harrington, Anne. *Reenchanted Science: Holism in German Culture from Wilhelm II to Hitler*. Princeton, NJ: Princeton University Press, 1996.

Harrison, Brian Howard. *Drink and the Victorians: The Temperance Question in England, 1815–1872*. London: Faber and Faber, 1971.

Hartley, David. *Observations on Man, His Frame, His Duty, and His Expectations*. London: S. Richardson, 1749.

Hartman, Geoffrey H. *The Fate of Reading*. Chicago: University of Chicago Press, 1975.

——. *The Unremarkable Wordsworth*. Minneapolis: University of Minnesota Press, 1987.

Hazlitt, William. "Review of *A Lay-Sermon on the Distresses of the Country, Addressed to the Middle and Higher Orders*. By S. T. Coleridge." *Examiner* 454 (8 September 1816): 571–73.

Hedley, Douglas. *Coleridge, Philosophy and Religion: "Aids to Reflection" and the Mirror of the Spirit*. Cambridge: Cambridge University Press, 2000.

Hegel, G. W. F. *Aesthetics: Lectures on Fine Art*. Translated by T. M. Knox. 2 vols. Oxford: Clarendon Press, 1998.

——. *Hegel's Philosophy of Nature: Being Part Two of the Encyclopaedia of the Philosophical Sciences (1830)*. Edited by Arnold V. Miller and Karl Ludwig Michelet. Oxford: Clarendon Press, 1970.

——. *Phenomenology of Spirit*. Edited by Arnold V. Miller and J. N. Findlay. Oxford: Clarendon Press, 1977.

——. *The Science of Logic*. Translated by George di Giovanni. Cambridge: Cambridge University Press, 2010.

——. *Werke in zwanzig Bänden, Theorie-Werkausgabe*. 20 vols. Frankfurt: Suhrkamp, 1970.

Heidegger, Martin. *The Fundamental Concepts of Metaphysics: World, Finitude, Solitude*. Translated by William McNeill and Nicholas Walker. Bloomington: Indiana University Press, 1995.

Heilmeyer, Marina. *The Language of Flowers: Symbols and Myths*. London: Prestel, 2001.

Helmreich, Stefan. *Silicon Second Nature: Culturing Artificial Life in a Digital World*. Berkeley: University of California Press, 1998.

Herrnstein, Richard J., and Charles A. Murray. *The Bell Curve: Intelligence and Class Structure in American Life*. New York: Free Press, 1994.

Hill, J. *Centaury, the Great Stomachic: Its Preference to All Other Bitters*. London: Printed for R. Baldwin, and J. Ridley, 1765.

Hoffman, Frederick. *A Treatise on the Nature of Aliments, or Foods, in General . . . To Which Is Added, an Essay on the Digestion*. London: Printed for L. Davis and C. Reymers, 1761.

Hollander, John. *Vision and Resonance: Two Senses of Poetic Form*. New York: Oxford University Press, 1975.

Holmes, Richard. *The Age of Wonder: How the Romantic Generation Discovered the Beauty and Terror of Science*. London: HarperPress, 2008.

Hooke, Robert. *Micrographia; or, Some Physiological Descriptions of Minute Bodies Made by Magnifying Glasses*. London: Printed by Jo. Martyn and Ja. Allestry, 1665.

Hoover, Suzanne R. "Coleridge, Humphry Davy, and Some Early Experiments with a Consciousness-Altering Drug." *Bulletin of Research in the Humanities* 81 (1978): 9–27.

Hunter, John. "A Case of Paralysis of the Muscles of Deglutination Cured by an Artificial Mode of Conveying Food and Medicines into the Stomach." In *Works of John Hunter*, 3:622–24.

——. "Experiments and Observations on Animals, with Respect to the Power of Producing Heat." In *Observations* (1792), 99–127.

——. "Experiments on Animals and Vegetables, with Respect to the Power of Producing Heat." *Philosophical Transactions of the Royal Society of London* 65 (1775): 446–58.

——. *Lectures on the Principles of Surgery.* With notes by James F. Palmer. Philadelphia: Haswell, Harrington, and Haswell, 1839.

——. *Observations on Certain Parts of the Animal Oeconomy.* London: No. 13, Castle-Street, Leicester-Square, 1786.

——. *Observations on Certain Parts of the Animal Oeconomy.* 2nd ed. London: No. 13, Castle-Street, Leicester-Square, 1792.

——. "Of the Heat, &c. of Animals and Vegetables." *Philosophical Transactions of the Royal Society of London* 68 (1778): 7–49.

——. "Of the Progress and Peculiarities of the Chick." In Hunter and Owen, *Essays and Observations,* 1:199–216.

——. "Proposals for the Recovery of People Apparently Drowned." *Philosophical Transactions of the Royal Society of London* 66 (1776): 412–25.

——. *A Treatise on the Blood, Inflammation, and Gun-Shot Wounds.* London: George Nicol, 1794. Reprint, Birmingham, AL: Classics of Medicine Library, 1982.

——. *The Works of John Hunter, F.R.S., with Notes.* Edited by James F. Palmer. 4 vols. London: Longman, Rees, Orme, Brown, Green, and Longman, 1835.

Hunter, John, and Richard Owen. *Essays and Observations on Natural History, Anatomy, Physiology, Psychology, and Geology.* 2 vols. London: J. Van Voorst, 1861.

Hyder, David. "Foucault, Cavaillès, and Husserl on the Historical Epistemology of the Sciences." *Perspectives on Science* 11.1 (2003): 107–29.

Ingenhousz, Jan. *Experiments upon Vegetables . . . Examining the Accurate Degree of Salubrity of the Atmosphere.* London: Printed for P. Elmsly and H. Payne, 1779.

Ingenhousz, Jan, and Howard S. Reed. *Jan Ingenhousz, Plant Physiologist: With a History of the Discovery of Photosynthesis.* Waltham, MA: Chronica Botanica, 1949.

Jackson, Noel. *Science and Sensation in Romantic Poetry.* Cambridge: Cambridge University Press, 2008.

Jacob, François. *The Logic of Life: A History of Heredity.* Translated by Betty E. Spillmann. Princeton, NJ: Princeton University Press, 1973.

Jacobus, Mary. *Tradition and Experiment in Wordsworth's Lyrical Ballads (1798).* Oxford: Clarendon Press, 1976.

Jakobson, Roman. "Closing Statement: Linguistics and Poetics." In *Style in Language,* edited by Thomas A. Sebeok, 350–77. Cambridge, MA: Technology Press of MIT, 1960.

Jenner, Edward. *An Inquiry into the Causes and Effects of the Variolæ Vaccinæ.* London: Printed for the author, 1798.

Johns, Adrian. *The Nature of the Book: Print and Knowledge in the Making.* Chicago: University of Chicago Press, 1998.

Jonsson, Fredrik Albritton. "The Physiology of Hypochondria in Eighteenth-Century Britain." In Forth and Carden-Coyne, *Cultures of the Abdomen,* 15–30.

Kahn, Douglas. *Noise, Water, Meat: A History of Sound in the Arts.* Cambridge, MA: MIT Press, 1999.

Kant, Immanuel. *Anthropology from a Pragmatic Point of View.* Translated by Manfred Kuehn. Edited by Robert B. Louden. New York: Cambridge University Press, 2006.

——. *Critique of Judgment.* Edited by Werner S. Pluhar. Indianapolis: Hackett, 1987.

——. *Critique of Pure Reason*. Translated by Norman Kemp Smith. Unabridged ed. New York: St. Martin's Press, 1965.

——. *Essays and Treatises on Moral, Political, and Various Philosophical Subjects*. 2 vols. London: Printed for the translator; and sold by William Richardson, 1798–99.

——. *Gesammelte Schriften*. 23 vols. Berlin: Hrsg. von der Koeniglich-Preussischen Akademie der Wissenschaften zu Berlin, 1902–.

——. *Kant: Political Writings*. Translated by H. B. Nisbet. Edited by Hans Siegbert Reiss. 2nd ed. Cambridge: Cambridge University Press, 1991.

Kaplan, Morton. "Fantasy of Paternity and the *Doppelgänger*: Mary Shelley's *Frankenstein*." In *The Unspoken Motive: A Guide to Psychoanalytic Literary Criticism*, edited by Morton Kaplan and Robert Kloss, 119–45. New York: Free Press, 1973.

Keach, William. *Shelley's Style*. New York: Methuen, 1984.

Keats, John. *Complete Poems*. Edited by Jack Stillinger. Cambridge, MA: Belknap Press of Harvard University Press, 1982.

——. *The Letters of John Keats, 1814–1821*. Edited by Hyder Edward Rollins. 2 vols. Cambridge, MA: Harvard University Press, 1958.

——. "Ode on a Grecian Urn." In *Complete Poems*, 282–83.

——. "To Autumn." In *Complete Poems*, 360–61.

Keller, Evelyn Fox. *Refiguring Life: Metaphors of Twentieth-Century Biology*. New York: Columbia University Press, 1995.

Kelley, Theresa M. "Restless Romantic Plants: Goethe Meets Hegel." *European Romantic Review* 20.2 (2009): 187–95.

Khalip, Jacques. *Anonymous Life: Romanticism and Dispossession*. Stanford, CA: Stanford University Press, 2009.

Kirschner, Marc, John Gerhart, and Tim Mitchison. "Molecular 'Vitalism.'" *Cell* 100 (2000): 79–88.

Kittler, Friedrich A. *Discourse Networks 1800/1900*. Translated by Michael Metteer, with Chris Cullens. Stanford, CA: Stanford University Press, 1990.

——. "Media and Drugs in Pynchon's Second World War." In *Literature, Media, Information Systems: Essays*, edited by John Johnston, 101–16. Amsterdam: G+B Arts International, 1997.

——. "Towards an Ontology of Media." *Theory, Culture & Society* 26.2–3 (2009): 23–31.

Klein, Naomi. *The Shock Doctrine: The Rise of Disaster Capitalism*. 1st ed. New York: Metropolitan Books / Henry Holt, 2007.

Knight, Philip. *Flower Poetics in Nineteenth-Century France*. New York: Clarendon Press, 1986.

Kolnai, Aurel. *On Disgust*. Translated by Carolyn Korsmeyer and Barry Smith. Chicago: Open Court, 2004.

Korshin, Paul J. "Types of Eighteenth-Century Literary Patronage." *Eighteenth-Century Studies* 7.4 (1974): 453–73.

Koyré, Alexandre. *Metaphysics and Measurement: Essays in Scientific Revolution*. Cambridge, MA: Harvard University Press, 1968.

Krell, David Farrell. *Contagion: Sexuality, Disease, and Death in German Idealism and Romanticism*. Bloomington: Indiana University Press, 1998.

Krieger, Murray. *The Play and Place of Criticism*. Baltimore: Johns Hopkins Press, 1967.

Krieger, Nancy, Diane L. Rowley, Allen A. Herman, Byllye Avery, and Mona T. Phillips.

"Racism, Sexism, and Social Class: Implications for Studies of Health, Disease, and Well-being." *American Journal of Preventive Medicine* 9.6 Supplement (1993): 82–122.

Kroeber, Karl. *Ecological Literary Criticism: Romantic Imagining and the Biology of the Mind.* New York: Columbia University Press, 1994.

Kuhn, Thomas S. *The Structure of Scientific Revolutions.* 3rd ed. Chicago: University of Chicago Press, 1996.

Lamarck, Jean-Baptiste. *Philosophie zoologique; ou exposition des considérations relatives à l'histoire naturelle des animaux.* Edited by Charles Martins. Paris: F. Savy, 1873.

———. *Zoological Philosophy: An Exposition with Regard to the Natural History of Animals.* Translated by Hugh Samuel Roger Elliot. Chicago: University of Chicago Press, 1984.

Langhorne, John. *The Fables of Flora, by Dr. Langhorne.* London: Printed for J. Murray, 1771.

Latour, Bruno. "For David Bloor . . . and Beyond: A Reply to David Bloor's 'Anti-Latour.'" *Studies in the History and Philosophy of Science* 30.1 (1999): 113–29.

———. *Pandora's Hope: Essays on the Reality of Science Studies.* Cambridge, MA: Harvard University Press, 1999.

———. *The Pasteurization of France.* Translated by Alan Sheridan and John Law. Cambridge, MA: Harvard University Press, 1988.

Latour, Bruno, and Steve Woolgar. *Laboratory Life: The Construction of Scientific Facts.* Princeton, NJ: Princeton University Press, 1988.

Lawrence, Christopher. "The Nervous System and Society in the Scottish Enlightenment." In *Natural Order: Historical Studies of Scientific Culture,* edited by Barry Barnes and Steven Shapin, 19–40. Beverly Hills, CA: Sage Publications, 1979.

Lawrence, William. *An Introduction to Comparative Anatomy and Physiology: Being the Two Introductory Lectures.* London: J. Callow, 1816.

———. *Lectures on Physiology, Zoology, and the Natural History of Man: Delivered at the Royal College of Surgeons.* London: Callow, 1819.

Leask, Nigel. *Curiosity and the Aesthetics of Travel Writing, 1770–1840: 'From an Antique Land.'* Oxford: Oxford University Press, 2002.

Leavis, F. R. *Revaluation: Tradition and Development in English Poetry.* New York: G. W. Stewart, 1947.

Leicester, Henry Marshall. *Development of Biochemical Concepts from Ancient to Modern Times.* Cambridge, MA: Harvard University Press, 1974.

Lenoir, Tim. *The Strategy of Life: Teleology and Mechanics in Nineteenth-Century German Biology.* Chicago: University of Chicago Press, 1982.

Lessing, Gotthold Ephraim. *Laocoön: An Essay on the Limits of Painting and Poetry.* Translated by Edward Allen McCormick. Indianapolis, IN: Bobbs-Merrill, 1962.

Levere, Trevor Harvey. *Poetry Realized in Nature: Samuel Taylor Coleridge and Early Nineteenth-Century Science.* Cambridge: Cambridge University Press, 1981.

Levine, George. "The Ambiguous Heritage of *Frankenstein.*" In Levine and Knoepflmacher, *Endurance of Frankenstein,* 3–30.

Levine, George Lewis, and U. C. Knoepflmacher, eds. *The Endurance of Frankenstein: Essays on Mary Shelley's Novel.* Berkeley: University of California Press, 1979.

Levinson, Marjorie. *Keats's Life of Allegory: The Origins of a Style.* Cambridge, MA: Blackwell, 1990.

Levy, Steven. *Artificial Life: The Quest for a New Creation*. New York: Pantheon Books, 1992.

Leys, Ruth. *From Sympathy to Reflex: Marshall Hall and His Opponents*. New York: Garland Publishing, 1991.

Lindenberger, Herbert. "Keats's 'To Autumn' and Our Knowledge of a Poem." *College English* 32.2 (1970): 123–34.

Liu, Alan. *Wordsworth: The Sense of History*. Stanford, CA: Stanford University Press, 1989.

Logan, Peter Melville. *Nerves and Narratives: A Cultural History of Hysteria in Nineteenth-Century British Prose*. Berkeley: University of California Press, 1997.

Lukács, György. *The Young Hegel: Studies in the Relations between Dialectics and Economics*. Translated by Rodney Livingstone. London: Merlin Press, 1975.

Luke, Tim. "The Dreams of Deep Ecology." *Telos* 76 (1988): 65–92.

Lynn, Michael R. *Popular Science and Public Opinion in Eighteenth-Century France*. New York: Palgrave, 2006.

Mackie, Erin Skye. *The Commerce of Everyday Life: Selections from "The Tatler" and "The Spectator."* Boston: Bedford / St. Martin's, 1998.

Maddison, Carol. *Apollo and the Nine: A History of the Ode*. London: Routledge and Kegan Paul, 1960.

Malachuk, Daniel S. "Coleridge's Republicanism and the Aphorism in 'Aids to Reflection.'" *Studies in Romanticism* 39.3 (2000): 397–417.

Maniquis, Robert. "The Puzzling Mimosa: Sensitivity and Plant Symbols in Romanticism." *Studies in Romanticism* 8.3 (1969): 129–55.

Margulis, Lynn, and Dorion Sagan. *Acquiring Genomes: A Theory of the Origins of Species*. New York: Basic Books, 2002.

———. *What Is Life?* Berkeley: University of California Press, 1995.

Marshall, David. *The Surprising Effects of Sympathy: Marivaux, Diderot, Rousseau, and Mary Shelley*. Chicago: University of Chicago Press, 1988.

Mason, Simon. *The Good and Bad Effects of Tea Consider'd*. London: Printed for M. Cooper, 1745.

Massumi, Brian. *Parables for the Virtual: Movement, Affect, Sensation*. Durham, NC: Duke University Press, 2002.

Mattelart, Armand. *The Invention of Communication*. Translated by Susan Emanuel. Minneapolis: University of Minnesota Press, 1996.

Mayo, Robert. "The Contemporaneity of the *Lyrical Ballads*." *PMLA* 69.3 (1954): 486–522.

Mayr, Ernst. "Darwin and the Evolutionary Theory in Biology." In *Evolution and Anthropology: A Centennial Appraisal*, edited by Betty J. Meggers, 1–10. Washington, DC: Anthropological Society of Washington, 1959.

McKendrick, Neil. "The Consumer Revolution of Eighteenth-Century England." In *The Birth of a Consumer Society: The Commercialization of Eighteenth-Century England*, edited by Neil McKendrick, John Brewer, and J. H. Plumb, 9–33. Bloomington: Indiana University Press, 1982.

McKusick, James C. *Coleridge's Philosophy of Language*. New Haven, CT: Yale University Press, 1986.

McLane, Maureen N. *Romanticism and the Human Sciences: Poetry, Population, and the Discourse of the Species*. Cambridge: Cambridge University Press, 2000.

McShea, Daniel W. "Complexity and Evolution: What Everybody Knows." *Biology and Philosophy* 6 (1991): 303–24.

Mendelsohn, Everett. *Heat and Life: The Development of the Theory of Animal Heat.* Cambridge, MA: Harvard University Press, 1964.

Mendelssohn, Moses. "Rhapsodie oder Zusätze zu den Briefen über die Empfindungen." In *Ästhetische Schriften in Auswahl*, edited by Otto F. Best, 127–65. Darmstadt: Wissenschaftliche Buchgesellschaft, 1986.

Menninghaus, Winfried. *Disgust: The Theory and History of a Strong Sensation.* Translated by Howard Eiland and Joel Golb. Albany, NY: SUNY Press, 2003.

Miller, J. Hillis. *The Linguistic Moment: From Wordsworth to Stevens.* Princeton, NJ: Princeton University Press, 1985.

Miller, William Ian. *The Anatomy of Disgust.* Cambridge, MA: Harvard University Press, 1997.

Milton, John. "Sonnet XIX: 'When I Consider . . .'" In *Complete Poems and Major Prose*, edited by Merritt Y. Hughes, 168. New York: Odyssey Press, 1957.

Mintz, Sidney. *Sweetness and Power: The Place of Sugar in Modern History.* New York: Viking, 1985.

Mirowski, Phillip. "What's Kuhn Got to Do with It?" In *The Effortless Economy of Science?* 85–96. Durham, NC: Duke University Press, 2004.

Mitchell, Robert. *Bioart and the Vitality of Media.* Seattle: University of Washington Press, 2010.

———. "Suspended Animation, Slow Time, and the Poetics of Trance." *PMLA* 126.1 (2011): 107–22.

———. *Sympathy and the State in the Romantic Era: Systems, State Finance, and the Shadows of Futurity.* New York: Routledge, 2007.

———. "The Transcendental: Deleuze, P. B. Shelley, and the Freedom of Immobility." In *Romanticism and the New Deleuze*, Romantic Circles Praxis Series (January 2008), www.rc.umd.edu/praxis/deleuze/mitchell/mitchell.html.

———. "U.S. Biobanking Strategies and Biomedical Immaterial Labor." *Biosocieties* 7.3 (2012): 224–44.

Mitchell, Robert, Helen Burgess, and Phillip Thurtle. *Biofutures: Owning Body Parts and Information.* Philadelphia: University of Pennsylvania Press, 2008.

Mitchell, W. J. T. *Picture Theory: Essays on Verbal and Visual Representation.* Chicago: University of Chicago Press, 1994.

———. *What Do Pictures Want?: The Lives and Loves of Images.* Chicago: University of Chicago Press, 2005.

Modiano, Raimonda. "Coleridge and Wordsworth: The Ethics of Gift Exchange and Literary Ownership." In *Coleridge's Theory of Imagination Today*, edited by Christine Gallant, 243–56. New York: AMS Press, 1989.

Moretti, Franco. *Graphs, Maps, Trees: Abstract Models for a Literary History.* New York: Verso, 2005.

———. "The Slaughterhouse of Literature." *Modern Language Quarterly* 61.1 (2000): 207–27.

———. *The Way of the World: The* Bildungsroman *in European Culture.* London: Verso, 1987.

Morton, Tim. "Blood Sugar." In Fulford and Kitson, *Romanticism and Colonialism*, 87–106.

———, ed. *Cultures of Taste / Theories of Appetite: Eating Romanticism*. New York: Palgrave Macmillan, 2004.

———. *Ecology without Nature: Rethinking Environmental Aesthetics*. Cambridge, MA: Harvard University Press, 2007.

Motte, Warren F. *Oulipo: A Primer of Potential Literature*. Lincoln: University of Nebraska Press, 1986.

Nadarajan, Gunalan. "Ornamental Biotechnology and Parergonal Aesthetics." In *Signs of Life: Bio Art and Beyond*, edited by Eduardo Kac, 43–56. Cambridge, MA: MIT Press, 2007.

Nemoianu, Virgil. "The Dialectics of Movement in Keats's 'To Autumn.'" *PMLA* 93.2 (1978): 205–14.

Newton, Isaac. *Optical Lectures Read in the Publick Schools of the University of Cambridge*. London: Printed for Francis Fayram, 1728.

Nichols, Ashton. "The Loves of Plants and Animals: Romantic Science and the Pleasures of Nature." In *Romanticism & Ecology*, Romantic Circles Praxis Series (November 2001), www.rc.umd.edu/praxis/ecology/nichols/nichols.html.

Nicholson, William. *The British Encyclopedia, or Dictionary of Arts and Sciences, Comprising an Accurate and Popular View of the Present Improved State of Human Knowledge*. 6 vols. London: Longman, Hurst, Rees, and Orme, 1809.

Nietzsche, Friedrich Wilhelm. *The Birth of Tragedy and the Case of Wagner*. Translated by Walter Kaufmann. New York: Vintage, 1967.

———. *The Portable Nietzsche*. Translated by Walter Kaufman. New York: Penguin Books, 1982.

Nollet, Jean Antoine. *Leçons de physique experimentale*. 6 vols. Amsterdam: Chez Arksteé and Merku, 1745–65.

Nyman, Michael. *Experimental Music: Cage and Beyond*. 2nd ed. Cambridge: Cambridge University Press, 1999.

Ogden, Bernard. *Plain Directions for the Treatment of Wounds in General . . . To Which Are Added Remarks on Suspended Animation*. Sunderland: Printed by T. Reed, 1797.

Olson, Charles. "Projective Verse." In *Collected Prose*, edited by D. Allen and B. Friedlander, with an introduction by Robert Creeley, 239–49. Berkeley: University of California Press, 1997.

O'Neill, Robert V. "Is It Time to Bury the Ecosystem Concept? (with Full Military Honors, of Course!)." *Ecology* 82.12 (2001): 3275–84.

Osborne, Thomas. "What Is a Problem?" *History of the Human Sciences* 16.1 (2003): 1–17.

Ousby, Ian. "My Servant Caleb: Godwin's Caleb Williams and the Political Trials of the 1790s." *University of Toronto Quarterly* 44 (1974): 47–55.

Paracelsus, and Arthur Edward Waite. *The Hermetic and Alchemical Writings of Aureolus Philippus Theophrastus Bombast, of Hohenheim, Called Paracelsus the Great*. 2 vols. London: James Elliott, 1894.

Paulli, Simon. *A Treatise on Tobacco, Tea, Coffee, and Chocolate*. London: T. Osborne, 1746.

Pearson, Keith Ansell. *Germinal Life: The Difference and Repetition of Deleuze*. New York: Routledge, 1999.

Pfau, Thomas. *Romantic Moods: Paranoia, Trauma, and Melancholy, 1790–1840*. Baltimore: Johns Hopkins University Press, 2005.

Pickering, Andrew. *The Mangle of Practice: Time, Agency, and Science*. Chicago: University of Chicago Press, 1995.

Poggioli, Renato. *The Theory of the Avant-Garde*. Cambridge, MA: Belknap Press of Harvard University Press, 1968.

Pollan, Michael. *The Botany of Desire: A Plant's Eye View of the World*. 1st ed. New York: Random House, 2001.

Pollmann, Inga. "Invisible Worlds, Visible: Uexküll's Umwelt, Film, and Film Theory." *Critical Inquiry* 39.4 (2013): 777–816.

Pope, Alexander. "An Essay on Man in Four Epistles to H. St John Lord Bolingbroke." In *Alexander Pope*, edited by Pat Rogers, 270–309. Oxford: Oxford University Press, 1993.

Popper, Karl. *The Logic of Scientific Discovery*. New York: Routledge, 2002.

Porter, Roy. "Addicted to Modernity: Nervousness in the Early Consumer Society." In *Culture in History: Production, Consumption and Values in Historical Perspective*, edited by Joseph Melling and Jonathan Berry, 180–94. Exeter: Exeter University Press, 1992.

Potter, George R. "Coleridge and the Idea of Evolution." *PMLA* 40.2 (1925): 379–97.

Pound, Ezra. "A Retrospect." In *Literary Essays of Ezra Pound*, edited by T. S. Eliot, 3–14. New York: New Directions, 1968.

Priestley, Joseph. *Experiments and Observations on Different Kinds of Air*. London: Printed for J. Johnson, 1774.

Punter, David. *The Literature of Terror: A History of Gothic Fictions from 1765 to the Present Day*. 2nd ed. 2 vols. London: Longman, 1996.

Pynchon, Thomas. *Gravity's Rainbow*. New York: Penguin Books, 1973.

Quincy, John. *Lexicon Physico-Medicum; or, a New Medicinal Dictionary*. London: Printed for J. Osborn and T. Longman, 1726.

Rabinow, Paul. *French Modern: Norms and Forms of the Social Environment*. Cambridge, MA: MIT Press, 1989.

Rabinowitch, Eugene. *Photosynthesis and Related Processes*. 3 vols. New York: Interscience Publishers, 1945.

Raithby, John. *Peace . . . Addressed to the Inhabitants of Great Britain, through the Medium of Their Representatives in Parliament*. London: Printed for Allen and West, and Owen, 1795.

Rajan, Tilottama. "The Encyclopedia and the University of Theory: Idealism and the Organization of Knowledge." *Textual Practice* 21.2 (2007): 335–58.

———. "First Outline of a System of Theory: Schelling and the Margins of Philosophy, 1799–1815." *Studies in Romanticism* 46.3 (2007): 311–35.

———. "(In)Digestible Material: Illness and Dialectic in Hegel's *the Philosophy of Nature*." In Morton, *Cultures of Taste / Theories of Appetite*, 217–36.

———. "Keats, Poetry, and 'The Absence of Work.'" *Modern Philology* 46.3 (1998): 334–51.

Rancière, Jacques. *The Politics of Aesthetics: The Distribution of the Sensible*. Translated and with an introduction by Gabriel Rockhill. New York: Continuum, 2004.

Reill, Peter Hanns. *Vitalizing Nature in the Enlightenment*. Berkeley: University of California Press, 2005.

Rheinberger, Hans-Jörg. "Experiment, Difference, and Writing: I. Tracing Protein Synthesis." *Studies in the History and Philosophy of Science* 23.2 (1992): 305–31.

———. "Gaston Bachelard and the Notion of 'Phenomenotechnique.'" *Perspectives on Science* 13.3 (2005): 313–28.

———. *Toward a History of Epistemic Things: Synthesizing Proteins in the Test Tube.* Stanford, CA: Stanford University Press, 1997.

Richards, Robert J. *The Romantic Conception of Life: Science and Philosophy in the Age of Goethe.* Chicago: University of Chicago Press, 2002.

Richardson, Alan. *British Romanticism and the Science of the Mind.* Cambridge: Cambridge University Press, 2001.

———. "Darkness Visible? Race and Representation in Bristol Abolitionist Poetry, 1770–1810." In Fulford and Kitson, *Romanticism and Colonialism,* 129–47.

Ritterbush, Philip C. *Overtures to Biology: The Speculations of Eighteenth-Century Naturalists.* New Haven, CT: Yale University Press, 1964.

Robinson, Charles E. "Mary Shelley and the Roger Dodsworth Hoax." *Keats-Shelley Journal* 24 (1975): 20–28.

Rogers, John. *The Matter of Revolution: Science, Poetry, and Politics in the Age of Milton.* Ithaca, NY: Cornell University Press, 1996.

Romantic Circles. "Mary Wollstonecraft Shelley Chronology and Resource Site." www .rc.umd.edu/reference/chronologies/mschronology/reviews.html.

Rose, Mark. *Authors and Owners: The Invention of Copyright.* Cambridge, MA: Harvard University Press, 1993.

Rose, Nikolas S. *The Politics of Life Itself: Biomedicine, Power, and Subjectivity in the Twenty-First Century.* Princeton, NJ: Princeton University Press, 2007.

Rousseau, George. "Coleridge's Dreaming Gut: Digestion, Genius, Hypochondria." In Forth and Carden-Coyne, *Cultures of the Abdomen,* 105–26.

Rousseau, Jean-Jacques. *Reveries of the Solitary Walker.* Translated by Peter France. New York: Penguin, 1979.

Ruston, Sharon. *Shelley and Vitality.* New York: Palgrave Macmillan, 2005.

Santner, Eric L. *On Creaturely Life: Rilke, Benjamin, Sebald.* Chicago: University of Chicago Press, 2006.

Saumarez, Richard. *A New System of Physiology, Comprehending . . . Several States of Health and Disease.* 2nd ed. 2 vols. London: Sold by J. Johnson, 1799.

Schaffer, Simon. "The Consuming Flame: Electrical Showmen and Tory Mystics in the World of Goods." In *Consumption and the World of Goods,* edited by John Brewer and Roy Porter, 489–526. London: Routledge, 1993.

———. "The Nebular Hypothesis and the Science of Progress." In *History, Humanity and Evolution: Essays for John C. Greene,* edited by James R. Moore, 131–64. New York: Cambridge University Press, 1989.

———. "Scientific Discoveries and the End of Natural Philosophy." *Social Studies of Science* 16:3 (1986): 387–420.

———. "Self Evidence." *Critical Inquiry* 18.2 (1992): 327–62.

Schelling, Friedrich Wilhelm Joseph von. *Erster Entwurf eines Systems der Naturphilosophie* [1798]. In *Schellings Werke,* 2:1–268.

———. *First Outline of a System of the Philosophy of Nature.* Translated by Keith R. Peterson. Albany: SUNY Press, 2004.

———. *The Philosophy of Art.* Translated by Douglas W. Stott. Minneapolis: University of Minnesota Press, 1989.

——. *Schellings Werke: Münchner Jubiläumsdruck.* Edited by Manfred Schröter. 12 vols. Munich: C. H. Beck und R. Didenbourg, 1927.

——. *System des Transcendentalen Idealismus.* Tübingen: J.G. Cotta, 1800.

——. *System of Transcendental Idealism (1800).* Translated by Peter Lauchlan Heath. Charlottesville: University Press of Virginia, 1978.

Schelling, Friedrich Wilhelm Joseph von, and Slavoj Žižek. *The Abyss of Freedom: Ages of the World.* Ann Arbor: University of Michigan Press, 1997.

Schiller, Friedrich. *On the Aesthetic Education of Man, in a Series of Letters.* Translated by Reginald Snell. New York: Frederick Ungar, 1965.

Schivelbusch, Wolfgang. *Tastes of Paradise: A Social History of Spices, Stimulants, and Intoxicants.* New York: Pantheon Books, 1992.

Schlegel, Johann Adolf. "Anmerkungen über Ekel." In *Einschränkung der schönen Künste auf einen einzigen Grundsatz*, 3rd ed., 106–20. Leipzig: Weidmann, 1770.

Scott, Grant F. *The Sculpted Word: Keats, Ekphrasis, and the Visual Arts.* Hanover, NH: University Press of New England, 1994.

Scott, Walter. "Remarks on *Frankenstein, or the Modern Prometheus.*" *Blackwood's Edinburgh Magazine* 2 (1818): 613–20.

Scurlock, R. G. *History and Origins of Cryogenics.* Oxford: Oxford University Press, 1992.

Secord, James A. *Victorian Sensation: The Extraordinary Publication, Reception, and Secret Authorship of Vestiges of the Natural History of Creation.* Chicago: University of Chicago Press, 2000.

Serres, Michel. *The Parasite.* Translated by Lawrence R. Schehr. Baltimore, MD: Johns Hopkins University Press, 1982.

Shapin, Steven. "The House of Experiment in Seventeenth-Century England." *Isis* 79.3 (1988): 373–404.

——. "Pump and Circumstance: Robert Boyle's Literary Technology." *Social Studies of Science* 14.4 (1984): 481–520.

——. *A Social History of Truth: Civility and Science in Seventeenth-Century England.* Chicago: University of Chicago Press, 1994.

Shapin, Steven, and Simon Schaffer. *Leviathan and the Air-Pump: Hobbes, Boyle, and the Experimental Life.* Princeton, NJ: Princeton University Press, 1985.

Shelley, Mary Wollstonecraft. *Frankenstein; or, the Modern Prometheus.* Edited by David Lorne Macdonald and Kathleen Dorothy Scherf. 2nd ed. Peterborough, ON: Broadview Press, 1999.

——. *Frankenstein; or, the Modern Prometheus: The 1818 Text.* Edited by Marilyn Butler. Oxford: Oxford University Press, 1998.

——. "Roger Dodsworth: The Reanimated Englishman." In *Collected Tales and Stories*, edited by Charles E. Robinson, 43–50. Baltimore: Johns Hopkins University Press, 1976.

Shelley, Mary Wollstonecraft, and Percy Bysshe Shelley. *History of a Six Weeks' Tour, 1817.* Oxford: Woodstock, 1989.

Shelley, Percy Bysshe. *A Defence of Poetry.* In *Shelley's Prose*, 498–513.

——. "Essay on Life." In *Shelley's Prose*, 171–75.

——. "The Flower That Smiles Today." In *Shelley's Poetry and Prose*, 441–42.

——. *Letters.* Edited by Frederick L. Jones. 2 vols. Oxford: Clarendon Press, 1964.

——. "Mont Blanc; Lines Written in the Vale of Chamouni." In *Shelley's Poetry and Prose*, 89–93.

———. "Ode to the West Wind." In *Shelley's Poetry and Prose*, 221–23.

———. *Queen Mab; a Philosophical Poem* [1813]. Edited by Jonathan Wordsworth. New York: Woodstock Books, 1990.

———. *The Sensitive-Plant*. In *Shelley's Poetry and Prose*, 201–19.

———. *Shelley's Poetry and Prose*. Edited by Donald H. Reiman and Sharon B. Powers. New York: Norton, 1977.

———. *Shelley's Prose; or, the Trumpet of a Prophecy*. Edited by David Lee Clark. New York: New Amsterdam, 1988.

Shiner, Larry. *The Invention of Art: A Cultural History*. Chicago: University of Chicago Press, 2001.

Short, Thomas. *A Dissertation upon Tea . . . the Various Effects It Has on Different Constitutions*. London: Printed by W. Bowyer, for Fletcher Gyles, 1730.

Silver, Lee. *Remaking Eden*. New York: Harper, 1998.

Siskin, Clifford. "1798: The Year of the System." In *1798: The Year of the Lyrical Ballads*, edited by Richard Cronin, 9–31. New York: St. Martin's Press, 1998.

———. *The Work of Writing: Literature and Social Change in Britain, 1700–1830*. Baltimore: Johns Hopkins University Press, 1998.

Siskin, Clifford, and William Warner, eds. *This Is Enlightenment*. Chicago: University of Chicago Press, 2010.

———. "This Is Enlightenment: An Invitation in the Form of an Argument." In Siskin and Warner, *This Is Enlightenment*, 1–33.

Smith, Adam. *An Inquiry into the Nature and Causes of the Wealth of Nations*. Edited by R. H. Campbell and A. S. Skinner. *The Glasgow Edition of the Works and Correspondence of Adam Smith*, vol. 2a–2b. Indianapolis, IN: Liberty Fund, 1981.

Smith, Barbara Herrnstein. *Scandalous Knowledge: Science, Truth, and the Human*. Durham, NC: Duke University Press, 2006.

Smith, Robyn. "'As Yet Unknown': Rat Feeding Experiments in Early Vitamin Research." In *Animal Encounters*, edited by Manuela Rossini and Tom Tyler, 99–114. Boston: Brill, 2009.

Solhdju, Katrin. *Selbstexperimente: Die Suche nach der Innenperspektive und ihre epistemologischen Folgen*. Munich: Wilhelm Fink, 2011.

Spallanzani, Lazzaro. *Dissertations Relative to the Natural History of Animals and Vegetables*. 2 vols. London: G. G. and J. Robinson, 1797.

Spears, André. "Evolution in Context: 'Deep Time,' Archaeology, and the Post-Romantic Paradigm." *Comparative Literature* 48.4 (1996): 343–58.

Sperry, Jr., Stuart M. "Keats and the Chemistry of Poetic Creation." *PMLA* 85.2 (1970): 268–77.

Spitzer, Leo. *Essays in Historical Semantics*. New York: S. F. Vanni, 1948.

Sprat, Thomas. *The History of the Royal Society of London, for the Improving of Natural Knowledge*. London: Printed for J. Knapton et al., 1734.

St. Clair, William. *The Reading Nation in the Romantic Period*. Cambridge: Cambridge University Press, 2004.

Stengers, Isabelle. "A Constructivist Reading of *Process and Reality*." *Theory, Culture, & Society* 25.4 (2008): 91–110.

———. *The Invention of Modern Science*. Translated by Daniel W. Smith. Minneapolis: University of Minnesota Press, 2000.

Stewart, Larry R. *The Rise of Public Science: Rhetoric, Technology, and Natural Philosophy in Newtonian Britain, 1660–1750*. New York: Cambridge University Press, 1992.

Sullivan, Garrett A., Jr. "'A Story to Be Hastily Gobbled up': *Caleb Williams* and Print Culture." *Studies in Romanticism* 32 (1993): 323–37.

Swift, Jonathan. *The Writings of Jonathan Swift*. Edited by Robert A. Greenberg and William B. Piper. New York: Norton, 1973.

Taminiaux, Jacques. *The Metamorphoses of Phenomenological Reduction*. Milwaukee, MI: Marquette University Press, 2004.

Thacker, Eugene. *Biomedia*. Minneapolis: University of Minnesota Press, 2004.

Thelwall, John. *An Essay Towards a Definition of Animal Vitality*. London: T. Rickaby, 1793.

Thompson, James. "Surveillance in William Godwin's *Caleb Williams*." In *Gothic Fictions: Prohibition/Transgression*, edited by Kenneth W. Graham, 173–98. New York: AMS Press, 1989.

Todd, Dennis. *Imagining Monsters: Miscreations of the Self in Eighteenth-Century England*. Chicago: University of Chicago Press, 1995.

Toews, John Edward. *Hegelianism: The Path toward Dialectical Humanism, 1805–1841*. Cambridge: Cambridge University Press, 1980.

Tomko, Michael. "Politics, Performance, and Coleridge's 'Suspension of Disbelief.'" *Victorian Studies* 49.2 (Winter 2007): 241–49.

Tresch, John. *The Romantic Machine: Utopian Science and Technology after Napoleon*. Chicago: University of Chicago Press, 2012.

Treviranus, Gottfried Reinhold. *Biologie, oder Philosophie der lebenden Natur für Naturforscher und Aerzte*. 6 vols. Göttingen: Johann Friedrich Röwer, 1802–22.

Trott, Nicola. "Wordsworth's Loves of the Plants." In *1800: The New Lyrical Ballads*, edited by Nicola Trott and Seamus Perry, 141–68. New York: Palgrave, 2001.

Trotter, David. *Paranoid Modernism: Literary Experiment, Psychosis, and the Professionalization of English Society*. Oxford: Oxford University Press, 2001.

Trotter, Thomas. *An Essay, Medical, Philosophical, and Chemical, on Drunkenness and Its Effects on the Human Body*. Edited by Roy Porter. London: Routledge, 1988.

———. *Medicina Nautica: An Essay on the Diseases of Seamen*. 3 vols. London: Longman, Hurst, Rees, and Orme, 1804.

———. *A View of the Nervous Temperament; Being a Practical Enquiry*. London: Longman, Hurst, Rees, and Orme, 1807.

Underwood, T. L. *The Work of the Sun: Literature, Science, and Political Economy, 1760–1860*. New York: Palgrave Macmillan, 2005.

Vila, Anne C. *Enlightenment and Pathology: Sensibility in the Literature and Medicine of Eighteenth-Century France*. Baltimore: Johns Hopkins University Press, 1998.

Waldby, Catherine, and Robert Mitchell. *Tissue Economies: Blood, Organs, and Cell Lines in Late Capitalism*. Durham, NC: Duke University Press, 2006.

Wallen, Martin. "Coleridge's Scrofulous Dejection." *Journal of English and Germanic Philology* 99.4 (2000): 555–75.

Walvin, James. *Fruits of Empire: Exotic Produce and British Taste, 1660–1800*. New York: New York University Press, 1997.

Wang, Orrin N. C. "Romantic Sobriety." *Modern Language Quarterly* 60.4 (1999): 469–93.

Wellbery, David E. *Lessing's Laocoon: Semiotics and Aesthetics in the Age of Reason*. Cambridge: Cambridge University Press, 1984.

Wheeler, L. Richmond. *Vitalism: Its History and Theory.* London: H. F. and G. Witherby, 1939.

White, Hayden V. *Metahistory: The Historical Imagination in Nineteenth-Century Europe.* Baltimore: Johns Hopkins University Press, 1975.

Whiter, Walter. *A Dissertation on the Disorder of Death; or, That State of the Frame under the Signs of Death Called Suspended Animation.* London: Printed for the author, 1819.

Williams, Carolyn D. "'The Luxury of Doing Good': Benevolence, Sensibility, and the Royal Humane Society." In *Pleasure in the Eighteenth Century,* edited by Roy Porter and Marie Mulvey Roberts, 77–107. New York: New York University Press, 1996.

Williams, Raymond. *Keywords: A Vocabulary of Culture and Society.* Rev. ed. New York: Oxford University Press, 1985.

Wilson, Elizabeth A. *Affect and Artificial Intelligence.* Seattle: University of Washington Press, 2010.

———. "Gut Feminism." *differences* 15.3 (2004): 66–94.

Wimsatt, William K. *The Verbal Icon: Studies in the Meaning of Poetry.* Lexington: University of Kentucky Press, 1954.

Winslow, C.-E. A., and R. R. Bellinger. "Hippocratic and Galenic Concepts of Metabolism." *Bulletin of the History of Medicine* 17 (1945): 127–37.

Winter, Alison. "Ethereal Epidemic: Mesmerism and the Introduction of Inhalation Anaesthesia to Early Victorian London." *Social History of Medicine* 4.1 (1991): 1–27.

Wolfe, Cary. *Before the Law: Humans and Other Animals in a Biopolitical Frame.* Chicago: University of Chicago Press, 2012.

Woodward, Kathleen. "Statistical Panic." *differences* 11.2 (1999): 177–203.

Wordsworth, William. *The Excursion, Being a Portion of the Recluse, a Poem.* London: Printed for Longman, Hurst, Rees, Orme, and Brown, 1814.

———. "Lines Written a Few Miles above Tintern Abbey." In Coleridge and Wordsworth, *Lyrical Ballads,* 142–47.

———. *Lyrical Ballads, with Pastoral and Other Poems.* London: Longman and Rees, 1802.

———. *The Poetical Works of William Wordsworth: Edited from the Manuscripts with Textual and Critical Notes.* Edited by Ernest De Selincourt and Helen Darbishire. 5 vols. Oxford: Clarendon Press, 1952.

———. *The Ruined Cottage.* In *Romanticism,* edited by Duncan Wu, 2nd ed., 277–89. Oxford: Blackwell Publishers, 1998.

———. "The Thorn." In Coleridge and Wordsworth, *Lyrical Ballads,* 103–10.

———. "Yew-Trees." In *Poetical Works,* 2:209–10.

Wordsworth, William, and Samuel Taylor Coleridge. *Lyrical Ballads: With a Few Other Poems.* 1st ed. London: Printed for J. and A. Arch, 1798.

Youngquist, Paul. "Rehabilitating Coleridge: Poetry, Philosophy, Excess." *ELH* 66.4 (1999): 885–909.

Zallen, Doris T. "The 'Light' Organism for the Job: Green Algae and Photosynthesis Research." *Journal of the History of Biology* 26.2 (1993): 269–79.

Zola, Émile. "The Experimental Novel." In *The Experimental Novel and Other Essays,* translated by Belle M. Sherman, 1–54. New York: Haskell House, 1964.

Zurawski, Magdalena. "The Mis-Education of Frankenstein." Lecture, Duke University, Durham, NC, 2011.

Index

abduction, 217

Abernethy, John, 10, 89–90, 92, 155, 247n27, 247n29

Abrams, M. H., 242n52

Académie Royal des Sciences, 28

actants, 21, 228

activity, 49; experiment as, 32–35, 40, 229; life as, 1, 62; mediation as, 160–63; and passivity, 15–16, 199, 207, 210, 268

Adams, George, 148–49, 184–85

addiction-to-synthesis, 69

Addison, Joseph, 109

Adorno, Theodor, 125, 253n41; on experiments, 12, 14–17, 22–27, 41–42, 235n24

"Advertisement" (Wordsworth and Coleridge). See *Lyrical Ballads*

aesthetics, 40, 56, 73, 141, 143, 233n4, 255n65; and atmospheres, 203, 205–6, 214, 217; experience and, 40, 139–40; judgment and, 192, 217; limits of, 108, 138–40; and science, 143

affect, 70, 72, 105, 178, 255n69; science and, 108, 194, 250n9

Agamben, Giorgio, 12, 219–20, 231n3

Aids to Reflection (Coleridge), 97, 99, 243n65

altered states, 3, 45, 54; and sensation, 54, 56, 62, 65–66

ambience, 204, 267n44. See also atmospheres

analogy, 193, 206, 209, 264n7, 268n46

anesthesia, 61, 71

animals, 56–57, 89–91, 149–50, 152–54, 164, 169, 183, 240n23, 247n29, 248n34; and the "abstract animal," 172–73; and animal heat, 9, 46–48, 50, 64–66, 70,

150, 173, 239n14; and animal rights, 220–21; fables and, 199–201; and plant life, 192–95, 264n7, 265n8, 266n33, 266n35; and territories, 202, 267n39

animism, 231n2

anti-slavery activism, 126–37

apostrophe, 67–69, 236n34

apprenticeship, 199, 207–8, 210, 268n48, 268n49, 268n57

Aristotle, 6, 49, 157, 164, 192–93, 232n9, 260n46. See also neo-Aristotelianism

art, 140–41, 233n4; and apprenticeship, 208; as compound, 72; and disgust, 139; "life" of, 2, 41–42; materiality of, 228; and nausea, 105; as network, 17, 39–40; and non-art, 39; as "organ" of philosophy, 169–70; and social conflict, 23–25; and temporality, 12, 38–42; as a unity contrasted with the arts, 12, 17, 24, 35, 38, 237n48

artificial life. See *under* life

artistic experimentation, 2–4, 11–12, 14–17, 22–27, 34–42, 227–29, 233n4, 235n24, 237n44. See also experiments

Art-Network, 39–42

assemblages, 18, 21, 183, 223, 235n19

associationism, 126–27

atmospheres, 65, 98–99, 229, 267n38; and beauty, 205, 207; Kant on, 203; local, 202, 267n37; medicated, 211, 267n40; ontology of, 202–4; risk as, 141; selective function of, 202; as territories, 202. See also mood; mutual atmospheres

audience, 61, 100, 105, 112–13, 118–19, 125–26, 128–29; and experiments, 15, 23–26, 28, 41